Geometry in Our
Three-Dimensional World

Problem Solving in Mathematics and Beyond

Print ISSN: 2591-7234
Online ISSN: 2591-7242

Series Editor: Dr. Alfred S. Posamentier
Distinguished Lecturer
New York City College of Technology - City University of New York

There are countless applications that would be considered problem solving in mathematics and beyond. One could even argue that most of mathematics in one way or another involves solving problems. However, this series is intended to be of interest to the general audience with the sole purpose of demonstrating the power and beauty of mathematics through clever problem-solving experiences.

Each of the books will be aimed at the general audience, which implies that the writing level will be such that it will not engulfed in technical language — rather the language will be simple everyday language so that the focus can remain on the content and not be distracted by unnecessarily sophiscated language. Again, the primary purpose of this series is to approach the topic of mathematics problem-solving in a most appealing and attractive way in order to win more of the general public to appreciate his most important subject rather than to fear it. At the same time we expect that professionals in the scientific community will also find these books attractive, as they will provide many entertaining surprises for the unsuspecting reader.

Published

Vol. 25 *Geometry in Our Three-Dimensional World*
by Alfred S Posamentier, Bernd Thaller, Christian Dorner, Robert Geretschläger, Guenter Maresch, Christian Spreitzer and David Stuhlpfarrer

Vol. 24 *Innovative Teaching: Best Practices from Business and Beyond for Mathematics Teachers*
by Denise H Sutton and Alfred S Posamentier

Vol. 23 *Learning Trigonometry by Problem Solving*
by Alexander Rozenblyum and Leonid Rozenblyum

Vol. 22 *Mathematical Labyrinths. Pathfinding*
by Boris Pritsker

For the complete list of volumes in this series, please visit www.worldscientific.com/series/psmb

Problem Solving in Mathematics and Beyond Volume **25**

Geometry in Our Three-Dimensional World

Alfred S. Posamentier
The City University of New York, USA

Bernd Thaller
University of Graz, Austria

Christian Dorner
University of Vienna, Austria

Robert Geretschläger
BRG Kepler H.S., Graz, Austria

Guenter Maresch
University of Salzburg, Austria

Christian Spreitzer
University College of Teacher Education Lower Austria, Austria

David Stuhlpfarrer
University College of Teacher Education Styria, Austria

World Scientific
NEW JERSEY · LONDON · SINGAPORE · BEIJING · SHANGHAI · HONG KONG · TAIPEI · CHENNAI · TOKYO

Published by

World Scientific Publishing Co. Pte. Ltd.

5 Toh Tuck Link, Singapore 596224

USA office: 27 Warren Street, Suite 401-402, Hackensack, NJ 07601

UK office: 57 Shelton Street, Covent Garden, London WC2H 9HE

Library of Congress Cataloging-in-Publication Data

Names: Posamentier, Alfred S., author.

Title: Geometry in our three-dimensional world / Alfred S. Posamentier,
The City University of New York, USA, Bernd Thaller, University of Graz, Austria,
Christian Dorner, University of Vienna, Austria, Robert Geretschläger, BRG Kepler H.S.,
Graz, Austria, Günter Maresch, University of Salzburg, Austria,
Christian Spreitzer, University College of Teacher Education Lower Austria, Austria,
David Stuhlpfarrer, University College of Teacher Education Styria, Austria.

Description: New Jersey : World Scientific, [2022] |
Series: Problem solving in mathematics and beyond, 2591-7234 ; Vol. 25

Identifiers: LCCN 2021042935 | ISBN 9789811237102 (hardcover) |
ISBN 9789811237744 (paperback) | ISBN 9789811237119 (ebook for institutions) |
ISBN 9789811237126 (ebook for individuals)

Subjects: LCSH: Geometry, Solid. | Polyhedra. | Three-dimensional modeling.

Classification: LCC QA491 .P68 2022 | DDC 516/.156--dc23/eng/202110005

LC record available at https://lccn.loc.gov/2021042935

British Library Cataloguing-in-Publication Data

A catalogue record for this book is available from the British Library.

For any available supplementary material, please visit
https://www.worldscientific.com/worldscibooks/10.1142/12283#t=suppl

Desk Editors: Vishnu Mohan/Tan Rok Ting

Typeset by Stallion Press
Email: enquiries@stallionpress.com

Preface

As we live in a three-dimensional world, it is surprising how little emphasis this aspect of our lives gets into the typical school curriculum. Although we are taught how to find the volume of a box or to measure three-dimensional objects, there is precious little attention given to the fact that we live on a sphere, namely, the Earth. When traveling by plane and tracking the flightpath on a video screen, people often wonder why a flight from New York to Vienna would be traveling along a curved path nearly touching Greenland when on a plane map that would seem not to be the shortest path. Or when asked which state of the continental United States is closest to the African continent, people are often shocked to learn that it is the State of Maine. These kinds of voids in our knowledge partly motivated this group of seven mathematicians and mathematics educators to collectively share their expertise to enlighten and enrich the general readership with a wide variety of aspects beyond plane geometry typically presented through secondary school.

Since ancient times, geometry has fascinated civilizations. Whether it was the famous artists such as Leonardo da Vinci or scientists such as Archimedes, the subject of geometry has always dominated human thinking. Therefore, it is no surprise that geometry has had a permanent place in academic education throughout the millennia. In fact, the word geometry is of Greek origin, "geo" meaning Earth and "metron" meaning measurement. The study of geometry in school has largely focused on plane geometry, namely, that which can be drawn on a piece of paper.

However, geometry unfolds its true usefulness largely through the knowledge that it provides for the three-dimensional world in which we live. Recently, much effort has been made to modernize geometry teaching by adding concepts of spatial geometry and virtual, computer generated reality. This recognizes the importance of points, lines, circles and polygons, as the tools by which to enrich our understanding of three-dimensional figures and their positional relationships.

One of the aims of this book is to raise awareness of the fascinating experiences of three-dimensional geometry. The book is intended for the general readership as it presents a broad overview of many different aspects of three-dimensional geometry in an intelligible fashion to motivate readers to investigate further topics of interest. Some readers will be amazed at the many aspects of three-dimensional geometry that are experienced on a daily basis, taken for granted, and that expose incredible nuances that can enrich our lives.

The first chapter contains a brief introduction to the concept of "dimension" and spatial coordinates, and also introduces some observations of everyday life in connection with spatial geometry that are often taken for granted. This also includes optical illusions that play tricks with our spatial perception.

Spatial understanding, visual perception and spatial skills, which are increasingly important in many professional areas, are the subject of Chapter 2. This aspect of human intelligence allows us to perceive our environment, understand shapes and sizes, estimate distances and be aware of our position in space and of its relationship to surrounding objects. With a consideration of some relevant aspects of psychological research, which explain how we see three-dimensional objects, we present some exercises that can be used to train one's spatial abilities — a skill which everyone should possess.

The third chapter is devoted to the question of how best to represent three-dimensional scenes in a two-dimensional graphic. We describe in some detail, how the art of perspective drawing has historically developed over time. You will learn how to create perspectivity with one or more vanishing points in order to create the illusion of space or distance. Taking this one step further to an investigation of projections — mapping points in three dimensions onto a plane — leads us to consider the principles of camera obscura and photography.

Although it is not part of our constant awareness, we do need to acknowledge that we live on the surface of a sphere. What appears to us to be a flat plane is, in fact, a small section of the spherical surface of the Earth when seen in three dimensions. This leads us in the fourth chapter to some interesting observations and modifications to two-dimensional school geometry. For example, it is well-known that in plane geometry a triangle has an angle sum is 180°, however, on a sphere the angle sum of a triangle ranges from 180° to 540°. One of the most important theorems of spherical geometry, which relates the angle sum of a spherical triangle to its area, will be presented with the expectation to provide unexpected amazement for the reader. Additionally, a discussion will follow about how a spherical surface can be mapped onto a plane. This leads us to the various methods that cartographers use to produce maps of the Earth that we typically see in geographical atlases. One of these methods is the Mercator projection, which is historically significant and has been brought back to new importance through the maps used in satellite navigation.

With the rapidly advancing technology and the activities in space exploration there are many aspects of three-dimensional observations, such as lunar phases, solar and lunar eclipses, and the various lengths of a day at various locations on Earth that can be explained with the geometric configurations of Earth, Sun, and Moon. Chapter 4 will also address the phenomena that convinced the ancient Greeks more than 2,000 years ago that the Earth was indeed spherical.

In Chapter 6, we enter the fascinating world of Platonic and Archimedean solids. These polyhedra are spatial equivalents of plane figures such as triangles, squares, and higher degree polygons. The five Platonic solids — whose faces are identical regular triangles, squares, or pentagons — will be presented and have their existence justified. These are merely the beginning aspects of a large number of increasingly complex bodies that have fascinating relationships with one another and yet prove to be quite aesthetically pleasing.

We tend to take the shapes of various sports' balls for granted. We show that these geometric shapes have come about through some curious geometric relationships which are presented in Chapter 7. The reader will subsequently look critically upon the various soccer balls, or basketballs, just to mention a few.

Chapter 8 describes some principles of arranging geometrical forms in two- and three-dimensional patterns. A first example is the regular pattern of a honeycomb in a beehive. Irregular patterns can also be analyzed mathematically. The typical structure created by a pile of soap bubbles can be described as a so-called Voronoi diagram. In this chapter, we describe the principles of Voronoi diagrams, some methods used to create them, and their application to economic questions, public health, art and architecture.

The honeycomb pattern shows how bee larvae can be accommodated in an optimal space-saving setting. This leads to a special case of so-called "packing problems," which are discussed more generally in Chapter 9. Packing problems are a class of optimization problems in mathematics that involve maximizing the density of packing variously shaped objects. Optimum packing problems continue to be relevant in a number of scientific disciplines and continue to receive significant attention. Interestingly, as early as the 16th century, the famous German mathematician Johannes Kepler postulated a solution for packing spheres, which then hundreds of years later was proved to be truly optimal.

Topology, a branch of geometry that deals with properties of surfaces that remain unchanged when the surfaces get bent or stretched without tearing is introduced in Chapter 10. This fascinating aspect of geometry considers, for example the single-sided surface of a Möbius strip and some of the spectacular unexpected results when the strip is cut in various ways. This and variants of this question put a significant strain to our spatial imagination and lead to some unexpected findings. We also discuss various realizations of the Klein bottle and other one-sided surfaces that, in three dimensions, can only be implemented with self-intersections.

The popular art of folding paper, referred to as origami, enables the creation of unexpected three-dimensional objects. In Chapter 11, we explore some fascinating three-dimensional tessellations and polyhedral, which can be crafted from paper or other two-dimensional materials with advanced origami techniques for purely aesthetic purposes. There are also numerous technological applications of these methods. Satellite technology, which is dependent on the transport of large flat or curved surfaces with minimal mass into space (think of such things as solar panel arrays or parabolic antennae) uses complex folding methods to pack and unpack such materials into deployable configurations. In medicine, stents can be folded by the methods of advanced origami to make them very small for human implantation, but at the same time stable and durable enough for long-term use.

The art of creating knots in three dimensions is the topic of Chapter 12. While inspired by knots which appear in everyday life, a mathematical knot differs from common knots in that the ends are joined together so that it cannot be undone. The simplest type of such knots is a ring. In this chapter, we show how to draw and classify knots. Any given knot can be represented in many different ways by knot diagrams. The fundamental problems in knot theory are determining when two descriptions represent the same knot and classifying the many different types of knots.

In Chapter 13, we take three-dimensional geometry one step further by considering the movement of mechanical machines. We discuss mechanisms, their degrees of freedom and geometric constraints, and look into advanced applications, such as robot arms, the Stewart platform, over-constrained mechanisms, as well as the Turbula® mixer. This leads to a consideration of various gears and how they can be made to transfer a twisting motion into another direction

Planning and designing mechanism requires engineers to use techniques called geometrical modeling, which is presented in Chapter 14. Mathematical principles underlying computer-aided design are presented along with the processes necessary to create virtual models, which then can be realized using three-dimensional printers. The construction of some fascinating examples will be described in some detail, such as the ambiguous cylinder and the Klein bottle.

To better understand aspects of three-dimensional geometry, we explore an outlook to topics involving four dimensions. This is not merely a topic of science fiction, since for example, machines with more than three degrees of freedom have to be described with higher-dimensional mathematics. In the fourth dimension, we look at the aspects of the assumption that we have one additional direction that is perpendicular to all previous spatial directions of three-dimensional space. This provides the stage for fascinating objects like the hypercube, which is the four-dimensional analog of the cube. As a conclusion, a discussion of the hypersphere in mathematics, physics, and art is presented.

The authors hope that through this book the reader will gain an understanding of geometrical principles that are helpful in a world in which computer-generated visualizations are becoming more ubiquitous and where good spatial awareness becomes important in more and more areas of everyday private and professional life.

About the Authors

Alfred S Posamentier is currently Distinguished Lecturer at New York City College of Technology of the City University of New York. He is also Professor Emeritus of Mathematics Education at The City College of the City University of New York, and former Dean of the School of Education, where he was tenured for 40 years. He is the author and co-author of more than 75 mathematics books for teachers, secondary school students, and the general readership. Dr Posamentier is also a frequent commentator in newspapers and journals on topics relating to mathematics and education. He is still involved in working with mathematics teachers and supervisors, nationally and internationally, to help them maximize their instructional effectiveness.

Bernd Thaller is Professor of Mathematics Education at the Institute of Mathematics and Scientific Computing of the University of Graz, Austria. After studying mathematics and physics in Graz at which he earned his doctorate, he did further studies at the University of Vienna and the Free University of Berlin. He became a mathematical physicist, making some important research contributions to relativistic quantum mechanics and quantum mechanical spectral and scattering theory. He has written a monograph about the Dirac equation and two books on Visual Quantum Mechanics.

He was coordinator of two large-scale European projects on research and development in educational math (Socrates 2005–2008 and Erasmus+ 2015–2017) and has been supervisor of more than 140 diploma theses and dissertations. In 2008, he founded and is currently head of the regional educational competence center for mathematics and geometry, whose main task is the coordination of activities of the teacher training institutions in Styria and the organization of in-service teacher training events, seminars, and workshops.

Christian Dorner is a member of the faculty in the Mathematics Department at the University of Vienna. His research focuses on stochastic thinking, financial mathematics for secondary school mathematics classes, procedural knowledge of students and research-based design of video-vignettes (short video clips) to be used for developing the practical didactic skills of prospective mathematics teachers. He has published research papers in peer-reviewed journals on mathematics education and has given several talks at international conferences on mathematics education.

Robert Geretschläger is currently a mathematics teacher at Bundesrealgymnasium Keplerstraße (BRG) in Graz, Austria, and a lecturer at the University of Graz. He is also actively involved in the production of challenging and entertaining mathematical problem-solving material for the *Mathematik macht Freu(n)de* project at the

University of Vienna, Austria. He is involved in the organization of many mathematics competitions on a local, regional, and national level. Among other duties, he has been the leader of the Austrian team at the International Mathematical Olympiad since 2007 and president of the organizing committee of the Mathematical Kangaroo in Austria since 1998.

Besides his interests in competition mathematics, his research interests lie in the areas of the geometry and mathematics of paper folding, polyhedra, higher elementary mathematics (especially geometry and inequalities), recreational mathematics and the didactics of teaching mathematically gifted students. He is a member of the didactics commission of the Austrian mathematical society, and has worked in curriculum development, the introduction of the Program for International Student Assessment (PISA) in Austria, and the development of mathematics standards for secondary school students.

Guenter J Maresch is Professor of Mathematics Education at the University of Salzburg (Austria). His research topics are spatial ability, spatial thinking, descriptive geometry, computer-aided design (CAD), digital media, didactic principles, and curriculum development. He is the editor-in-chief of two professional journals, has published 16 books, has developed different models for spatial thinking, and is the head of the research group for mathematics education at the University of Salzburg. From 2021–2024 he will be one of the partners of the international EU-project SellSTEM.

In 2012, Prof. Maresch was named "Educator of the Year" at the worldwide competition of Bentley Systems, and in 2020, he was one of the winners of the "Digital Teaching Award" of the University of Salzburg.

Christian Spreitzer is a professor of mathematics education at the University College of Teacher Education in Lower Austria and a lecturer at the University of Vienna. He teaches mathematics and science in the teacher education program for primary school teachers and mathematics and physics for future high-school teachers. In addition, he is involved in various in-service teacher training programs for mathematics teachers.

Besides his career as a university teacher and scientist, he has coauthored several books on popular mathematics, notably *Math Makers* (with Alfred S Posamentier), *The Mathematics of Everyday Life* (with Alfred S Posamentier), and *The Joy of Mathematics* (with Alfred S Posamentier, Robert Geretschläger, and Charles Li).

David Stuhlpfarrer is on the faculty of the Department of Mathematics Education, in the Institute for Secondary Teacher Education of the University College for Teacher Education, Styria, Graz.

Besides his keen interest in geometry, he is also very interested in everything concerning new DIY technologies and computational thinking, like 3D-printing, laser-cutting, and the use of single-board computers like the Arduino, the Raspberry Pi, the Micro:bit, or the Calliope for DIY projects.

Contents

Chapter 1

Our World is Three-Dimensional

When we speak of geometry, we typically think of the geometric figures that we can draw on a piece of paper, like triangles, squares, and circles. This is two-dimensional geometry and has for a long-time dominated school mathematics, and indeed many were fascinated by the intriguing wonders revealed in the relationships of two-dimensional figures. However, we live in a world with three spatial dimensions — length, width, depth — that is even richer in intricate geometric facts and relations. Sometimes, in everyday life, we encounter phenomena that are hard to explain unless one has a good understanding of the three-dimensional nature of the world in which we live. One example, where one is easily fooled by a two-dimensional world view, is the following phenomenon, which we call the flight-path reality.

1.1 Shortest Distance on the Globe

One of the authors recently took a flight from Frankfurt, Germany to Vancouver in western Canada, which are two cities approximately on same latitude. Vancouver is even a bit further south than Frankfurt. So, you might expect the plane to fly west from Frankfurt, until it eventually reaches Vancouver. To the surprise of many on board the flight, the aircraft flew from Frankfurt straight to the north. It swerved slightly to the west via Denmark, and then flew, further north, across

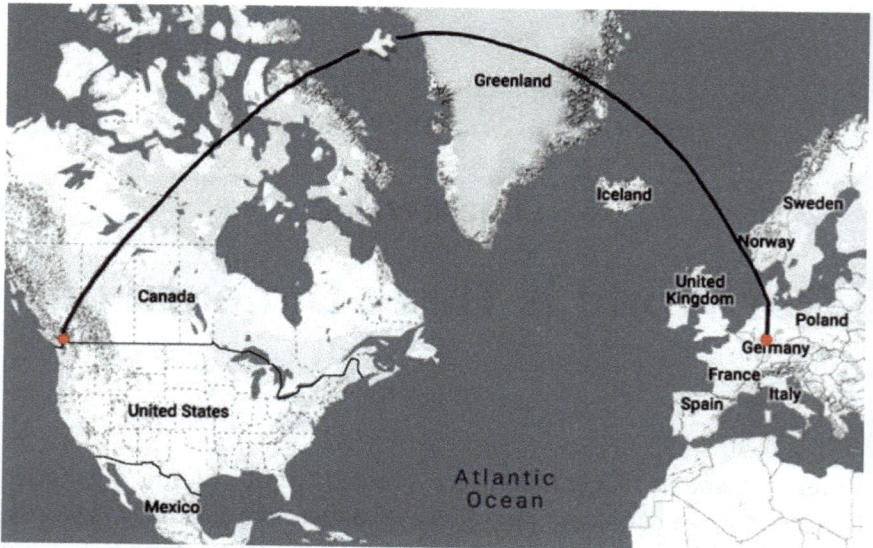

Fig. 1.1.   Path of flight LH492 from Frankfurt to Vancouver (generated using data from flightradar24.com and maps.google.com).

Greenland and the northern part of Canada, finally approaching Vancouver from the north-east. Figure 1.1 shows the aircraft's route generated by the actual flight data as shown on a flat screen.

If the map depicted in Figure 1.1 were a true image of the world, this would seem to be a very remarkable detour. One would think that the pilot is trying to avoid something, since he is not traveling in a straight line as envisioned on the two-dimensional world map (or on the video screen in the plane). Yet in fact, the pilot chooses a path that is rather close to the actual shortest possible path in three dimensions (see Figure 1.2).

What makes this unexpected is that we are dealing with geometry on a sphere, in this case the Earth's surface, which very easily explains this aspect of our three-dimensional world. If you happen to own a globe — in fact a scale model of our planet — you can easily determine the shortest path between two points on Earth's surface by stretching a rubber band between these two points on the globe. As the elastic band contracts, it automatically determines the shortest

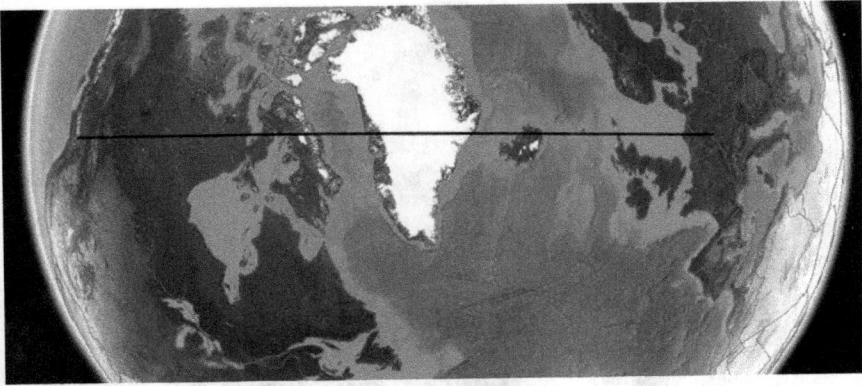

Fig. 1.2. Shortest flight path between Frankfurt and Vancouver.

path between its endpoints. This shortest path is actually a part of a so-called great circle — which is a circle placed on the surface of the sphere in such a way that the center of the circle coincides with the center of the sphere. The most prominent great circle on Earth's surface is the Equator — the plane of this special circle is perpendicular to the Earth's axis of rotation.

Another example, where our intuition about distances on the surface of a sphere can go wrong is the following: Of the 48 contiguous United States, which state is closest to the African continent? We typically think that the southern coast of Florida is closest to Africa, however, in actual fact, the tip of Maine is considerably closer to the coast of Africa: The closest distance (along a great circle) between the coast of West Sahara to Maine is about 5,100 km, the closest distance from Africa to Florida is about 6,400 km.

1.2 The Wrong Tilt of the Moon

Another, albeit related, phenomenon concerning straight lines on a spherical surface is called the wrong-tilt-of-the-Moon puzzle.

This photo (Figure 1.3) of the Waxing Moon was taken in New York City shortly after sunset. Among the people watching this beautiful scene, a discussion arose about why the Moon (a globe illuminated by the Sun) appeared to be illuminated from slightly above,

Fig. 1.3. The wrong-tilt of the Moon puzzle.

although the Sun was much lower in the sky at that time — in fact it was already below the horizon. Should not the Moon rather appear illuminated from diagonally below?

A similar effect occurs with the Waning Moon in the morning. The photo in Figure 1.4 was taken during sunrise. It is a cylindrical panorama composed of several shots. The large field of vision makes it possible to see the rising Sun and the still-visible Waning Moon on the same image, although both appear as very small dots. The enlarged detail clearly shows that the Moon is illuminated from above. How can that be when the Sun is still on the horizon and is, therefore, much lower in the sky than the Moon?

The angle between the Sun and the Moon is 180° for the Full Moon and 90° for the Half Moon. As shown in Figure 1.4 it is at about 140°. In this extremely wide-angle photo, straight lines necessarily appear as curved, only straight lines through the center of the image appear as straight lines (horizon). Included in the picture is also a vapor trail of a plane — a straight line that appears curved, because it is near the border of the image. A light-ray between Sun and Moon is

Fig. 1.4.   Bastei-Bridge, Germany, at sunrise (cylindrical panorama).

a straight line connecting the Sun and the Moon, but — similar to the vapor trail — it would appear as a curved line in the wide-angle panorama.

The phenomenon of the "wrong" tilt of the Moon is related to how we perceive straight lines on a spherical surface. In this case, the sphere is the celestial sphere and the straight lines are the rays of light from the Sun to the Moon. The straight connections between two distant points in the sky are no longer perceived as straight lines, but as curves that follow the curvature of the sky. This happens, in particular, when the two points are so far apart that one has to turn one's head to look from one point to the other. In the situation depicted in Figure 1.4, most people would imagine the straight line between the Moon and the Sun as an oblique line that gradually approaches the horizon, and not as a great circle that runs high across the celestial sphere.

The aspect of geometry, which deals with lines and figures on a spherical surface, is called *spherical geometry*, which will be discussed further on. The spherical surface is embedded in a three-dimensional space, yet its geometry is still very similar to the geometry of two-dimensional plane — in particular, if the sphere is very large.

1.3 Coordinates, Dimension, and Space

What do mathematicians actually mean when they speak of "dimension"? In colloquial language, the word "dimension" usually means the extent or size of a particular object or building. In physics, the "dimension" of a physical quantity describes the basic quantities (and, therefore, the appropriate units) that are necessary for its measurement. The dimension of "velocity" is, for example, "distance per time". That means, that in every system of units, the unit of speed is the unit of distance divided by the unit of time (e.g., miles per hour or meters per second). These uses of the word "dimension" are very different from its use in geometry.

In geometry, the dimension of a mathematical space or an object is always the number of independent coordinates that are required to describe its position in it. For example, a line has dimension 1 because a single coordinate is required to describe a position on the line. A coordinate is just a number (more precisely, a real number), but one has to specify a coordinate origin and a unit of length (i.e., a "coordinate system") in order to interpret the coordinate (see Figure 1.5). One often denotes the coordinate of an arbitrary point by the letter x, which stands for any real number (positive or negative), and gives the position of a point on the line with respect to a chosen coordinate system, as shown in Figure 1.5 for the coordinate value $x = 2.5$.

A plane has the dimension 2. In order to specify a location in a plane, one has to choose a coordinate system with two perpendicular "coordinate axes", as shown in Figure 1.6. Now we need two numbers, x and y, to describe the location of a point, and we usually write them as a pair in parenthesis, (x, y). The interpretation is as follows: In your chosen coordinate system, start at the origin, move horizontally to the

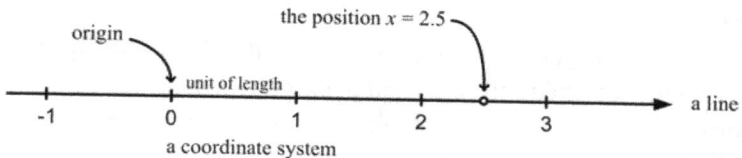

Fig. 1.5.   A coordinate system on a one-dimensional line.

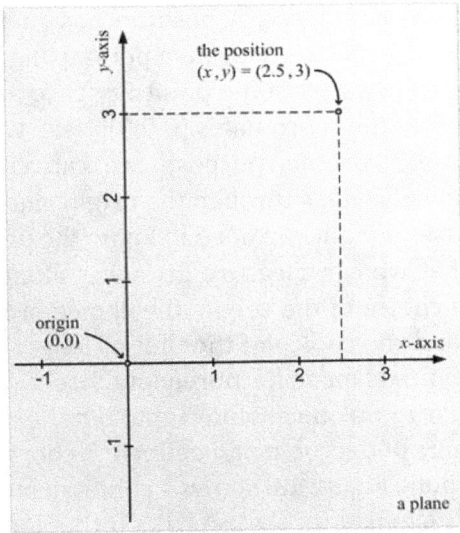

Fig. 1.6.   A coordinate system on a two-dimensional plane.

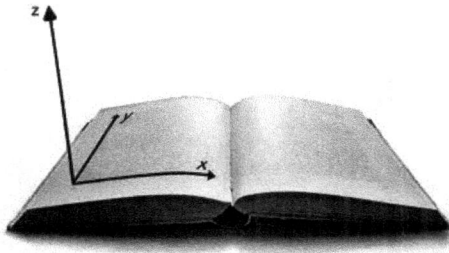

Fig. 1.7.   While the pages in a book are two-dimensional, our world has three dimensions.

point on the x-axis with coordinate x, then move vertically on a line parallel to the y-axis until you reach the position y.

The page of the book you are reading is an example of a two-dimensional object. You can move left or right along the x-axis (parallel to the lines), or up or down along the y-axis. In order to specify an arbitrary point on the page one would need two coordinates. The third dimension is entered by leaving the book page and moving vertically away from the page along the z-axis (see Figure 1.7).

In three-dimensional space, a position has to be described by three coordinates, x, y, and z. To locate a point using its three coordinates, we first need to select some point in space as the coordinate origin, which receives the coordinates $(0,0,0)$. Next, we have to specify the coordinate axes. For this purpose, we can choose any three mutually-perpendicular lines through the origin, and designate them as the x, y, and z axes. Finally, we need to know the unit of length (e.g., inch or cm) so that we can measure distances along the coordinate axes. As with the choice of the origin, the choice of the directions of coordinate axes and the choice of the unit of length are arbitrary. In most cases, the choice is made for purely practical reasons. For example, in a rectangular room one might want to put the coordinate origin in one of the corners of the room and define the coordinate axes along the edges of the room. Figure 1.8 shows a common method of displaying a coordinate system, with the x-axis pointing forward, the y-axis pointing to the right, and the z-axis pointing up. It is often convenient to add arrowheads to the coordinate axes, in order to visualize the

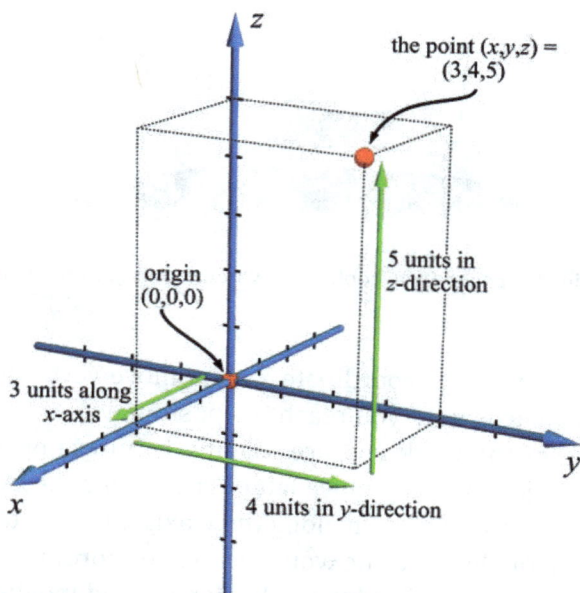

Fig. 1.8.   A coordinate system and instructions to find the point $(3,4,5)$.

positive direction (the direction, in which the coordinate-value increases). With respect to the selected coordinate system, each point in space is uniquely given by a triple of coordinates (x, y, z), which are three real numbers $x, y,$ and z. Figure 1.8 shows how to locate a particular point with coordinates $(3,4,5)$. It is easy: Starting from the coordinate origin, one goes three units along the x-axis, then 4 units in the direction parallel to the y-axis, and finally 5 units in the z-direction, then one arrives at the point with the coordinates $(3,4,5)$. If one of the coordinates happens to be negative, one just counts the corresponding number of units in the opposite direction. In fact, the position of any point in three-dimensional space can be described in this way by specifying three real numbers. In mathematics, three-dimensional space is therefore identified with the set of all possible coordinate triples (x,y,z), where $x, y,$ and z are arbitrary real numbers.

The concept of coordinates dates back to the French mathematician René Descartes (1596–1650). At that time, it was a rather revolutionary idea that made it possible to use the methods of arithmetic and algebra for investigations in geometry. Later, this idea also made it possible to generalize geometry to higher dimensions — to mathematical spaces in which we need four, five, or more coordinates to indicate a location. A four-dimensional space can thus be described as the set of all "points" of the form (x,y,z,w), where the four letters represent arbitrary numbers that are interpreted as coordinates. While this book is primarily devoted to a study of geometry of three-dimensional space, the last chapter of this book will take us on a brief excursion into the fascinating realm of higher-dimensional mathematics.

1.4 Philosophy of Space

While the definition of "dimension" as the minimum number of coordinates that are required to indicate a position is from modern times, the concept of space as a geometrical object is much older. Ancient Greek geometers introduced the mathematical concept of space as a model of the cosmos. It is in the books of Euclid of Alexandria, who

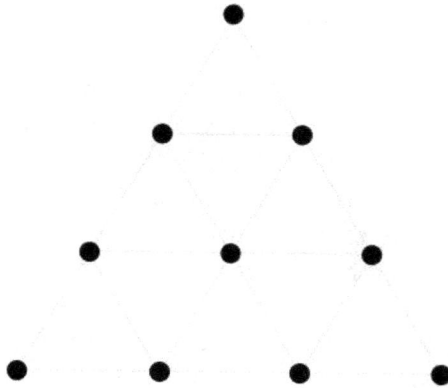

Fig. 1.9.   The Tetractys, the ancient Greek symbol for the hierarchical dimensionality of the cosmos.

lived around the year 300 BCE, where the fundamental properties of space are described in terms of a small set of axioms or postulates. Even today, the mathematical space that serves to describe the space of our intuition is called Euclidean space. Still earlier than Euclid, Pythagoras and his followers had ideas relating to a structure of the cosmos hierarchically organized by dimensions. This structure was symbolized by the tetractys, one of the mystical symbols of the Pythagoreans. The tetractys is a triangular figure composed of 10 points in a regular array, as shown in Figure 1.9.

This arrangement of points is nothing less than a symbol of the cosmos. The 10 points are arranged in four rows. The first row contains just one point. A point, according to Euclid, is an object that has no parts — an object of zero dimensions. The second row contains two points. Two points in space can be used to represent a straight line, which is a one-dimensional object. The line through the given points is even uniquely defined, provided that the two points do not coincide.

The next stage of the dimensional hierarchy is in the third row, containing three points. Three points are always in a plane, as long as they are not collinear, that is, if they are not on the same line. A plane is a two-dimensional object. Finally, with the four points in the fourth row we enter the world of three dimensions, the world we live in. If the four points are not coplanar (that is, if they are not in the same plane), they form the vertices of a three-sided pyramid — a three-dimensional object. This completes the hierarchical description of the cosmos and classifies the basic elements of Euclidean geometry according to their dimension: point — line — plane — space.

1.5 Optical Illusions and the Limits of Spatial Imagination

Three-dimensional geometry is an inexhaustible source of interesting effects and amazing surprises. The Dutch artist M.C. Escher (1898–1972) is famous for his naturalistic drawings of apparently impossible objects and irregular perspectives. In some of his drawings he made artistic use of the fascinating optical illusions shown in Figure 1.10, which are known as the Penrose triangle and the Penrose staircase, respectively.

The personal irritation felt by considering these images shows that our brain automatically tries to construct a three-dimensional impression even from two-dimensional images. It is a result of our evolution, where we perceive our environment as three-dimensional and try to interpret visual impressions as generated by a three-dimensional scene. Our brain's adaption to three dimensions and our ability to accurately assess one's own position in space in relation to other objects has always been of importance for survival. We need

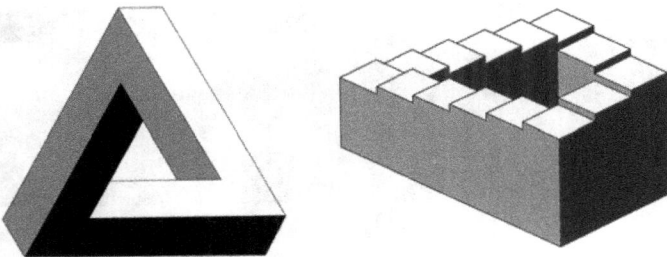

Fig. 1.10.   Penrose triangle and Penrose staircase.

to be able to sense depth and distance in order to behave accordingly and to avert damage. The importance of depth perception is probably the main reason for having two eyes. The sensory input from two eyes capturing a scene from two different angles allows our brain to create a three-dimensional impression of the environment. In the following chapters, we will deal with the psychological mechanisms of spatial perception and with the methods of representing three-dimensional scenes in two-dimensional images.

Spatial perception and spatial understanding are certainly fundamental components of human intelligence. Yet, spatial thinking is challenging for various people in different ways. The ability in three-dimensional thinking involving a variety of exercises can vary with the individual. Some examples of difficult tasks engaging different components of spatial intelligence can be found in Chapter 2.

In this context, it is interesting (and also fun) to explore the limits of our spatial imagination. There are many examples on the Internet that show how easily our spatial perception can be fooled. The Neural Correlate Society hosts an annual competition, the Best Illusion of the Year Contest (see http://illusionoftheyear.com), which has produced several great examples of misguided spatial perception. The ambiguous objects with seemingly impossible mirror images by Kōkichi Sugihara are particularly striking (see Figure 1.11). In Chapter 14, we will explore the ambiguous cylinder illusion (Figure 1.11(b)) in more detail.

We have seen some examples showing the importance of three-dimensional geometry for interpreting our perception of the world.

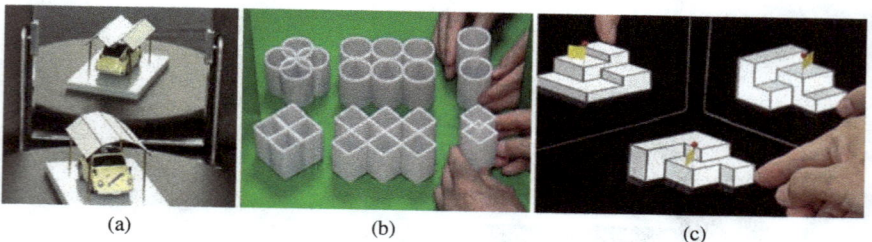

(a) (b) (c)

Fig. 1.11. Ambiguous objects by Kokichi Sugihara.

With virtual reality beginning to enter our consciousness, we are all the more exposed to looking at our environment more critically. This book is intended to help you to see the world in three dimensions and to discover some of the fascinating relationships that are the subject of higher-dimensional geometry.

Chapter 2

The Eight Basic Elements of Spatial Thinking

The talent of spatial understanding is a prerequisite for developing and understanding spatial geometry. It is, therefore, worthwhile to analyze in more detail this ability, which is partly innate and partly acquired. Extensive psychological research in recent years has shown that visual perception and spatial awareness consist of several components, each of which represents a very specialized skill. This chapter describes the components of spatial thinking in detail and illustrates them with a collection of sample problems that readers can use to test their own skills.

2.1 Spatial Skills — How Can They Help Us?

Among the most significant developments that will affect our lives in the future is the increasing digitalization of all areas of life, such as the use of *big data* and *artificial intelligence* to analyze and influence our behavior, and the growing importance of visualizations that permeate all types of communication. All of these trends will certainly accelerate in the next few years. While digitalization and big data often make the headlines, the trend of visualizing our world is a rather gradual process and often goes unnoticed. Nevertheless, this trend plays a growing role in our private and professional lives and we are increasingly confronted with the task of correctly interpreting a visual representation and drawing the right conclusions from it. In this chapter, we will consider which special skills and which components of our intelligence can help us process visual information, and, furthermore, how these skills can be trained.

A well-developed visual perception and a good ability to capture and process spatial information is certainly useful for many everyday tasks. These include refilling a refrigerator, driving and parking a car, and estimating distances between objects as well as the speed of vehicles. We need to think spatially whenever we read a map or use a compass. At first glance, navigation systems seem to make it very easy to find the way from point A to point B. But in many situations it is not always that easy to translate the information of the moving map on the screen of the navigation system into the real world (see Figure 2.1) and thereupon make correct decisions as a driver.

First we need to check our speed and then check to make sure that the navigation system actually shows our current position on the right street and that it really makes sense to turn right when our navigation system indicates to make a right turn. If it were very easy to follow the information from modern navigation systems, we never

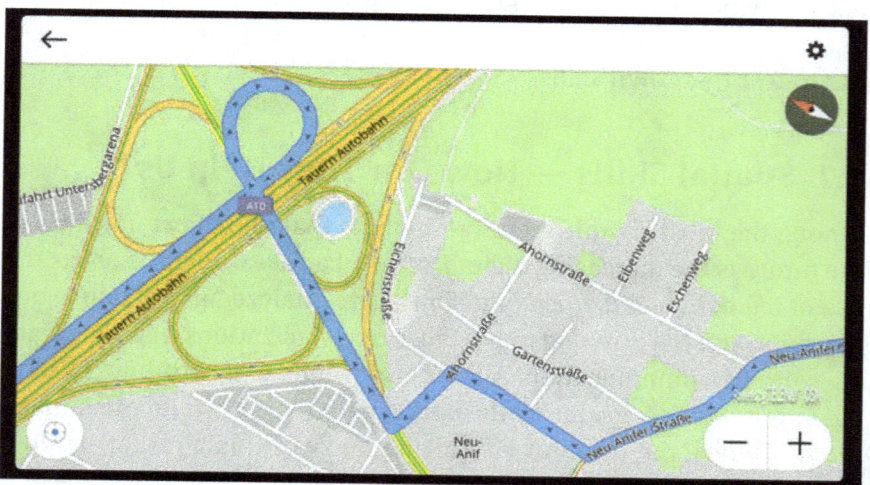

Fig. 2.1.   Navigation system.

Fig. 2.2.   Car accident caused by misinterpretation of a navigation system.

would see pictures of accidents like the one in Figure 2.2, where a driver thought that the next turn to the right would lead to a street. Instead it ended in a flight of stairs.

Here is another example. We need increasingly more spatial skills to use modern technical tools such as a three-dimensional printer. When using a traditional two-dimensional printer, we usually just have to push the print button and (in most cases) the sheet of paper will be printed correctly. Printing a three-dimensional object on a three-dimensional printer requires spatial skills and numerous preparations. First, the three-dimensional object must be moved (translated and rotated) in a way that a flat side surface of the object is correctly positioned on the base plate of the printer.

In Figure 2.3, we see that the three-dimensional object is *not* yet properly aligned with the base plate of the three-dimensional printer. Once the virtual three-dimensional object is correctly positioned, one has to scale the object and check whether it is properly placed on the three-dimensional-printer surface so that all parts of the surfaces will be printed in acceptable positions. A position is not acceptable if one or more surfaces would be floating in space unsupported. In this case, the three-dimensional object must either be repositioned, or the three-dimensional printer must receive the information that it has to print a so-called support structure. Support structures are additional

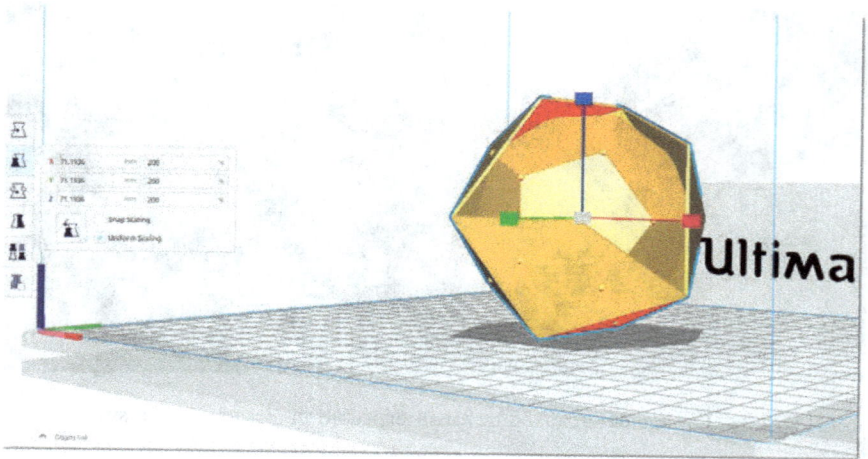

Fig. 2.3.   Preparing a three-dimensional model for printing with a three-dimensional printer (screenshot).

three-dimensional objects that are automatically generated by the three-dimensional printer and enable printing of the originally not-acceptable floating surfaces and that can be removed very easily after printing from the three-dimensional object.

Another example of our modern, visual world, that is, flying a drone (see Figure 2.4) appears to be a fairly simple activity at first. However, let's compare the following situations: When a drone flies away from the pilot, and he steers it to the right, the drone will turn to the right as seen from the pilot's vantage point. However, if the drone flies directly towards the pilot, the drone would turn to his left if he steered it to the right as before. The pilot must, therefore, always think of himself as a virtual pilot who flies with drone. This is not an easy task, and we see that flying a drone requires well-developed spatial thinking.

In addition, many modern-day professions require good visual perception and spatial ability. For example, a taxi driver uses these skills to find the best route from points A to point B. A crane operator also has well-developed spatial skills when he places a load gently and precisely on a construction site. A doctor must have excellent spatial imagination when she works with medical devices in a

Fig. 2.4. Drone (Image: Sven Teschke; Licence: CC BY-SA 3.0-de via Wikimedia Commons).

patient's body laparoscopically and controls the movements of the devices on a screen. And an architect needs these skills when constructing three-dimensional objects such as houses or bridges while working on a flat screen.

In a world where technological achievements play an increasingly important role, increasingly more professional careers are based on the fields of science, technology, engineering, and mathematics, which are known as STEM fields. All these subjects have in common that one needs very well-trained spatial skills to be successful. In 2009, American researcher Jonathan Wai and his colleagues compared the results of several long-term research projects from the past 50 years and found consistent evidence that people with successful careers in STEM fields already displayed high spatial abilities as adolescents. For example, in the late 1950s in a project called "TALENT", about 400,000 randomly selected high-school students were tested for their spatial skills and were then tracked for over 11 years through their subsequent professional careers. Almost all of those who earned a Ph.D. degree in a STEM field were among the top performers on the spatial ability test at school. As students 45% of these people even were among the best 4% in the spatial abilities test. These results demonstrate, in a very convincing way, that especially for all who want to work in one of the STEM fields, it is of significant importance to have well-developed spatial skills. The training and improvement of spatial skills is an essential prerequisite for STEM jobs, and thus, an excellent pathway to science, technology, engineering, and mathematics.

This leads to the question of whether spatial skills can be trained, and how that training should be designed to improve these important skills. There is, indeed, evidence that the ability of spatial thinking is not only a genetic predisposition but can also be trained and improved.

In the past, some folks had the notion that traits, which are genetically determined, are necessarily immutable, and that because some traits are modifiable, they must be caused by experiences after birth. None of these complementary positions are true. Today, we know that some have a better genetic predisposition than others. But everyone can train and improve their spatial skills enormously. The ability to think spatially comprises several components, each of which must be developed individually. We will first discuss the following two parts of spatial thinking that are commonly called "visual perception" and "spatial ability".

2.2 Visual System

Visual perception is the ability to recognize and identify spatial objects in the visual field. A necessary prerequisite is the processing of visual information through the brain's visual system. Visual information enters the brain through the eyes and is then processed by neural structures collectively referred to as the visual pathway. A schematic representation of the brain's visual pathway is shown in Figure 2.5.

Each eye casts a two-dimensional image of a scene onto the retina. These images are, in fact, upside-down and show the viewed scene from slightly shifted angles, which ultimately enables stereoscopic viewing. The retina is a layer of tissue on the back of the eye containing light sensitive cells. These cells are named rods and cones. Rods are only able the see black and white but no colors. The 120 million rods in our eye are very light sensitive, which means that we see contours of objects (in black/white) even when it is quite dark.

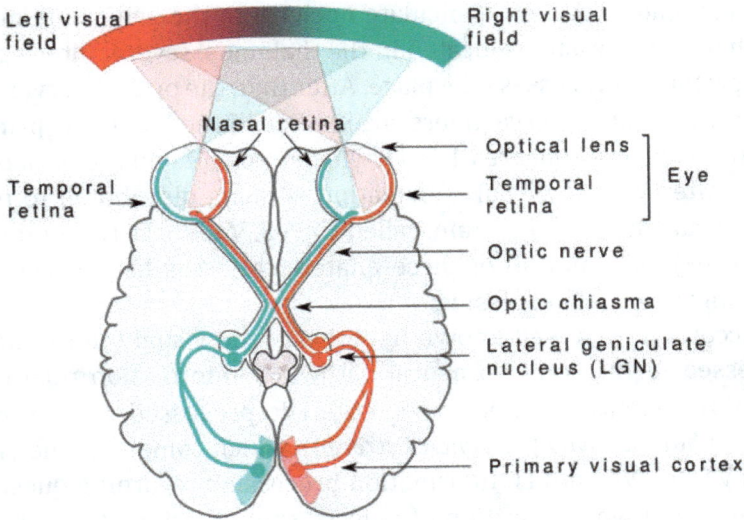

Fig. 2.5. Anatomy of the optical pathway in the brain (Miquel Perello Nieto/CC BY-SA) (https://creativecommons.org/licenses/by-sa/4.0).

To complement the vision that the rods provide, the cones are able to see colors. We have about 6 million cones in each eye. Most of them are located in the Fovea (centralis), which is a small area in our eyes where we have the best vision. When illuminated with light, these photoreceptor cells send out neuronal signals. As there are several layers of interconnected neuronal cells in the retina, some important steps of information processing already take place here, for example, enhancement of contrast and detection of color and motion.

The nerve fibers, or *axons,* carrying the neural signals from the retina are bundled in the *optic nerve,* which covers the eye's backside. The optic nerve transmits electrical signals from the retina into the brain for further processing. The two optic nerves from each eye cross each other only a few centimeters behind the eyes at the *optic chiasma.* As shown in Figure 2.5, each optic nerve is sub-divided in such a way that the axons carrying information from the left visual field of each eye are collected on the right side of the brain and forwarded to the *right geniculate nucleus.* All nerve fibers coming from the right visual field of each eye are collected after the optic chiasma on the left side and lead to the left geniculate nucleus. In the geniculate nuclei, which are relay systems located in the thalamus, several filtering and interpretation processes take place. After that, the optical nerves with about 1.5 million nerve fibers are routed directly to the primary visual cortex or visual area 1 (V1), which is an area on the back of the brain. After V1, the visual information is being distributed to many other visual areas of the brain, called V2, V3, V4, V5, IT (inferotemporal cortex), etc., and all of these anatomical areas have specialized functions in visual processing.

According to a well-known hypothesis, the visual information is processed further on two neural pathways. Information relating to shape and identity (object recognition) is processed along the so-called *What pathway* (or *ventral stream*), which comprises the visual areas V1, V2, V4, and IT. Information processing regarding questions of where and when (location of objects, spatial vision, direction and speed of motion) follow the *Where/When pathway* or *dorsal stream* (V1, V2, V3, V5, etc.). However, much research has yet to be done to learn more about what exactly happens after visual information enters the brain via the primary visual cortex.

2.3 Visual Perception

With the processing of information along the visual system, there is an increasing abstraction of the properties of what is seen. On the other hand, numerous connections are made with other information processing sub-networks of the brain. This enables us, when the information finally reaches our consciousness, to correctly interpret visual information, such as recognizing objects. The ability to recognize and identify spatial objects in one's field of vision is called "visual perception" which is a very complex process involving large parts of the brain and is still not fully understood on the level of neural processes. Psychological and behavioral research can still give us an idea of which functions and processes have to be performed by the information processing system so that visual perception can work.

The Austrian social worker, educator and psychologist Marianne Frostig (1906–1985) has dealt extensively with visual perception and examined it in detail. In her *Developmental Program in Visual Perception*, Frostig distinguishes five areas of visual perception: (1) *eye-motor coordination* — the coordinated control of eye movement and body movement (e.g., eye-hand coordination when one reaches for something), (2) *perception of figure-background* — the ability to distinguish a shape from the background, (3) *perception of form constancy* — knowing that a shape remains the same, even if it is moved or rotated, (4) *perception of position in space* — recognizing an object's position in its surroundings and the direction in which it is turned, and (5) *perception of spatial relationships* — perceiving the position of several objects in relation to each other (e.g., person A is to the left of person B).

These five areas are the key components of visual perception and are also particularly important for children's general learning ability (e.g., writing and reading). They are shown in Figure 2.6 as an overview.

2.4 Spatial Ability

The optical stimuli received by the brain are evaluated and processed in numerous brain areas entrusted with specific tasks. The recognition of spatial objects in the visual field in relation to their

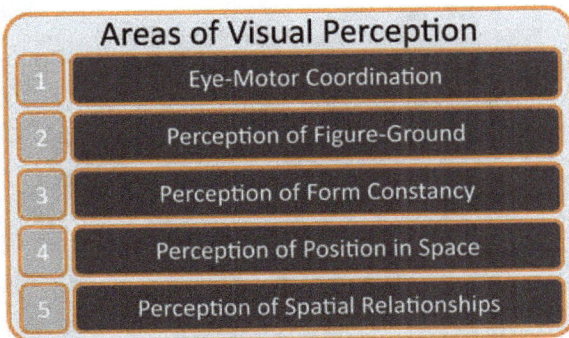

Areas of Visual Perception

1. Eye-Motor Coordination
2. Perception of Figure-Ground
3. Perception of Form Constancy
4. Perception of Position in Space
5. Perception of Spatial Relationships

Fig. 2.6.   The five components of visual perception according to Marianne Frostig.

surroundings is part of the visual perception. One usually considers this as a prerequisite for the next step, which is about the ability to keep objects in one's mind and perform operations with them in one's imagination. This is called "spatial ability" and can be defined as follows:

> *"Spatial ability" is the ability to create and transform spatial objects in one's imagination, to recognize and to establish relations between several of these mental objects, and to mentally put oneself in different spatial positions.*

Currently, numerous research projects are based on a psychological model where spatial ability is assumed to consist of the following four components or "sub-skills": (1) *Spatial Visualization* — the ability to create and hold a mental representation of an object or scene in one's mind, (2) *Spatial Relation* — the ability to mentally set objects into a spatial relation to each other, (3) *Mental Rotation* — the ability to mentally manipulate and rotate objects, and (4) *Spatial Orientation* — the ability to mentally put oneself in different spatial positions (see Figure 2.7).

Of concern is that the two component models just presented (Figures 2.6 and 2.7) do not take into account the many possible relationships between the sub-components of visual perception and spatial ability. A weakness in a component of visual perception is very

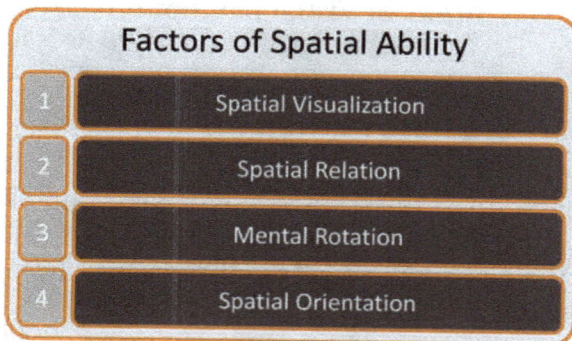

Factors of Spatial Ability

1	Spatial Visualization
2	Spatial Relation
3	Mental Rotation
4	Spatial Orientation

Fig. 2.7.   The four components of spatial ability.

likely to affect one or more components of spatial ability. These two areas are very closely related and should, therefore, be considered together in the diagnosis and training of spatial skills.

To overcome these current challenges, one of the authors (Guenter Maresch) developed a new structured model of spatial thinking that combines into a single model the two areas of visual perception and spatial ability. This model, which will be described in detail in the following section, is intended to clarify open scientific questions and to enable everyone (especially educators at all levels) to comprehensively diagnose, train, and foster one's visual perception and spatial abilities.

2.5 The Eight Basic Elements of Spatial Thinking

In this chapter, we present our model of spatial thinking. It combines and rearranges the sub-skills of visual perception and spatial ability into eight new basic elements, as shown in Figure 2.8. Each of the eight elements will be described in detail and will be accompanied by

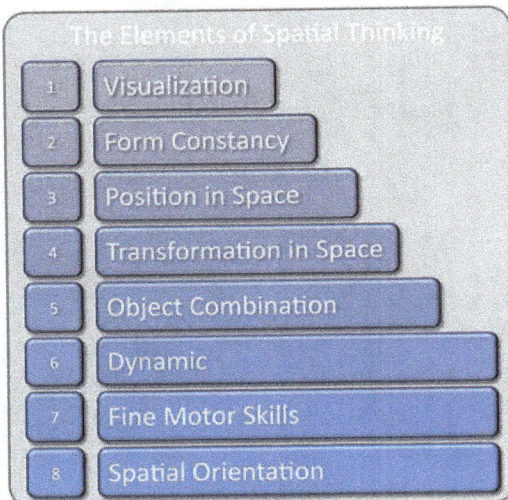

The Elements of Spatial Thinking

1. Visualization
2. Form Constancy
3. Position in Space
4. Transformation in Space
5. Object Combination
6. Dynamic
7. Fine Motor Skills
8. Spatial Orientation

Fig. 2.8.   A model with eight elements describing the ability of spatial thinking.

some examples of typical tasks along with training items that illustrate the special skills related to each element. We believe that these new items are to a higher degree independent of each other than the classical components of visual perception and spatial ability. Moreover, it should be easier to test (or improve) them individually without interference from other elements.

The elements 1–5 of the model shown in Figure 2.8, *Visualization, Form Constancy, Position in Space, Transformation in Space,* and *Object Combination,* are successive steps that build on one another, the most basic one being *Visualization.* Figure 2.7 shows them in ascending order. In general, children and students gradually develop these five spatial skills. The other three basic components, which are *Dynamics, Fine Motor Skills,* and *Spatial Orientation,* are mostly independent of the first five levels, and therefore, develop at an individual pace.

The first years of life bear the highest potential for the development of all basic skills. Up to the age of 13 or 14 years, the basic elements 3, 4, and 5 are developed on average up to about 80%. Results from current research show that the ability of *Spatial Orientation* grows slower than the elements 3, 4, and 5. In Figure 2.9, we show the

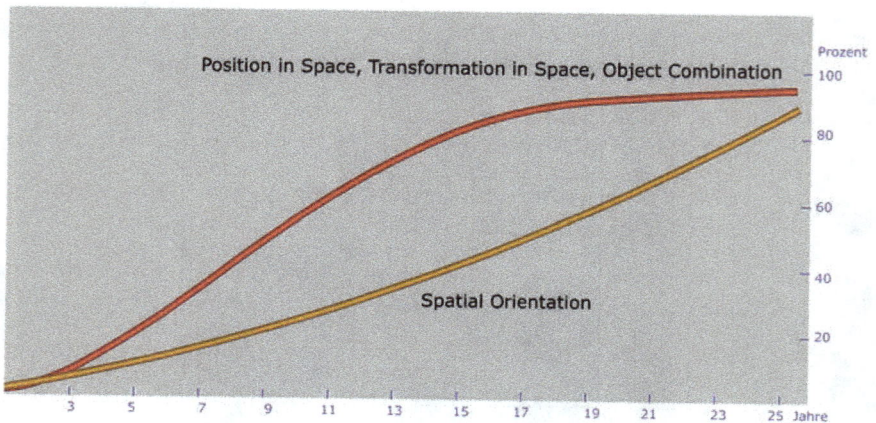

Fig. 2.9. The development of *Spatial Orientation* and other elements of spatial thinking.

development of the ability of *Spatial Orientation* in comparison to three other elements of spatial thinking.

We now describe the special skills by which each of the eight basic elements of spatial thinking are defined. This is best understood by considering some typical geometrical problems whose solutions require these special skills. The selection of problems presented below constitutes just a few representative examples, which by no means cover the entire area of competence of the respective element, which would require at least 20 different problems for each element.

Currently, further test items and training units are being developed, which should enable educators in all educational areas to diagnose and train the skills of students of all ages in the area of spatial thinking. All eight areas of competence can be addressed specifically and individually, using test items, tasks and problems similar to those presented below. The best success is guaranteed if this training is carried out continuously throughout the K-12 school years.

2.6 Element 1: Visualization

The most basic element in the model is here called *Visualization*, which is the ability to recognize and focus attention on relevant aspects of a presented image. Spatial objects can be recognized within a pictorial scene by their contours, colors and specific patterns. At this stage, individuals master the separation of background and shape and are able to separate relevant from irrelevant information. Spatial objects can be recognized in different forms of representation and from different vantage points. The element *Visualization* is related to Frostig's figure-background component of visual perception. At this first stage, neither objects nor viewers are in motion. The ability of Visualization can be tested with tasks similar to those presented in Figures 2.10–2.13. Solutions are given in Section A.1.

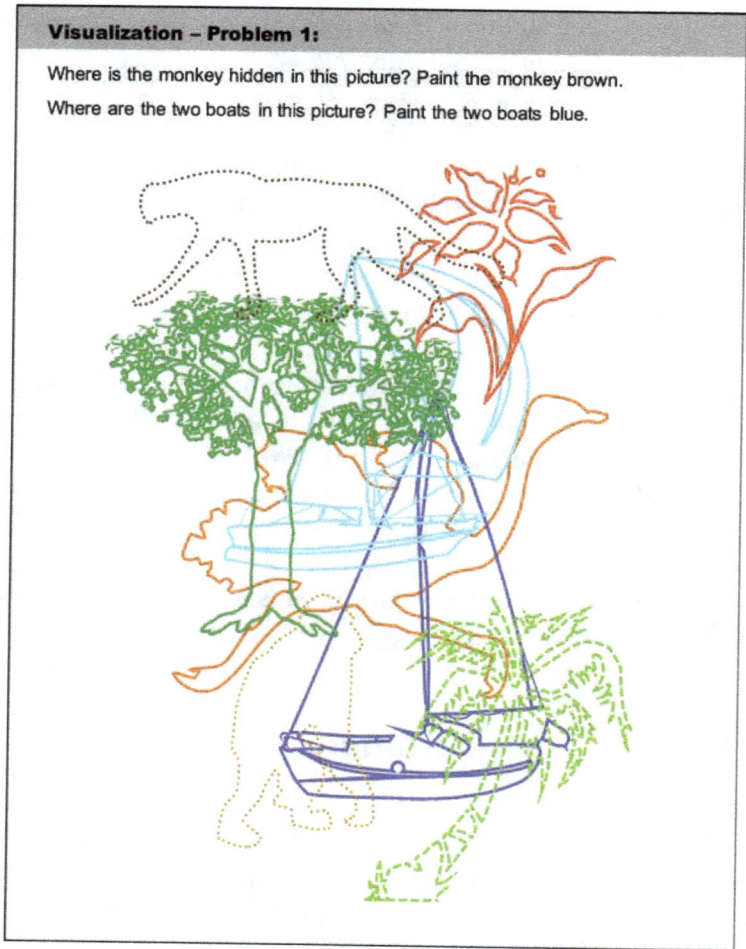

Visualization – Problem 1:

Where is the monkey hidden in this picture? Paint the monkey brown.

Where are the two boats in this picture? Paint the two boats blue.

Fig. 2.10.   Problem 1 illustrating the ability *Visualization.*

2.7 Element 2: Form Constancy

The ability *Form Constancy* means that simple planar and spatial objects (e.g., triangles, circles or spheres) can be identified by their characteristics and can be reliably recognized in different forms of representation, different views and variations, and in different

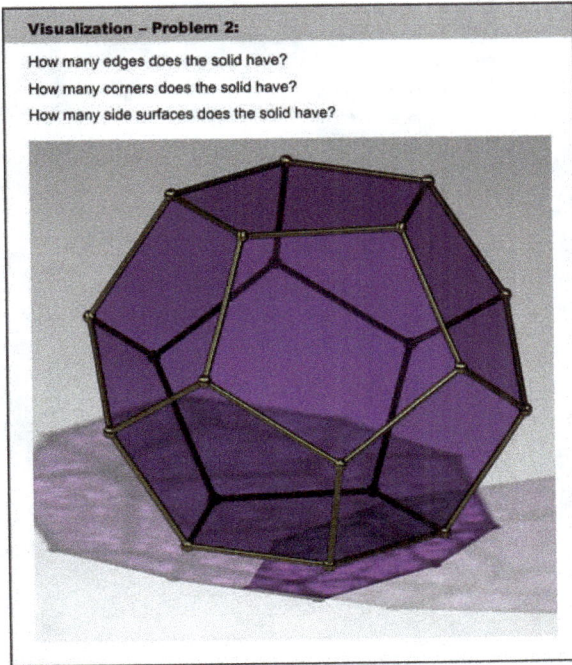

Visualization – Problem 2:

How many edges does the solid have?
How many corners does the solid have?
How many side surfaces does the solid have?

Fig. 2.11.   Problem 2 illustrating the ability *Visualization*.

environments. Different forms of presentation, for example, are free-hand sketches, ruler and compass designs, photos, computer images, or real models, but may also differ in brightness or with a different background from one another. Views may differ in the viewing angle from which an object is viewed (e.g., top/bottom, side, closer/further). Different variations of an object can, for example, be characterized by different sizes, colors and patterns. The ability *Form Constancy* means that one can not only recognize similar objects, but can also draw and describe them verbally. This element builds on the visualization ability and is similar to the corresponding component of Frostig's five-level model of visual perception. Some tasks for testing the ability Form Constancy are shown in Figures 2.14–2.16. Solutions are given in Section A.2.

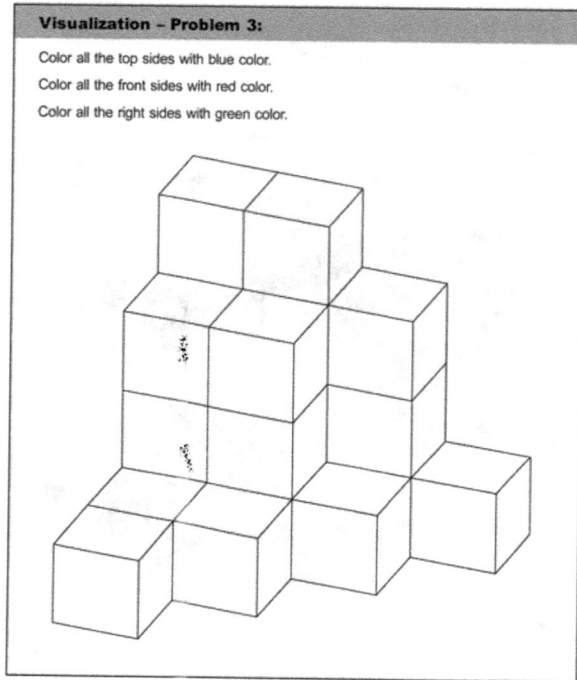

Visualization – Problem 3:

Color all the top sides with blue color.

Color all the front sides with red color.

Color all the right sides with green color.

Fig. 2.12. Problem 3 illustrating the ability *Visualization*.

2.8 Element 3: Position in Space

The ability to identify different positions in space is the third basic element. Positions of objects relative to each other (such as top/bottom, front/back, left/right, or horizontal/vertical) can be detected. In addition, the positioning of objects with given spatial relationships is considered mastered, if one can position objects in different views and in different forms of presentation at appropriate positions, or to generate them in one or more view(s), or to transfer a specific position of a spatial object to another view. The objects that may be fully or partially visible at position-in-space problems may not be moving and usually must not even be moved mentally by the viewer. The ability Position in Space is illustrated in Figures 2.17 and 2.18. Solutions are given in Section A.3.

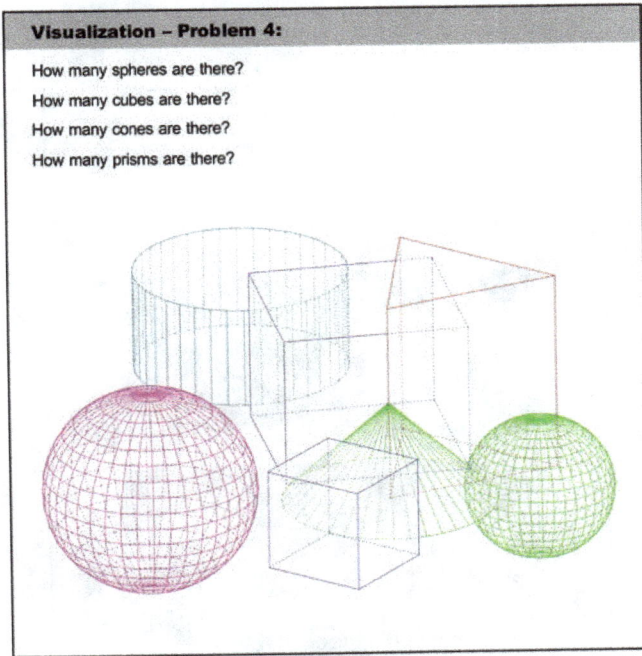

Fig. 2.13.   Problem 4 illustrating the ability *Visualization*.

2.9 Element 4: Transformation in Space

Mastering different basic *Transformations in Space* is characteristic of the fourth basic element of spatial thinking. References between the source object and the target object (before and after spatial transformations) can be detected. The basic spatial transformations of this routine include rotation, translation, mirroring, and scaling of two-dimensional and three-dimensional objects. Problem solvers and objects are — as in the examples before — at rest and not moving. But in contrast to the three examples before, here the problem solver must mentally move the given objects. The objects are either completely or partially visible in the start and end positions. This fourth basic element has references to Frostig's factor position in space and to the factors of visualization, mental rotation, and spatial relation of

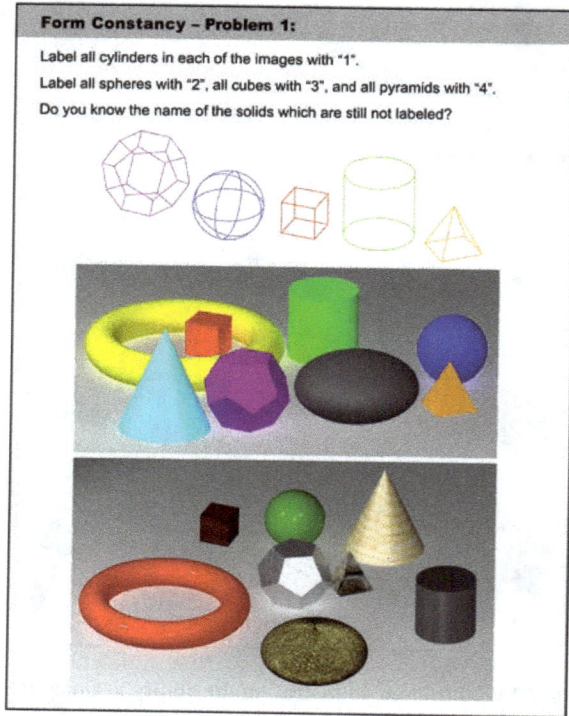

Form Constancy – Problem 1:

Label all cylinders in each of the images with "1".

Label all spheres with "2", all cubes with "3", and all pyramids with "4".

Do you know the name of the solids which are still not labeled?

Fig. 2.14.   Problem 1 illustrating the ability *Form Constancy*.

the spatial ability factors. In particular, it focuses on the ability to understand and then physically and mentally perform fundamental spatial transformations. Some tasks illustrating the ability to perform transformations in space are illustrated in Figures 2.19–2.21. Solutions are given in Section A.4.

2.10 Element 5: Object Combination

The fifth basic element *Object Combination* describes people's ability to mentally recognize and create cuts of two or more spatial objects, and mentally recognize and create supplementary parts of given objects and Boolean operations (difference, union, and intersection).

Form Constancy – Problem 2:

Each row presents four figures. One of the figures is slightly different from the others. Can you find the one in each row, that differs from the others?

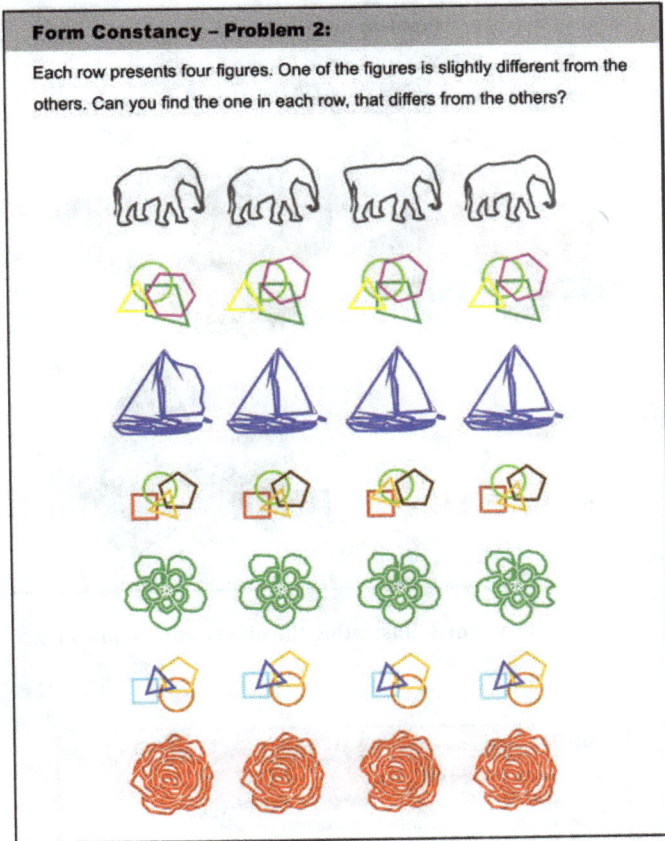

Fig. 2.15.   Problem 2 illustrating the ability *Form Constancy.*

It is not about how the individual parts can be transformed in order to be combined into a total object, but about recognizing spatial objects even if they are cut off or show further combinations (intersection, differences, union) with other objects. The basic routine transformation in space is primarily concerned with the recognition and control of spatial transformations, whereby, this component identifies an at-least mental dynamic facet. In the basic element *Object Combinations*, the analysis and consideration of spatial objects are in the foreground, which are cut off, or penetrate, and can, therefore, be

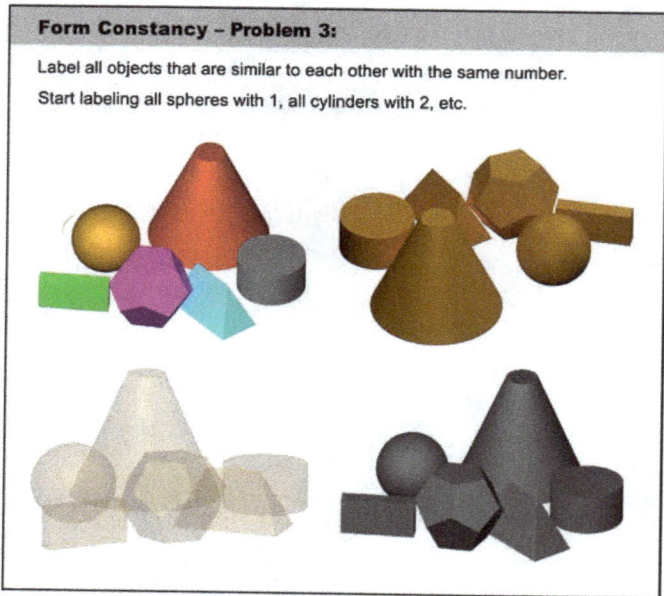

Fig. 2.16.   Problem 3 illustrating the ability *Form Constancy.*

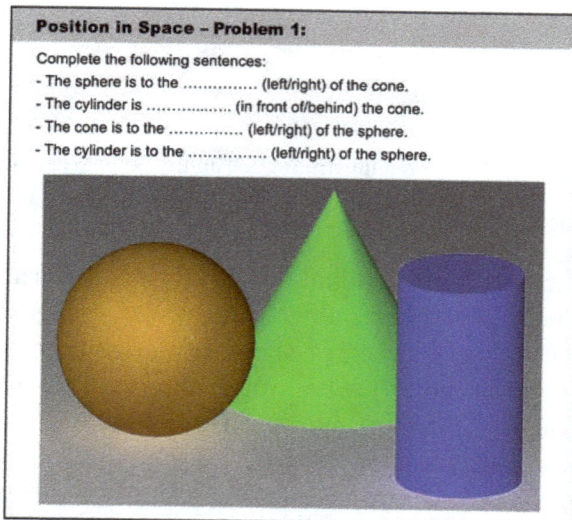

Fig. 2.17.   Problem 1 illustrating the ability *Position in Space.*

Position in Space – Problem 2:

Color all the sides of the given object in the center of the image with different colors.

Color the *front view* (the object as seen from the front) in the coordinate system (y", z").

Then color the *side view* (the object as seen from the right side) in the coordinate system (x"', z"').

Finally, color the *top view* (the object as seen from the top) in the coordinate system (x', y') in the lower part of the figure.

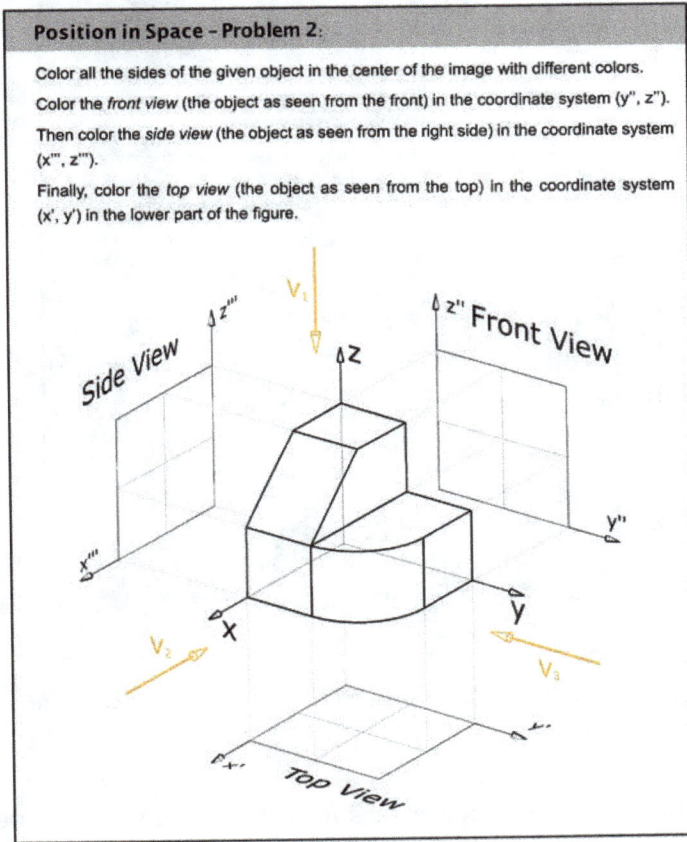

Fig. 2.18.   Problem 2 illustrating the ability *Position in Space*.

regarded as a predominantly static aspect. The considered objects are mostly only partly present or visible. Tasks illustrating this ability are presented in Figures 2.22 and 2.23. Solutions are given in Section A.5.

2.11 Element 6: Dynamics

The first five basic elements mentioned above (*Visualization, Form Constancy, Position in Space, Transformation in Space,* and *Object Combination*) tend to develop sequentially in humans. Only mastering

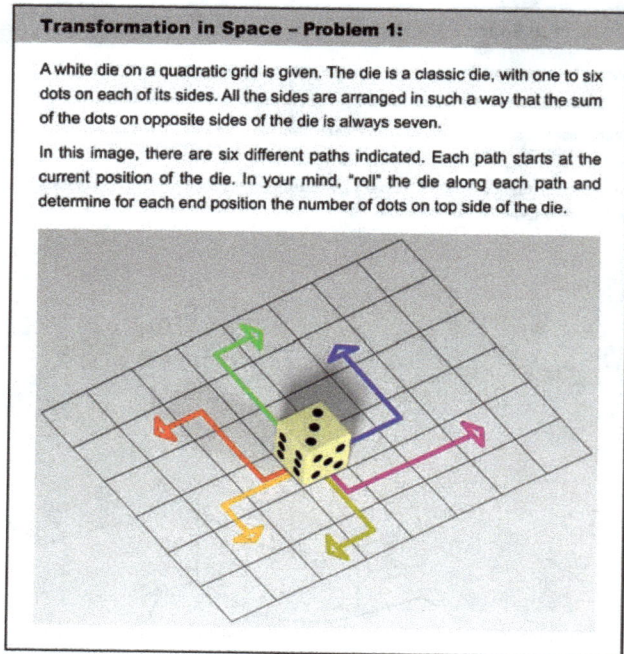

Fig. 2.19.   Problem 1 illustrating the ability *Transformation in Space.*

the abilities of one element, makes it possible and easier, to develop and strengthen the skills of a subsequent element. The three other basic elements (*Dynamics, Fine Motor Skills,* and *Spatial Orientation*) are a bit more independent in terms of development, as they address different continuing capabilities of spatial thinking.

The element *Dynamics* focuses on the recognition of movements and speeds. The first five basic elements, without exception, deal with abilities which presuppose resting problem solvers and stationary spatial objects. The basic element *Dynamics* describes the ability of people to be able to follow moving objects while the problem solver is at rest and to be able to anticipate different courses of motion. Other aspects of the element *Dynamics* are, for example, the ability to bring sequences of images of moving objects into the correct order (such as images of a spinning carousel), the ability of the

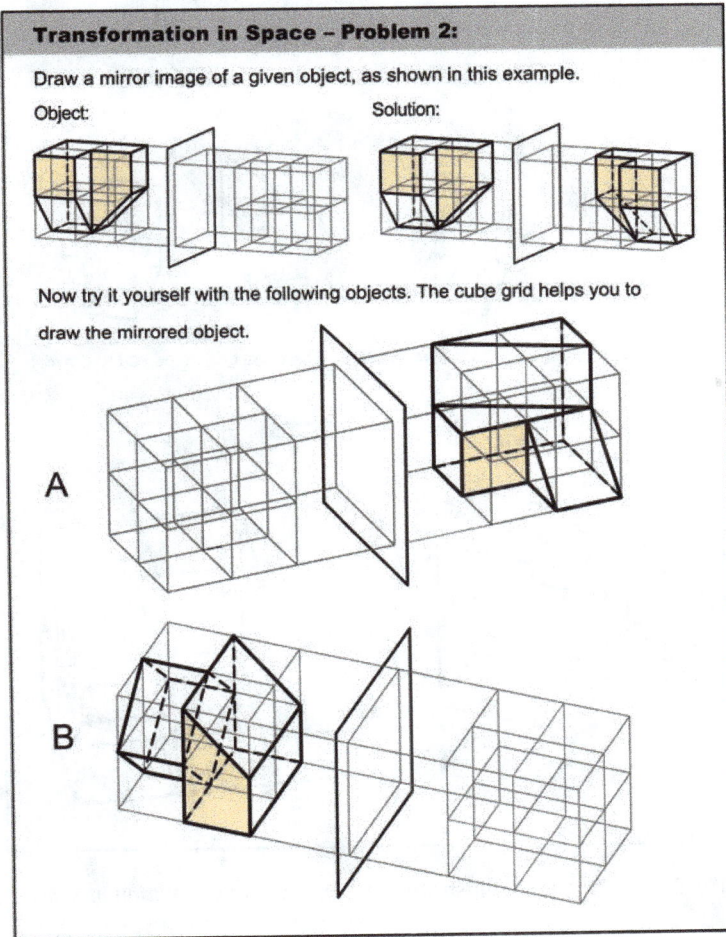

Fig. 2.20.   Problem 2 illustrating the ability *Transformation in Space.*

recognition and appreciation of motion patterns with or without a scenic background, and the ability to estimate, if and when two moving objects are colliding. It also includes the ability to move objects so that they can reach a goal at a certain time. Some tasks for testing the ability Dynamics are shown in Figures 2.24 and 2.25. Solutions are given in Section A.6.

Transformation in Space – Problem 3:

Draw a rotated image of a given object, as shown in this example.

Object: Solution:

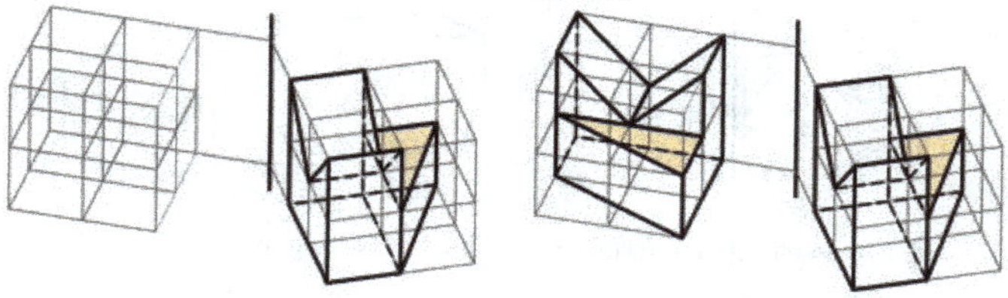

Now try it yourself with the following object. The cube grid helps you to draw the rotated object.

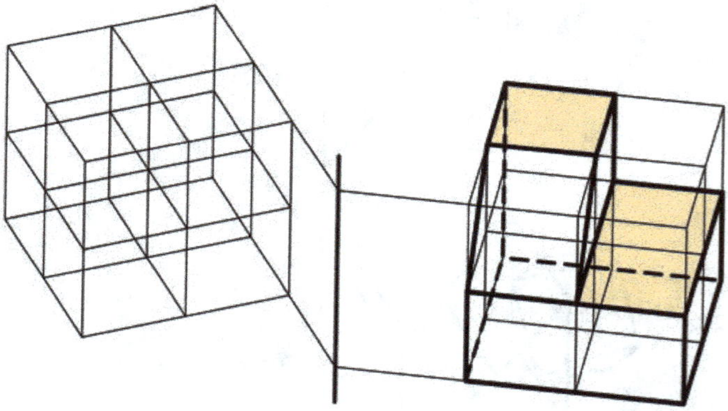

Fig. 2.21. Problem 3 illustrating the ability *Transformation in Space.*

2.12 Element 7: Fine Motor Skills

The *Fine Motor Skills* routine monitors whether the visual stimuli and information after their processing in the brain are passed on to the (fine) motor brain areas and lead to appropriate fine motor activities. Many everyday activities require a well-developed visual perception, which is the ability to be able to record and filter visual information appropriately through our optical apparatus, and to be able to relay

Object Combination – Problem 1

The object on top consists of 19 small cubes. Which of the six objects shown below can be used to complete the object on top so that the result is a cube made of 3 × 3 × 3 small cubes? At least one or more of the objects can provide a correct solution. The objects can be moved and/or rotated as needed.

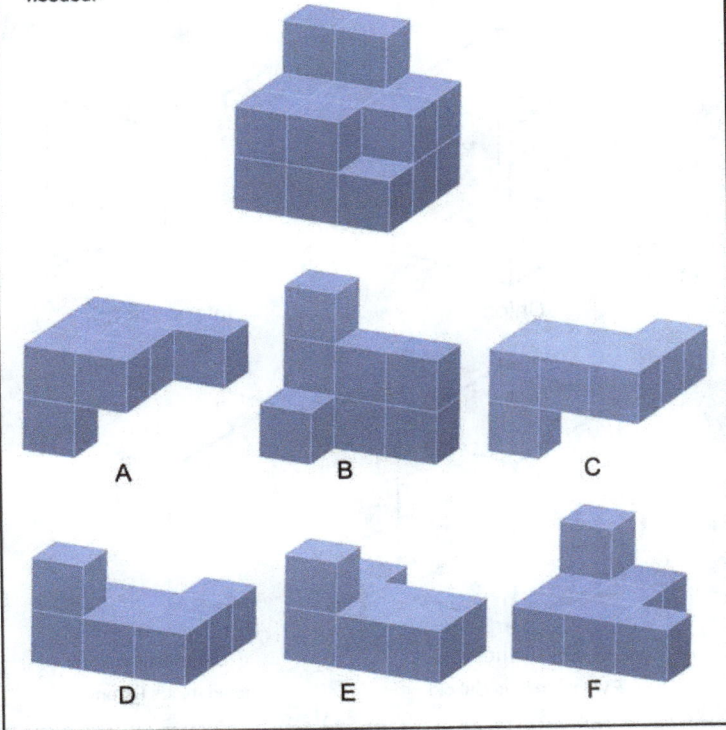

Fig. 2.22. Problem 1 illustrating the ability *Object Combination.*

it to the brain in a stable manner, and also to enable the recognition of spatial objects. They also require well-trained spatial ability, which is the ability to process spatial objects in a purely mental way and to be able to imagine themselves in scenes from other perspectives. Finally, a well-coordinated (fine) motor apparatus is needed that enables us to realize mentally planned movements and to be able to

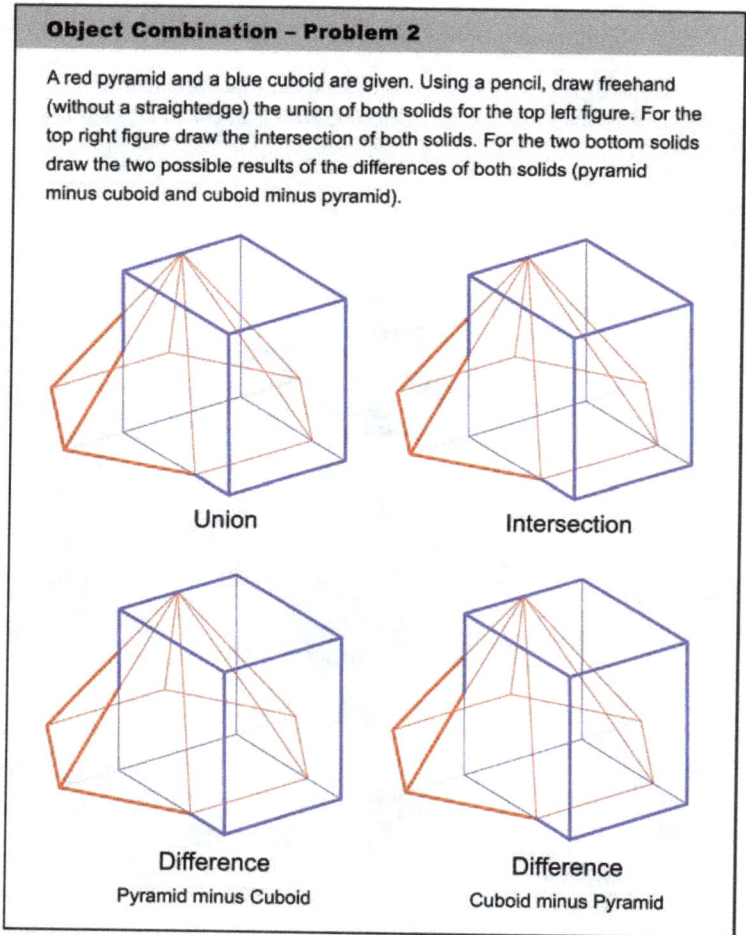

Fig. 2.23.   Problem 2 illustrating the ability *Object Combination.*

make drawings and constructions, as well as to be able to write or hammer a nail into a wooden board. The basic routine of *Fine Motor Skills* shows an affinity for the so-called eye-hand coordination of Marianne Frostig, and therefore, observes the ability of humans to match visual and (fine) motor stimuli and, thus, the ability to follow certain visual information with appropriate real movements (such as

Dynamics – Problem 1

Anne, who stands beside the platform, takes three consecutive pictures with her camera from the same direction, while the platform with some people and a dog on it is rotating around its center in a clockwise direction. Unfortunately, the three pictures get mixed up and Anne wants to place them in the correct order. Please help Anne place the pictures 1, 2, and 3 in the correct order by keeping in mind that the platform was rotating in a clockwise direction, which is indicated by the arrow in the first image.

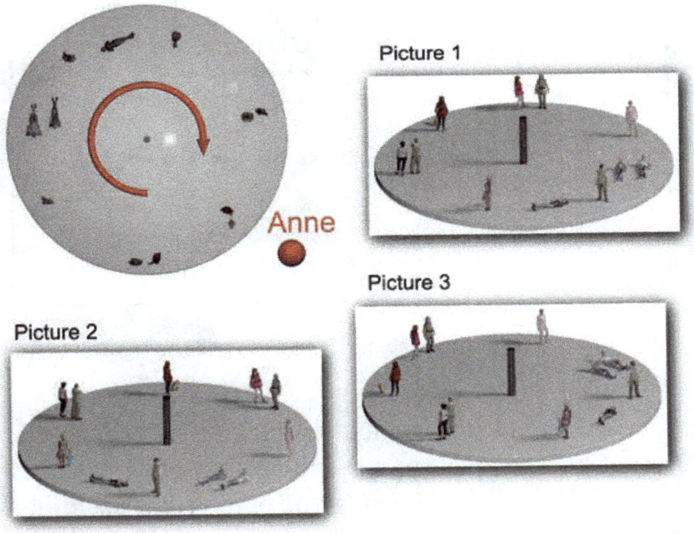

Anne

Picture 1

Picture 2

Picture 3

Fig. 2.24. Problem 1 illustrating the ability *Dynamics.*

draw tracks with a pencil, draw with both hands and move real objects to specific positions). This ability is illustrated by the tasks shown in Figures 2.26 and 2.27. Solutions are given in Section A.7.

2.13 Element 8: Spatial Orientation

Finally, the ability *Spatial Orientation* is mainly aimed at being able to mentally switch into other positions of a spatial scene. *Spatial*

Dynamics – Problem 2

In this figure, part B that makes the other part move is called the driver. The black circle in the center of part B represents the axle, which can turn but cannot move from where it is shown. Part A can only move horizontally. Now answer the questions below.

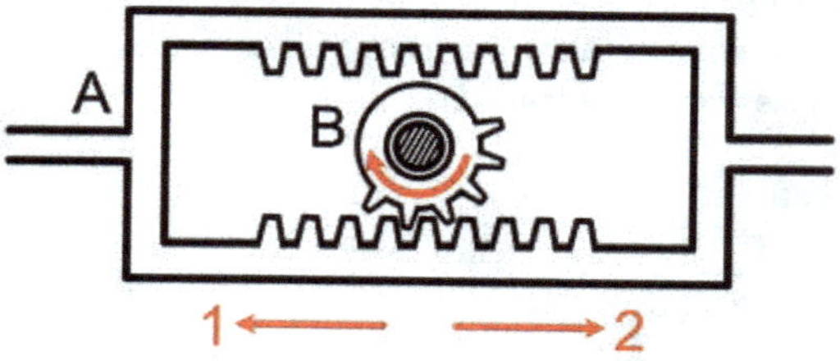

1.   If part B starts moving in the direction shown, which way will part A move? In the direction 1 or 2?

2.   In which direction will part A be moving when part B has turned halfway around from where it is now? 1 or 2?

Fig. 2.25.   Problem 2 illustrating the ability *Dynamics*. Based on Figure 114 from Brown.[1]

Orientation is the ability to be able to imagine certain spatial scenes from different perspectives, and then mentally move into other positions in a spatial scene. It is the only one of the eight basic elements of spatial thinking in which the problem solver mentally moves in the given spatial scene. The basic element of Spatial Orientation also includes the ability to determine (after moving in a real scene) specific distances and angles to other objects in the scene. The ability of *Spatial Orientation*, which is processed neurologically in the hippocampus and in the adjacent entorhinal cortex, is a decisive prerequisite for a successful practical life-management. The ability Spatial Orientation is needed to complete the tasks shown in Figures 2.28–2.30. Solutions are given in Section A.8.[1]

---

[1] Brown, Henry T. (1868). *Five Hundred and Seven Mechanical Movements*: Embracing all those which are most important in dynamics, hydraulics, hydrostatics, pneumatics, steam engines, mill and other gearing ... and including many movements never before published, and several which have only recently come into use. New-York: Coombs & Co.

Fine Motor Skills – Problem 1

Do you see the car standing at the "Start" point of the street? A person who is sitting in the car wants to drive home to the house, which is labeled with the word "Finish". The person wants to drive very carefully along the street, which has red colored borders. Use a pencil to trace the path that the person in the car drives right in the middle of the street to the house. Remain exactly in the middle of the street with the pencil and do not touch the border lines of the street. Try to draw the line without a break.

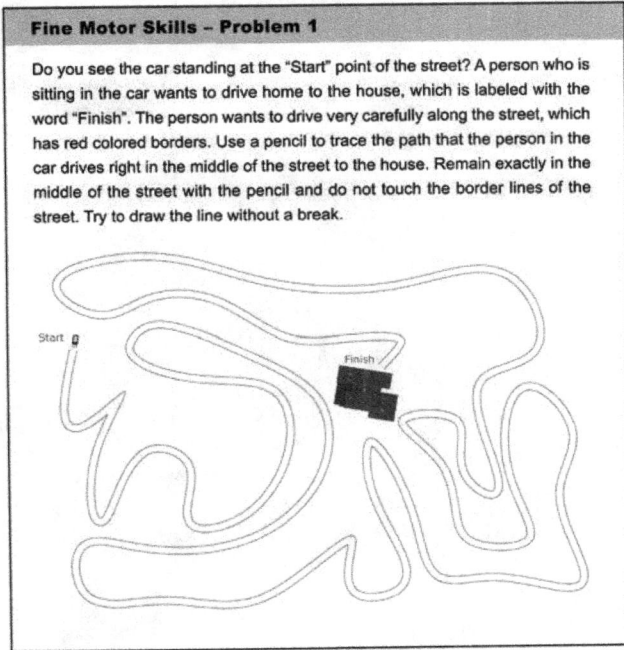

Fig. 2.26.   Problem 1 illustrating *Fine Motor Skills*.

Fine Motor Skills – Problem 2

Wire loop game: Guide the metal loop carefully along the curved wire without touching the wire. If you touch the wire with the loop, then a light will go on or a sound will be heard.

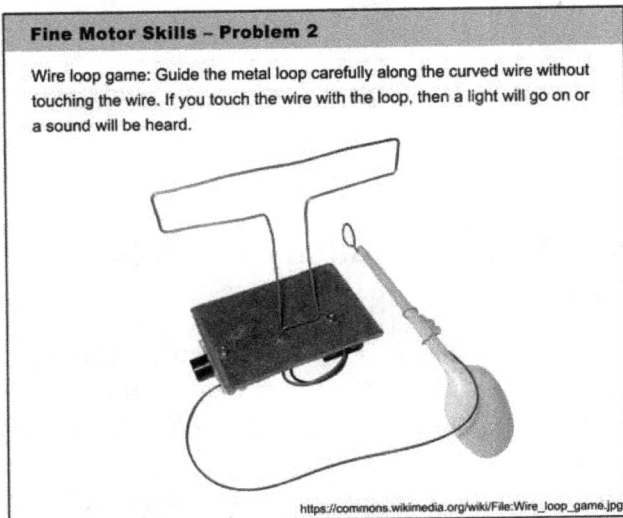

https://commons.wikimedia.org/wiki/File:Wire_loop_game.jpg

Fig. 2.27.   Problem 2 illustrating *Fine Motor Skills*.

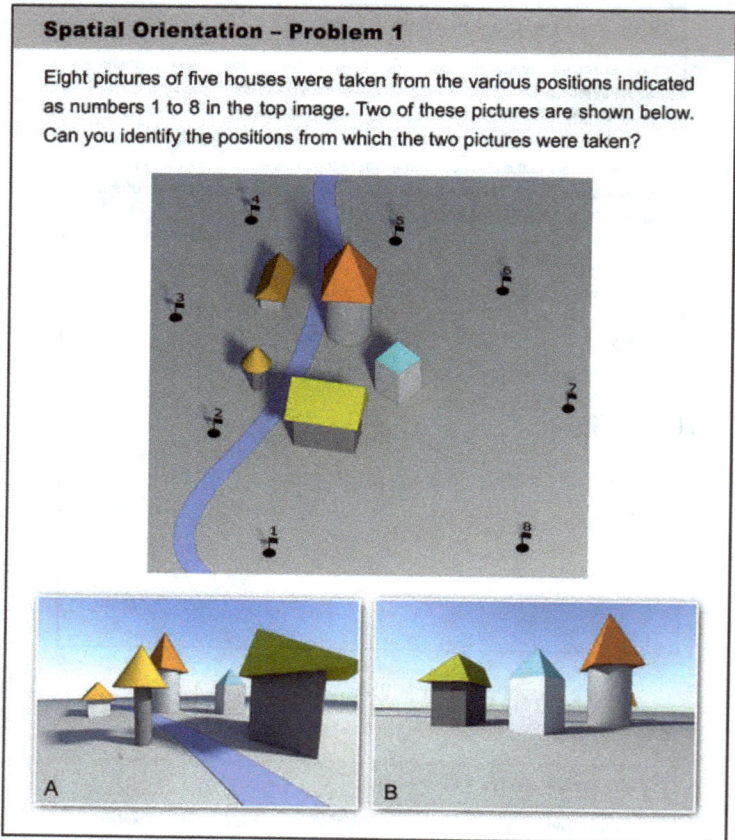

Fig. 2.28. Problem 1 illustrating the ability *Spatial Orientation*.

2.14 Crosslinks to Mathematics and Writing/Reading

The model of the eight basic elements of spatial thinking reflects the present competence models of geometry, takes into account current neurological findings, shows clear links to the five levels of visual perception of Marianne Frostig, and integrates the relevant factor-related results of spatial ability research. Cross connections of spatial ability and visual perception to other abilities and skills, such as

Spatial Orientation – Problem 2

Imagine that you are flying in a helicopter once around a small village. Here is a map of the village and the flight path of the helicopter around the village, indicating the direction of the flight (clockwise) and the starting point.

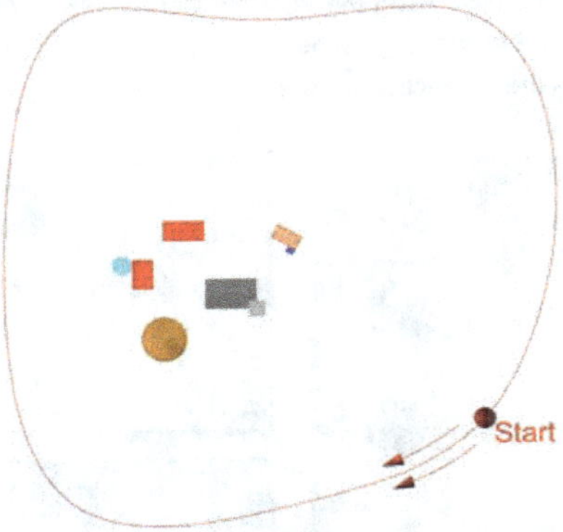

Start

During the flight, a photographer in the helicopter takes three pictures of the village. Unfortunately, the order of the pictures gets mixed up.
Your task is to arrange the pictures A, B, and C in chronological order

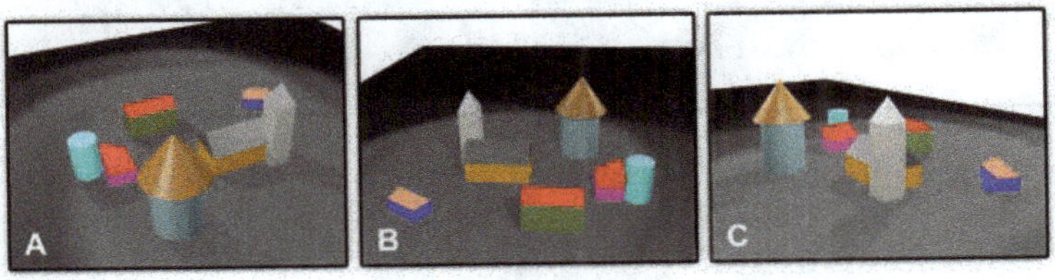

Fig. 2.29.   Problem 2 illustrating the ability *Spatial Orientation*.

mathematical performance or the ability to write and read are examined in various studies. In 2015, Zachary Hawes and colleagues from the University of Toronto discovered that students with well-developed spatial thinking are better in mathematics than others. This fact is most evident on missing term problems, such as $5 + \_\_ = 7$. It seems that if children are good in spatial thinking, they are also able to approach this kind of mathematical problem through spatially reorganizing the problems, for example, $5 + \_\_ = 7$ becomes $\_\_ = 7 - 5$. This – according to Hawes – is an important finding, as it is the first

---

Spatial Orientation – Problem 3

The image at the left shows the top view of a parking area where nine cars are parked. In addition to the cars and the trees, there is a large orange ball at the bottom of the image. A photographer takes two pictures of the parking area. For each of the pictures at the right, he stands next to one of the nine cars and focuses on the big sphere. Determine from which car the photographer took the respective picture.

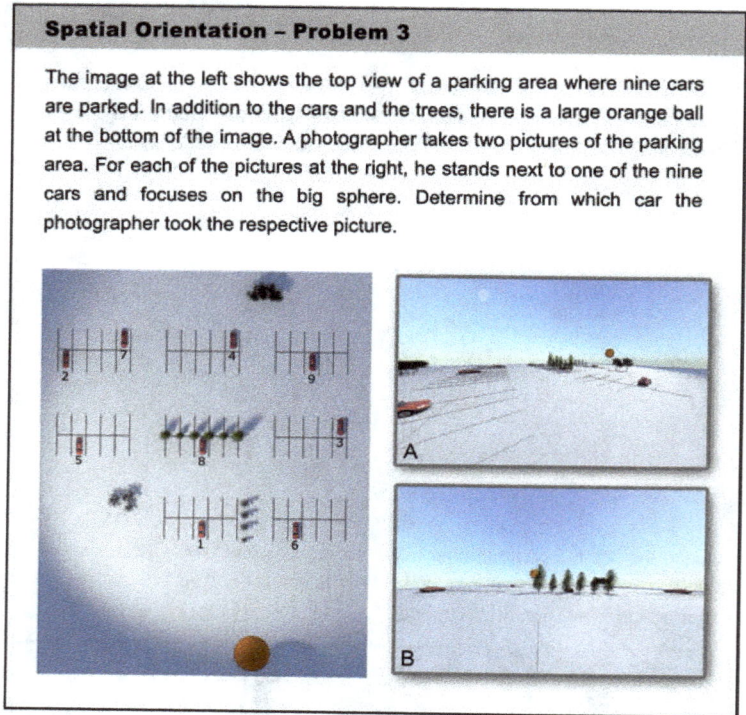

Fig. 2.30.   Problem 3 illustrating the ability *Spatial Orientation*.

empirical study to demonstrate the potential of spatial training as a means to facilitate calculation performance.[2]

Expected examples of the basic elements of spatial thinking with writing and reading are formulated in Table 2.1. Here, we exemplify the sometimes-complex and close links between spatial-visual abilities and other basic abilities as well as the skills of humans on the connections of spatial thinking to writing and reading.

---

[2] Hawes, Z., Tepylo, D., & Moss, J. (2015). Developing spatial thinking: Implications for early mathematics education In B. Davis and Spatial Reasoning Study Group (Eds.), *Spatial Reasoning in the Early Years: Principles, Assertions and Speculations* (pp. 29–44). New York, NY: Routledge.

Table 2.1.   Relationships of the eight basic elements of spatial thinking with writing and reading.

Basic elements	Exemplary relations to writing and reading
Visualization	Recognize letters even if they are printed on colored paper or paper with patterns (e.g., on posters, or graph paper)
Form constancy	Among other things for recognizing letters in general and letters in different styles and fonts
Position in space	Recognize order of letters in words
Transformation in space	Among other things important for the distinction between the letters b and d, m and w, or p and q
Object combination	Recognize word blocks (e.g., abbreviations, connections)
Dynamics	Read moving tickers (e.g., on screens of digital devices, railway stations, cinema, public transport)
Fine motor skills	Basic precondition for the activity of writing
Spatial orientation	Read mirrored written words correctly

A. Appendix — Solutions of the Problems

A.1 *Solution of the Visualization Problems*

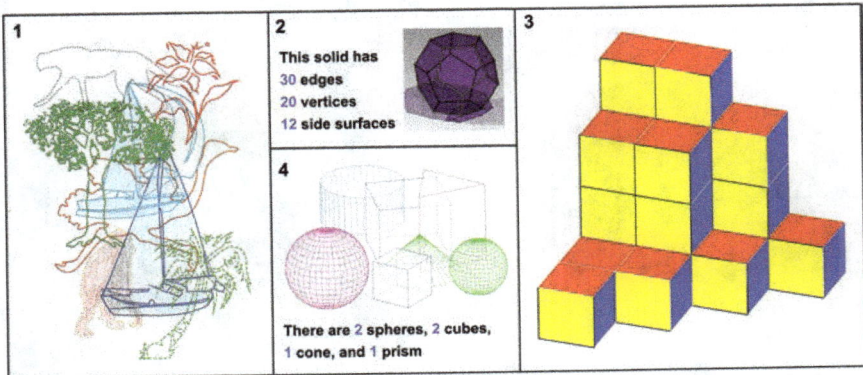

Fig. A.1.   Solution of the four problems illustrating the ability Visualization.

A.2 *Solution of the Form Constancy Problems*

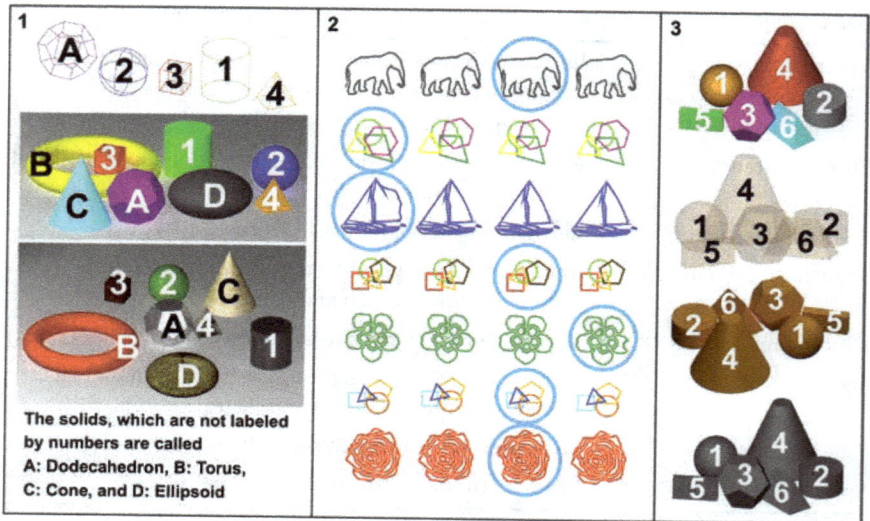

The solids, which are not labeled
by numbers are called
A: Dodecahedron, B: Torus,
C: Cone, and D: Ellipsoid

Fig. A.2.   Solution of the three problems illustrating the ability Form Constancy.

A.3 *Solution of the Position-in-Space Problems*

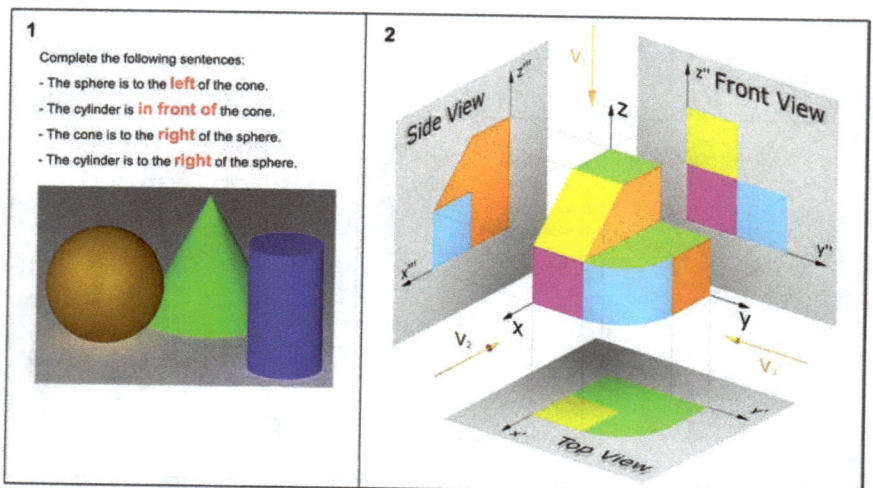

Complete the following sentences:
- The sphere is to the left of the cone.
- The cylinder is in front of the cone.
- The cone is to the right of the sphere.
- The cylinder is to the right of the sphere.

Fig. A.3.   Solution of the two problems illustrating the ability Position in Space.

A.4 *Solution of the Transformation-in-Space Problems*

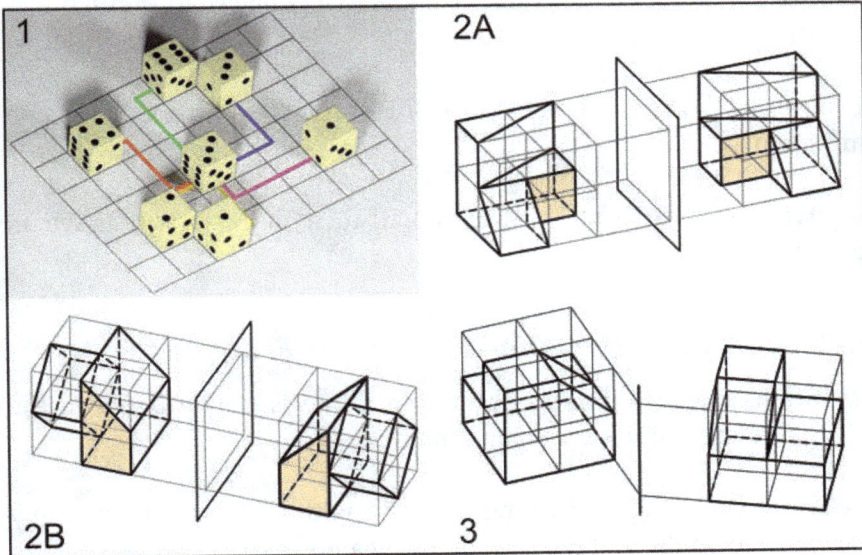

Fig. A.4. Solution of the three problems illustrating the ability Transformation in Space.

A.5 *Solution of the Object-Combination Problems*

Problem 1: The parts that can be used to complete the cube are **B** and **D**.

Problem 2: See Figure below.

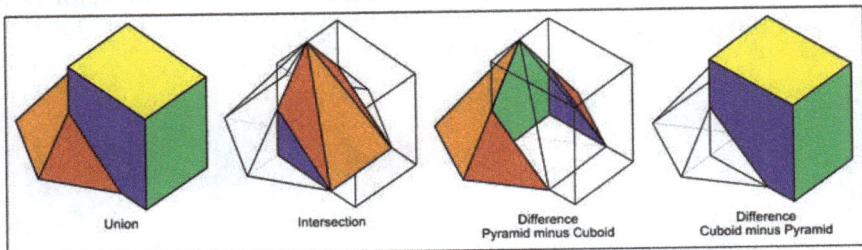

Fig. A.5. Solution of problem 2 illustrating the ability Object Combination.

A.6 *Solution of the Dynamics Problems*

Problem 1: Anne took the photos in the following order: **Picture 3 — Picture 1 — Picture 2**.

Problem 2: Answer to question 1: If B starts moving, A will move in direction **1**.

Answer to question 2: After a half-turn of B, A will move in direction **2**.

A.7 *Solution of the Fine-Motor-Skills Problems*

Problem 1: The solution is correct if a line is drawn between the given red lines without touching them.

Problem 2: When performing this task with the actual device, any error will be indicated by the light or by a noise.

A.8 *Solution of the Spatial-Orientation Problems*

Problem 1: Picture A was taken from **position 2** and the picture B was taken from **position 7**.

Problem 2: The photographer took the **picture C** first, then **picture A** and finally **picture B**.

Problem 3: For picture A, the photographer was next to **car 2**, for the picture B next to **car 4**.

Chapter 3

Spatial Representation

Now that we have seen how important spatial perception is for our lives and how the brain processes spatial sensations, we will ask how these spatial impressions can be conveyed from person to person. One then can ask how a spatial scene can be represented on a piece of paper so that the viewer can correctly interpret the scene. This question has played an important role in art history, since it has long been regarded as the task of a draftsman or painter to present on paper a real or imagined scene in a realistic way. This is by no means an easy task. After all, one has to convince the viewer to imagine a three-dimensional scene by looking at a two-dimensional flat piece of paper or canvas. An understanding of how this can be achieved has developed slowly and gradually over a period of several centuries.

3.1 Parallel Projection

The easiest way to obtain a two-dimensional representation of a three-dimensional object is by a so-called *parallel projection*, which is still used today in technical drawing. We illustrate this method in Figure 3.1 using a wireframe model of a cube. Let us assume that the model is illuminated by parallel light rays from a distant source such as the Sun. The model casts an image onto the flat screen, this is the *projection plane* (the vertical plane at the left in Figure 3.1, also called *image plane*). The image is the two-dimensional picture of the cube.

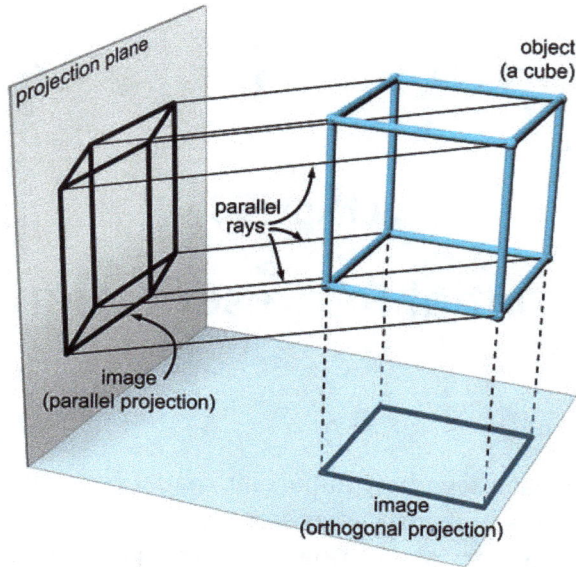

Fig. 3.1. Parallel projection of a wire-frame model of a cube.

In the illustration, we indicated the parallel light rays with oblique lines connecting the corners of the cube with the corresponding points on the two-dimensional projection. Note that the appearance of a cube in the projection can be very different, depending on the angle of incidence of the light rays and the orientation of the cube in space. For example, the projection onto the ground through vertical light rays from above is simply a square, which is also shown in Figure 3.1. A parallel projection, where the projecting rays are perpendicular to the projection surface, is called an *orthogonal projection*.

In Figure 3.2, we show the image of the cube on the projection plane, which is how one would actually see this image when one looks straight on the vertical projection plane in Figure 3.1. In this arrangement, two sides of the cube are parallel to the projection plane, and therefore, their shadow image, shown in Figure 3.2, is a square the same size as the side of the cube.

Note that the perception of this image has a certain ambiguity. One can see it in two ways, depending on which of the "interior

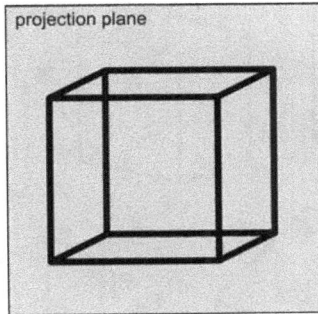

Fig. 3.2.   Two-dimensional image of the cube on the projection plane.

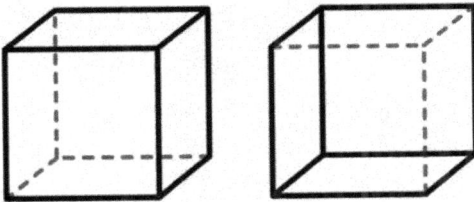

Fig. 3.3.   Two possible interpretations of Figure 3.2.

corners" in the image one perceives to be at the front. Figure 3.3 shows the two possible interpretations. Here we use dashed lines to indicate the edges at the back, which are further away from the viewer.

If you rotate the cube in space, eventually all sides in the parallel projection get distorted into parallelograms. All the images of cubes in Figure 3.4 can be interpreted as shadows of wire-frame models cast onto a flat surface (projection plane). Depending on the angle between the projection rays and the projection plane, some edges of the cube's image can become greatly elongated while others appear rather short. Therefore, parallel projections do not always give realistically-looking images.

While some of the views obtained via a parallel projection are acceptable, one cannot claim that a parallel projection gives a realistic impression of the three-dimensional world on a two-dimensional surface. If the projection is made from the side and the rays hit the projection plane at a very oblique angle, as in Figure 3.5, the image is

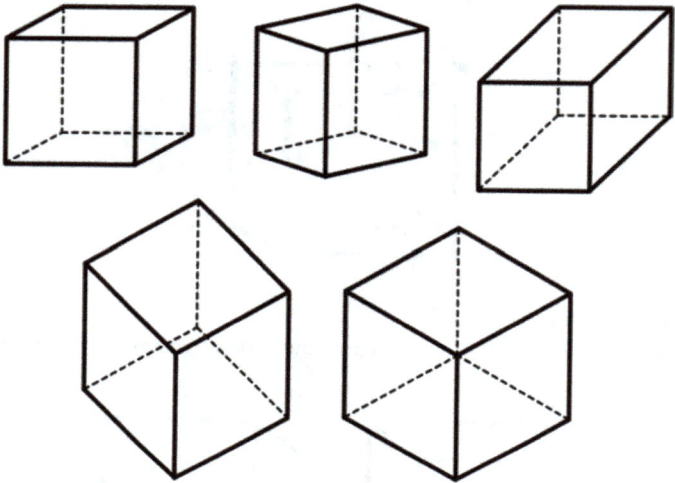

Fig. 3.4.   Various views of a cube, obtained through parallel projection.

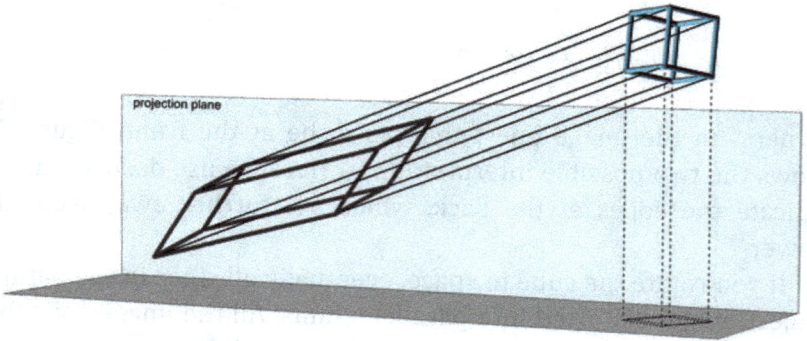

Fig. 3.5.   Projection with very obliquely incident light.

drawn in a very elongated fashion and the object is difficult, if not impossible, to recognize.

Figure 3.6(a) shows the result of the projection in Figure 3.5. The result is barely recognizable as a silhouette of a cube. But if one looks at the image from about the same direction from which the projection was made (that is, obliquely from the top right), one can see a clear, undistorted image again, as in Figure 3.6(b).

projection plane

(a)   (b)

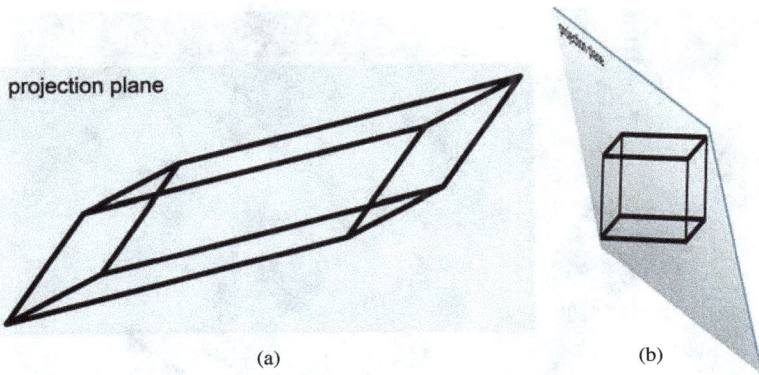

Fig. 3.6.   Distorted image of a cube generated by an oblique parallel projection as in Figure 3.5.

A distorted projection where the viewer is required to occupy a specific vantage point in order to see the image undistorted, is called anamorphosis. The German renaissance painter Hans Holbein the Younger (ca. 1497–1543) demonstrated anamorphosis in his famous image "Die Gesandten" ("The Ambassadors"), shown in Figure 3.7. Among the many symbols contained in this image, the strange shape in the foreground is particularly striking. It is a distorted image of a skull, clearly a reference to mortality. Holbein created this distortion by a sophisticated method of "oblique projection", which is quite similar to the method used for Figure 3.6. We can also look at both images from approximately the same direction to get a more realistic view of the object. In Figure 3.8, we show Holbein's image from the side (from a vantage point to the right of the image and close to the image plane). Here, we can recognize the nature of the object in the foreground, further highlighted by the enlarged view in Figure 3.8.

The main problem that renders the method of parallel projection unsuitable for realistic imaging is that distant objects are shown in the same size as nearby ones. This makes it sometimes difficult to interpret a scene. We illustrate this problem in Figure 3.9(a), where the difference in height of the location of the two pawns cannot be seen, which leads to profound irritation. The perspective illustration on the right brings clarity as to how to interpret this scene.

Fig. 3.7. "The Ambassadors" (1533) by Hans Holbein the Younger, The National Gallery, London.

The same effect can be used to obtain three-dimensional interpretations of the impossible triangle and the Penrose staircase shown earlier (see Figure 1.10). The Dutch graphic artist Maurits Cornelis Escher (1898–1972) made use of this illusion in several of his lithographs, for example in *Ascending and Descending* (1960) or *Waterfall* (1961).

3.2 Central Projection and Perspective

An important breakthrough in the art of creating realistic representations of three-dimensional objects and scenes was the invention of *perspectivity* in medieval European art. This invention did not happen

Fig. 3.8. Holbein's image from the side with an undistorted view of the skull.

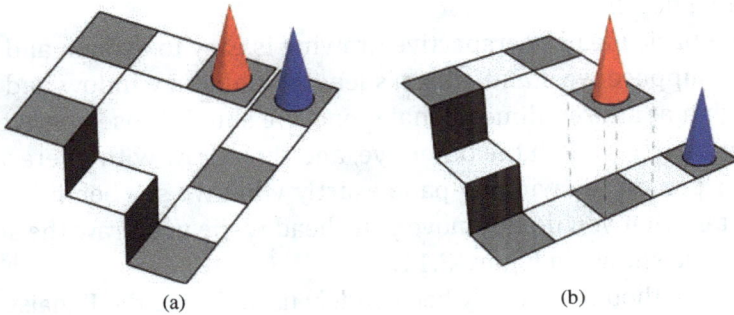

(a)   (b)

Fig. 3.9. Parallel projection of a scene (a) and perspective view from another vantage point (b).

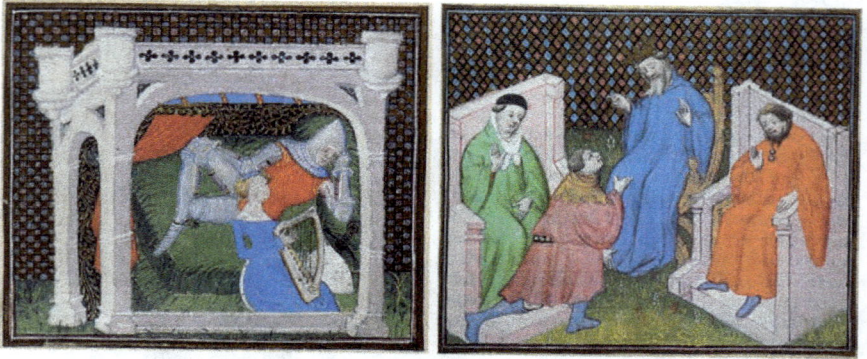

Fig. 3.10.   Incorrect use of perspective. Codex 2537, ÖNB Vienna, fol.103 and 268, early 15th century.

spontaneously, but took place in several incremental steps. The artists understood very early on that objects farther away have to be drawn smaller and closer objects have to be drawn larger. Our brain automatically interprets the size reduction of an object as a greater distance in the third dimension. But from there, the way forward is less clear. It is clear that horizontal lines leading away from the viewer have to be somehow drawn obliquely, but to what extent, is not easy to say. Figure 3.10 shows an example of book illustrations by an unknown artist from the early 15th century, in which the artist tried to place people in a three-dimensional environment. However, the use of perspective appears inconsistent and does not really give the illusion of spatial depth.

The basic idea of perspective drawing is easy to explain and easy to use. Suppose we stand at arm's length behind a window and look through it at a three-dimensional scene. We should close one eye and look at the scene with the other eye, and then draw with a (erasable) marker pen on the window-pane exactly what we see behind it. It is important that you do not move your head while you draw the scene, as we have shown in Figure 3.11.

This method has already been widely used during the Renaissance. Figure 3.12 shows an illustration by famous German artist Albrecht Dürer (1471–1528), where the artist fixes the eye with a visor and

Fig. 3.11.   Drawing on a window. Mapping a three-dimensional scene onto a two-dimensional surface.

paints an outline of the figure to be portrayed on a glass plate clamped into a frame. The title of the illustration is "A draughtsman establishing details for a portrait, using a perspective apparatus for drawing on glass".

A schematic representation of this situation is shown in Figure 3.13. Here, the "window" shown in Dürer's illustration corresponds to the xz-plane. The three-dimensional scene is represented by a simple geometric object — a cube. The two-dimensional image of the three-dimensional object is created in the xz-plane. Its points on the projection plane are created by intersecting the lines of sight from the eye to the points of the object.

Figure 3.14 shows the result, a two-dimensional view of the cube obtained from the projection method described in Figure 3.13. It gives a very realistic impression of the three-dimensional object, quite similar to the impression that you get when you take a photograph of the cube.

Fig. 3.12.   Illustration by Albrecht Dürer (1471–1528), for a "Course in the Art of Drawing", published in Nuremberg 1525.

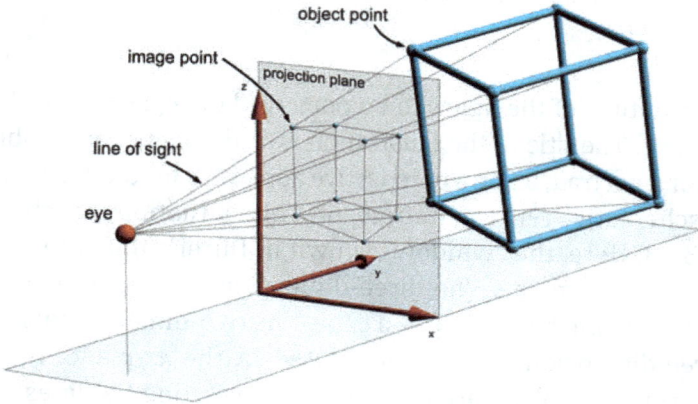

Fig. 3.13.   Creating a two-dimensional image of a three-dimensional scene.

This method of creating a perspective view is called "central projection". The idea behind this method is very old and had already been described by the Greek mathematician Euclid, who lived around 300 BCE, and is famous for his monumental multi-volume work *Elements,*

Fig. 3.14. Perspective image of a cube (as it appears in the projection plane of Figure 3.13).

which is a comprehensive treatment of the mathematics of his time. Euclid also wrote a book entitled *Optics* (Greek: Ὀπτικά) that dealt with the geometry of vision. He postulated that in order to see an object, the eye sends out visual rays, which are straight lines connecting the eye with points on the object, very much like the lines shown in Figure 3.13. Indeed, there seem to have been systematic efforts to develop a system of perspective in ancient Greek art, especially for the purpose of designing a theater stage. The name "scenography" denotes the art of scenic painting, especially for the purpose of creating the illusion of depth on a stage. This art had been developed to a somewhat limited extent prior to Euclid's work by Agatharchus of Samos, who lived in the 5th century BCE, and is said to have used graphical perspective in his work. Around 465 BCE, the tragedian Aeschylus (525–456 BCE) commissioned Agatharchus to set the stage for one of his dramatic works. The result was undoubtedly a great success, because of the lifelike and convincing visual effect in the representation of spatial depth, which had been given great praise. Some later commentators and historians even regarded Agatharchus as the inventor of central projection. However, very little of the visual art from that time has been preserved (except for some vase paintings and murals), and so we cannot really know which methods were used in antiquity to achieve a spatial effect in painting.

In any case, accurate perspective methods of drawing in the modern sense are an invention of the early Renaissance. Italian architect Filippo Brunelleschi (1377–1446), who designed the magnificent dome of the Florence cathedral, is usually credited to have developed the first systematic and precise method of using perspective in drawing — the well-known method of using one or more "vanishing points". This method helps us to create a correct perspective view, even if there is no glass pane or similar device (as in Figure 3.12) available to copy a three-dimensional scene.

3.3 Perspective Drawing Using Vanishing Points

We shall begin using the method with one vanishing point, which is quite simple and works well for creating a perspective view of special types of spatial scenes. Let us consider the image obtained by the projection method shown in Figure 3.13, when applied to a model room. The setup is shown in Figure 3.15, and the result in Figure 3.16 shows the image on the projection plane, that is, the view obtained when looking straight into the room.

Let us focus on the lines that are perpendicular to the image plane (the long edges of the room). In the three-dimensional scene (Figure 3.15), these are parallel lines leading away from the viewer.

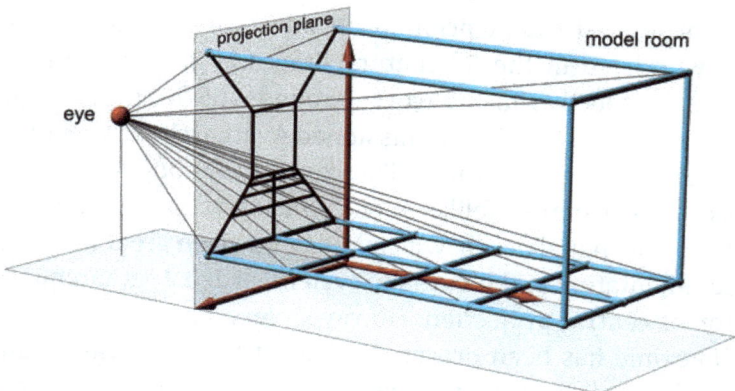

Fig. 3.15.   Applying a central projection to a wire-frame model of a room.

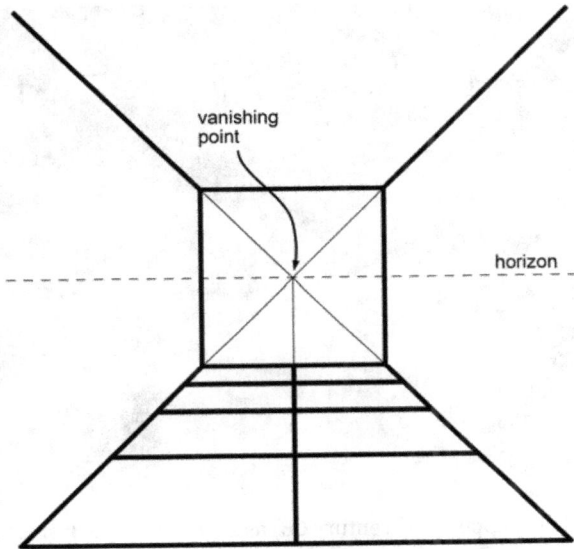

Fig. 3.16. Perspective image of the room on the projection plane of Figure 3.15.

In the two-dimensional image, Figure 3.16, these lines are no longer parallel, instead they appear to converge to a particular point in the center of the image. This point is called the "vanishing point". In everyday life, the vanishing point is where parallel railroad tracks seem to merge at an infinite distance on the horizon.

Among the most noteworthy pieces of art that use this method are those by some of the famous Italian artists such as: Leonardo da Vinci's: (1452–1519) "The Last Supper" (Figure 3.18), and Raphael's (1483–1520) "The School of Athens" (Figure 3.20). In the case of da Vinci's "The Last Supper", which is located on the wall of an empty room in the former church Santa Maria delle Grazie in Milan, Italy, the observer on the other side of the room would have the feeling that the table actually exists with its characters as shown. The vanishing point is just behind Jesus' head, drawing the viewer's attention to the main character (Figure 3.19). It is the first time after many renditions of Last Supper paintings in previous centuries that true depth perception was achieved. One of these earlier versions of the Last Supper is shown in Figure 3.17. About 14 years after

Fig. 3.17. The Last Supper (13th century CE) as a mosaic in the Basilica di San Marco, Venice.

Fig. 3.18. Leonardo da Vinci, The Last Supper (1495–1498), fresco in the Santa Maria delle Grazie, Milan, Italy.

Leonardo da Vinci, Raphael, in his "School of Athens" also uses the method of placing the main characters Plato (left) and Aristotle (right) together with the architecture's vanishing point at the center of the image (Figure 3.20).

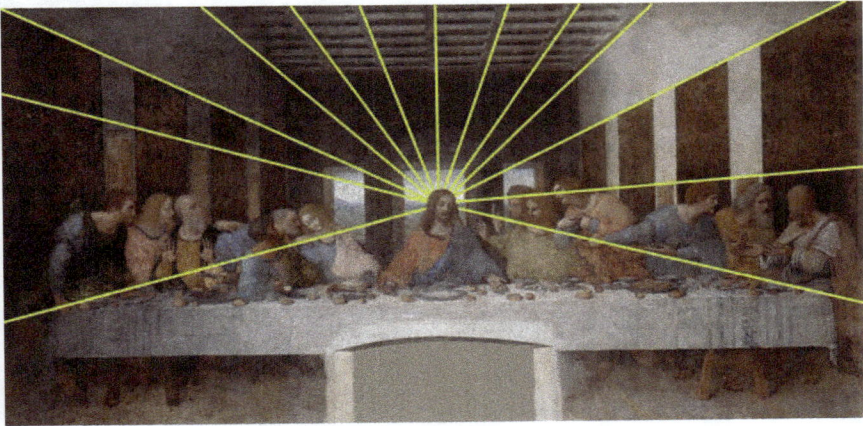

Fig. 3.19.   The Last Supper's vanishing point.

Fig. 3.20.   Raphael (1483–1520), School of Athens (1509–1511), fresco in the Apostolic Palace, Vatican.

3.4 Two-Point Perspective

Linear perspective becomes more complicated, when more groups of parallel lines are involved.

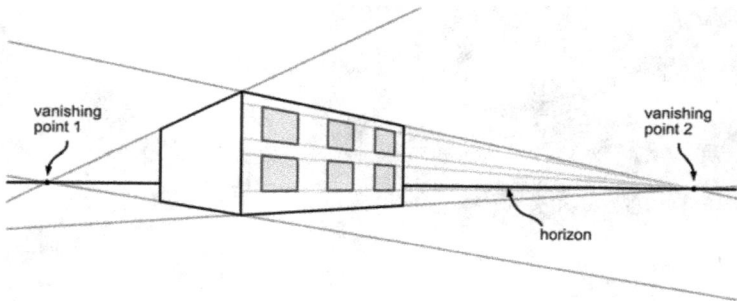

Fig. 3.21.   Two-point perspective.

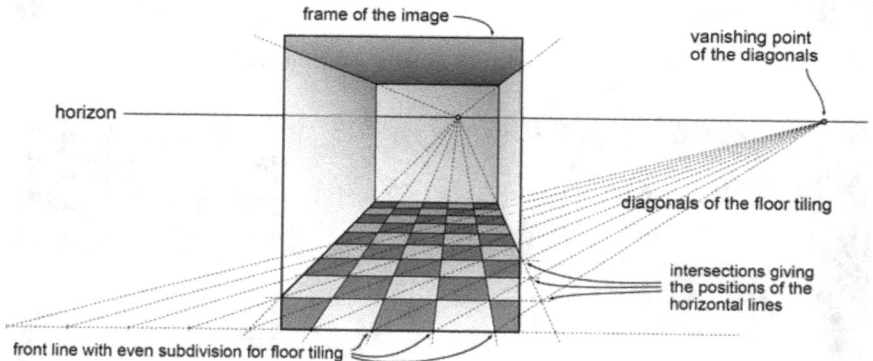

Fig. 3.22.   Illusion of depth in a checkerboard pattern.

Figure 3.21 shows an example of a perspective image with two vanishing points. The two sides of the building force us to consider two groups of parallel lines, which converge to two different vanishing points in the image plane. All vanishing points of the straight lines that are parallel to the ground plane are situated at the horizon. In this case, the horizon is determined by the horizontal line representing the eye-level of the observer.

A brilliant trick using the two-point perspective already enabled Renaissance artists to position objects in a scene so as to provide the convincing illusion of spatial depth. Consider Figure 3.22, showing an image of a room created with the technique using a single vanishing point, as described earlier and as illustrated in Figure 3.15.

The special issue with this picture is that the floor of the room is paved in a checkerboard pattern. In the final image, the horizontal rows of this tiling must, of course, be drawn closer and closer together in order to simulate the increasing distance from the viewer. In order to determine the spacing of the horizontal lines in the checkerboard pattern, one could use the following construction with two vanishing points.

The artist starts by drawing the lines pointing away from the observer using the vanishing point in the middle of the room. These lines start at the horizontal front line of the image, which, for this purpose, is subdivided into parts of equal length. Next the artist creates the first row of squares at the bottom of the image, by drawing the first horizontal line at a position that looks as realistic as possible. The artist proceeds with determining the vanishing point of the diagonals of all the squares, which is shown in Figure 3.22, as the second vanishing point to the right, and far outside the final image. The diagonals will not appear in the final picture, they are just auxiliary lines in the construction. Of particular interest are the intersections of these diagonals with the lower right edge of the room. These intersections are, of course, the corner points of the square tiles laid along this edge. Therefore, they also mark the positions of the horizontal lines that determine the horizontal rows of the checkerboard pattern. The depth of the room is then determined by the number of rows of the floor tiling, in this case, the room is nine rows deep. The appearance of the picture depends on how wide the first row of floor tiles was drawn at the beginning of the construction. If this were drawn very narrowly, the room would appear as if photographed with a telephoto lens, while a large width would give the impression of a wide-angle lens.

3.5 The Pinhole Camera

In the mathematical theory of central projection, it is possible to change the place between the "eye" and image plane, as in Figure 3.23. This figure shows the principle of a "pinhole camera". A pinhole camera (or "camera obscura") is a dark box with a little hole

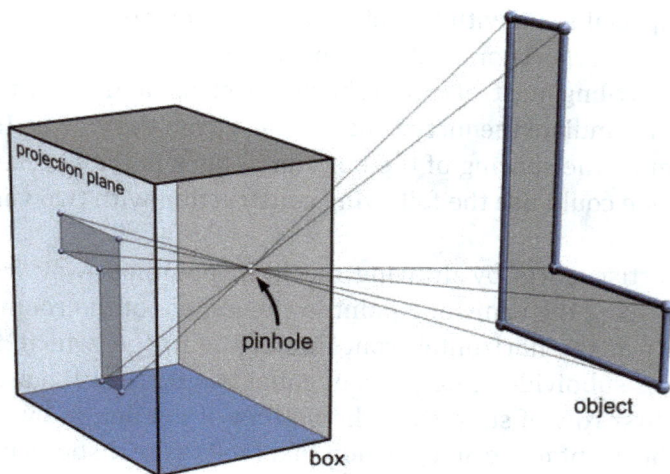

Fig. 3.23.   Pinhole camera.

on one side. Light enters through this hole and creates an upside-down image of the scene on the backside of the box. Using translucent paper as a screen on the back of the box, one can easily view the projected image, provided that a bright light from outside is prevented from shining onto the screen.

The pinhole camera creates a perspective image of the three-dimensional scene in front of the hole. The image, which is upside-down, has no distortion (straight lines are always mapped onto straight lines) and everything is in focus (the depth of field extends almost from the front of the camera to infinity). An example can be seen in Figure 3.24, created with a modern camera, where the lens has been replaced by a 0.2-mm hole in aluminum foil attached to the camera body.

The sharpness is not ideal. A point-like source of light illuminates a small disc on the image plane that is slightly larger than the hole itself (see Figure 3.25). Hence, every point of the real scene is mapped onto a small "disk of confusion", reducing the sharpness of the image. The sharpness and resolution of the image is, therefore, determined by the size of the hole. The image tends to get sharper as the hole gets smaller, but when the hole becomes too small, the wave nature of light

Fig. 3.24.   Image of a pinhole camera.

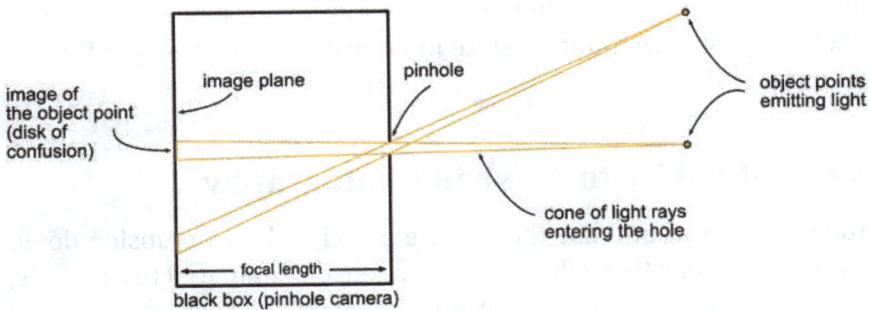

Fig. 3.25.   Disk of confusion: Points have unsharp images.

causes diffraction, thus, blurring the image once again. Moreover, as less light enters through a smaller hole, the picture will soon become too dark, especially if one makes the hole far too small. While a circular hole provides the sharpest images, the shape of the hole is not that critical. People have long been puzzled by the fact, that the image of the Sun is still circular, even if it shines through a rectangular peep hole. But the shape of the hole just influences the shape of the "area of confusion," and might thus, reduce the sharpness, but it does not influence the shape of the Sun's image.

The pinhole-principle of creating images has been known for a long time. People observed that in a dark room, where light enters through a small hole in the wall, an upside-down picture of the outside world is cast on the wall opposite the hole, analogous to the pinhole camera in Figure 3.23. This was described by the Han Chinese philosopher Mozi (墨子, ca. 470–391 BCE) and later by the Greek philosopher Aristotle (384–322 BCE). The phenomenon of the camera obscura was studied in a number of systematic experiments and described mathematically by the Arab physicist Ibn al-Haitham (965–1039 CE) in his book *On the Shape of the Eclipse*. He was the first to explain that the crescent shape that appears in a camera obscura during a partial solar eclipse was actually an image of the Sun, created by light rays travelling along straight lines.

The first photograph ever taken was made with a pinhole camera in 1826, by the French inventor Joseph Nicephore Niépce (1765–1833), who created a method for coating a pewter plate to make it photosensitive. He placed it in a pinhole camera and took a photograph on his country estate in France with an exposure time of 8 hours.

3.6 Optical Distortions in Photography

Today's cameras still use the pinhole principle to drop upside-down images of a scene through a lens onto a sensor. Compared to a pinhole, a lens collects much more light through a much larger hole and still produces images with pinpoint sharpness. The refraction of light in the lens also makes it possible to intervene in the light path in such a way that certain optical distortions arise in the image. But why would that be desirable?

Suppose we consider the following situation. Let's say we are very close to a large house and want to take a picture of it. If it does not fit in the picture, we have to take several photos, as in Figure 3.26. These pictures appear distortion-free through a good lens, and even if we use a standard wide-angle lens, the pictures appear as if they were taken with a pinhole camera. In the middle picture, the facade of the building is parallel to the picture plane, and therefore, appears undistorted. One can see the effects of the perspective in the right and left picture (central projection). The edges of the building converge to a vanishing point at the horizon.

If we want to combine these three images into a single image, we cannot just stitch them together, as this would cause corners to appear on the edges of the house. A smooth transition between the right and the left image of Figure 3.26 can only be achieved, if the originally straight lines become curved, as shown in Figure 3.27.

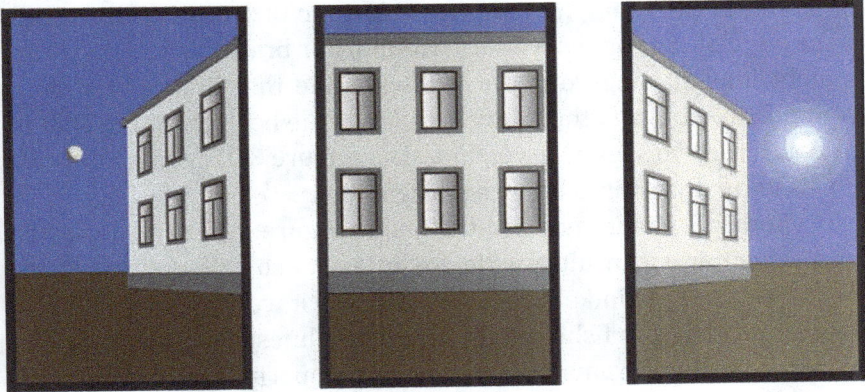

Fig. 3.26.   Several images of a house.

Fig. 3.27.   Cylindrical panorama of the house in Figure 3.26.

This shows a so-called cylindrical panorama, where vertical lines remain vertical, but horizontal lines appear curved (except for the line chosen as the horizon).

A smaller section of the image still gives an approximately-correct impression. In reality, one cannot see the whole building without turning one's head. If we were to extend the roof edge to infinity in both directions, it would connect one vanishing point on the horizon (the one to the right of the Sun) with the other vanishing point exactly opposite it (that is, the one to the left of the Moon). If we look in both directions at the same time, we would have to turn around. If we look at the edge of the roof as we turn around, our brain recognizes it as a straight line, although our eye follows a line that first moves away from the horizon, and then lowers towards the horizon again. That is, in reality, the eye sees a curved line, as in Figure 3.27, while our brain lets us know that it is actually a straight line.

Cylindrical panoramas like the one in Figure 3.27 can be created with a computer. An ultra-wide-angle lens, also called *fisheye* lens, creates a different kind of distortion, because it would also bend vertical lines. With a true fisheye lens, all straight lines that go through the center of the image remain straight, while all peripheral straight lines get curved (barrel distortion). This is the price for achieving viewing angles of 180° or more, which would be impossible with a pinhole camera.

Figures 3.26 and 3.27 also give us a better understanding of the puzzle with the wrong tilt of the Moon mentioned in the first chapter (see Section 1.2, Figure 1.4). A straight beam of light between the Sun and the Moon would run along the edge of the roof, which is pointing up from the Sun in the photograph to the right, and pointing down towards the Moon on the other side (left photograph). Therefore, the Moon appears in the left photograph as if it were illuminated from above, although the Sun is, approximately, the same height in the sky as the Moon.

Chapter 4

The Geometry on a Sphere

The geometry that can be done on the surface of the sphere is often referred to as *spherical geometry* as opposed to the geometry in a plane, which is called plane geometry, or Euclidean geometry. Although there are some similarities between spherical and plane geometry, there are also important differences as we will see as we navigate this very interesting topic. Spherical geometry is a special case of Riemannian geometry, named after the German mathematician Bernhard Riemann (1826–1866) who departed from Euclidean geometry to develop a geometry of various surfaces in higher dimensions.

4.1 History of Spherical Geometry

Earlier explorations in the field of non-Euclidean geometry should be attributed to the Hungarian mathematician Farkas Bolyai (1775–1856) and to the German mathematician Carl Friedrich Gauss (1777–1855). In 1818, Gauss carried out a geodetic survey of the Kingdom of Hanover, which aroused his interest in (non-Euclidean) differential geometry and led him to investigate the curvature of surfaces. He lectured extensively on these topics as evidenced by an announcement shown in Figure 4.1. In his *Theorema Egregium* (Latin for "Remarkable Theorem") he showed how a curvature can be determined just by measuring distances and angles on the surface. A special case of this theorem will be presented later in this chapter.

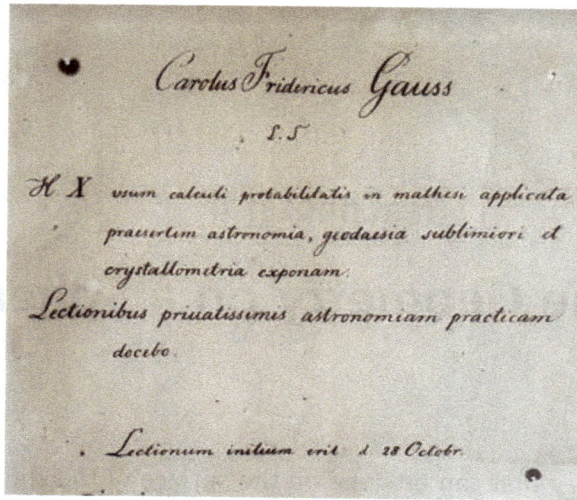

Fig. 4.1.   Announcement of a lecture of Gauss in his own handwriting.

Note: A translation of Figure 4.1 is as follows: Carl Friedrich Gauss: At 10 o'clock I will explain the use of probability calculus in applied mathematics, especially astronomy, advanced geodesy, and crystalometry.
I will teach practical astronomy in most sessions.
The first lecture will be on October 28.

It was Gauss who urged his student Riemann in 1853 to write a treatise on the foundations of (non-Euclidean) geometry. Riemann's paper, entitled "On the hypotheses which underlie geometry," was published posthumously in 1868, and is still regarded as one of the most important contributions to geometry.

Important practical applications of spherical geometry can be found in geodesy, geography and navigation as well as in astronomy. And it is its generalization, non-Euclidean geometry, that shapes our physical picture of the space–time continuum and the large-scale geometric structure of the universe.

4.2 Basic Properties of the Sphere: Surface and Volume

We shall start with some important basic properties of the sphere. These results were first published by the Greek mathematician and scientist Archimedes of Syracuse (approximately 287–212 BCE) using ingenious methods that anticipated much of modern analysis. He published his results in 225 BCE in a two-volume work entitled *On the Sphere and Cylinder*.

The circumference of a sphere is typically measured along a "great circle" on the sphere's surface — which is a circle whose center coincides with the center of the sphere and which has the same radius as the sphere. Therefore, if r is the radius of the sphere, the circumference of the sphere is $2\pi r$. The surface area of the sphere is, according to Archimedes' well-known formula, given by $4\pi r^2$, which equals the area of four great circles of the sphere. We can also interpret this area as the product of the diameter (or $2r$) and the circumference of a great circle. Symbolically, that would be: $(2r) \cdot (2\pi r) = 4\pi r^2$.

We also note that the surface of the sphere is the same as the surface of a cylinder whose radius at the base and whose height are both the same as the radius of the sphere (see Figure 4.2).

Similarly, the volume of the sphere is exactly 4 times the volume of a cone whose radius at the base and whose height are both the same as the radius of the sphere. Symbolically, the formula for the volume of a sphere is $\frac{4}{3}r^3$. Archimedes was the first to compare the volumes of a sphere, a cylinder, and a cone. Figure 4.3 shows a sphere, a cylinder with diameter and height both given by the sphere's diameter $2r$, and a double-cone with the same data, namely, the diameter and height of the double-cone are equal to the diameter of the sphere.

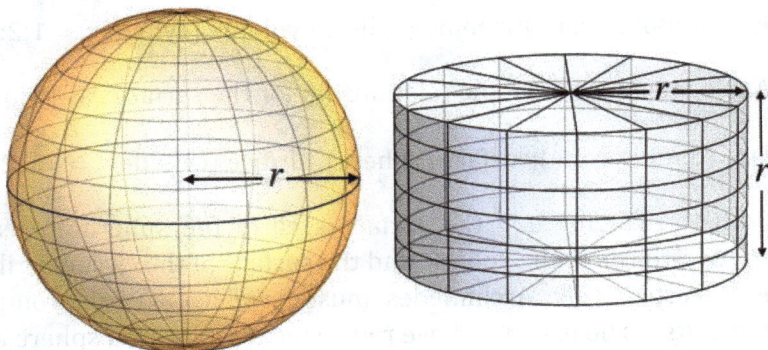

Fig. 4.2.   Sphere and cylinder with the same surface area $4\pi r^2$.

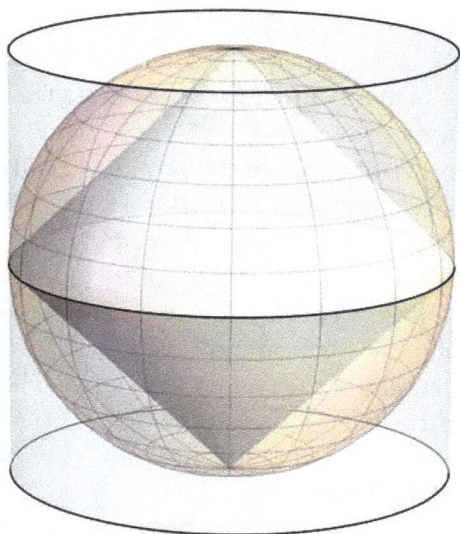

Fig. 4.3.   Comparing cone, sphere, and cylinder.

Let's recall how to compute the volumes of these shapes. The volume of a cone is equal to the area of base (πr^2) times one-third the height, r, which in this case gives us $\frac{1}{3}\pi r^3$, and for the double cone, we need to take twice this quantity to get $\frac{2}{3}\pi r^3$. The formula for the volume of a sphere, as mentioned earlier, is $\frac{4}{3}\pi r^3$. And the volume of a cylinder is equal to the area of the base (πr^2) times the height $2r$, which is $2\pi r^3 = \frac{6}{3}\pi r^3$. This then provides a beautiful resulting relationship that the ratio between the volumes of these solids is:

volume (double-cone):volume (sphere):volume (cylinder) $= 1{:}2{:}3$.

Analogously, we have the relationship for the surfaces as follows:

surface (double-cone):surface (sphere):surface (cylinder) $= \sqrt{2}{:}2{:}3$.

Both the volume *and* the surface area of the sphere are each exactly two-thirds of the volume and the surface of the circumscribed cylinder, respectively. Archimedes must have been very proud of these results, as he is said to have requested a sculptured sphere and cylinder placed on his tomb.

4.3 Great Circles

As we have seen, great circles are important for the interpretation of the formulas for the surface and volume of a sphere. In fact, they form the very foundation of spherical geometry, because a great circle on the sphere is the analog of a straight line in plane geometry. In other words, just as we work with straight lines on a plane, we work with great circle lines on the surface of a sphere. Recall that a great circle is defined as the intersection of a plane through the center of the sphere with the sphere, as we show in Figure 4.4. It is essentially a circle on the sphere whose center coincides with the center C of the sphere. Among all circles that one can draw on the surface of the sphere, a great circle is the largest and shares the radius of the sphere. A natural example of a great circle on Earth is the equator. All meridians (lines of longitude) are halves of great circles connecting the North Pole and the South Pole. The circles of latitude other than the equator are smaller circles, because they lie in planes that do not contain the center of the Earth, and are not used in spherical geometry.

Every great circle cuts the sphere in two halves — for example, the equator bisects Earth into a Northern Hemisphere and a Southern Hemisphere. A great circle that contains a point P also contains the *antipodal point P'*, which is exactly diametrically on other side of the sphere, as we can see in Figure 4.4. Two points on the sphere are antipodal, if they are the endpoints of a diameter of the sphere.

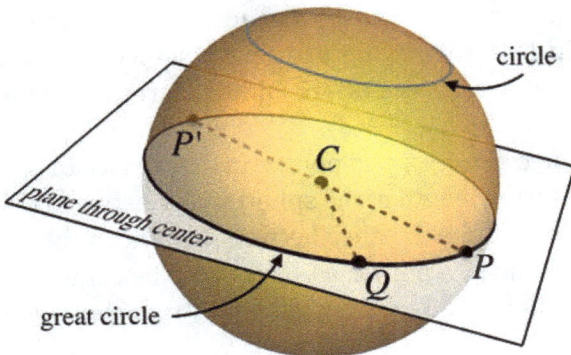

Fig. 4.4. Great circle.

4.4 Great Circles as Straight Lines on the Sphere

Remember, that any three non-collinear points determine a unique plane. Given any two points on the sphere that are not the endpoints of a diameter, a unique plane is defined by these two points and the center of the sphere as a third point. The intersection of this plane with the sphere is the unique great circle through the two given points. We, therefore, conclude that through any two points on a sphere (excluding the endpoints of a diameter) we can draw one and only one great circle. This is analogous to two points in a plane determining a unique line.

Thus, two antipodal points divide this great circle into two arcs, one is longer and one is shorter than a semi-circle of the great circle. The shorter arc will be referred to as the *spherical arc* or *great circle arc* joining these two points. The spherical arc is actually the shortest distance between these two points on the sphere. This would make it analogous to a segment of a straight line in a plane, which is also the shortest distance between two points in a plane. In mathematics, the curves that make the shortest connection between two points in a surface are called the *geodesics*. Great circles are the geodesics of spheres.

4.5 Measuring Distances and Angles

As we mentioned in Chapter 1, it is sometimes questioned why an airplane flight from Europe to Canada typically goes north over Greenland. In fact, this is the shortest distance — the great circle route — following the circular arc joining starting point and destination (see Figure 4.5).

The distance between two points on the surface of the sphere can be measured by the length of the spherical arc between the points. We could also describe the distance between points on the sphere by the angle between the two points, as seen from the center of the sphere. In Figure 4.4, this would be the angle $\angle QCP$ between the dashed lines with vertex at C.

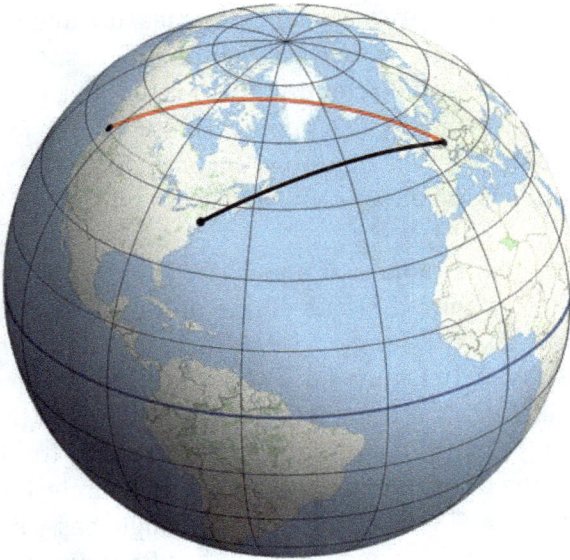

Fig. 4.5. Spherical arc between Frankfurt and Vancouver (red) and between Frankfurt and New York (black).

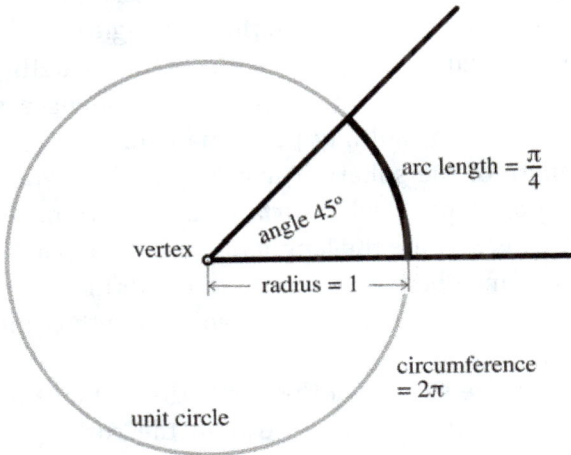

arc length = $\frac{\pi}{4}$

angle 45°

vertex

radius = 1

circumference = 2π

unit circle

Fig. 4.6. Angle in degrees and as an arc length.

In everyday life, we are accustomed to measure angles in degrees. For example, the angle shown in Figure 4.6 has 45°. But going forward, we shall use the scientific method of measuring angles in terms of *radians*, which is sometimes more useful for the kinds of measurements we will be doing. In particular, we will need to understand radians, when we come to discuss the area of triangles on the sphere.

In order to define radians as a measure for an angle, we need the unit circle (a circle with a radius of one-unit length) drawn with its center at the vertex of the angle. The given angle cuts an arc out of the unit circle, as shown in Figure 4.6. The scientific measure of the angle is just the length of this arc and it is called the *radian* of the angle. For example, an angle of 45° cuts $\frac{1}{8}$ out of the unit circle. Its arc length is, therefore, $\frac{1}{8}$ of the circumference (2π) of the unit circle, hence, the arc length of 45° is $\frac{2\pi}{8} = \frac{\pi}{4}$. We then say that this angle has $\frac{\pi}{4}$ radians.

We notice that the full angle of 360° has 2π radians, because the arc length of the full angle is the circumference of the unit circle. Dividing 360° by 2π ($\approx$6.2832) gives the number of degrees per radian, which is about 57.296°. This angle with a size of one radian, together with a few other very common angles, is shown in Figure 4.7.

The big advantage of measuring angles in terms of radians is that it is easy to determine the arc length that the angle cuts out of a circle with an arbitrary radius. That is, to find the arc length joining two points on the surface of a sphere (obviously, along a great circle route), one just has to multiply the radians of the arc measure by the radius length of the sphere. For example, if we know the angle between two points on Earth's surface, as seen from the center of the Earth, we can compute the length of the great circle arc (which is the shortest distance between these two points) by multiplying the angle's radian measure with Earth's radius length, which is about 6,371 km or 3,958 miles.

As we live on the surface of the Earth, the distance between two points seems to be more accessible than the corresponding angle measure at the center of the sphere. If we measure the distance between two points on the surface of the Earth, we can determine the corresponding angle by dividing the measured distance by

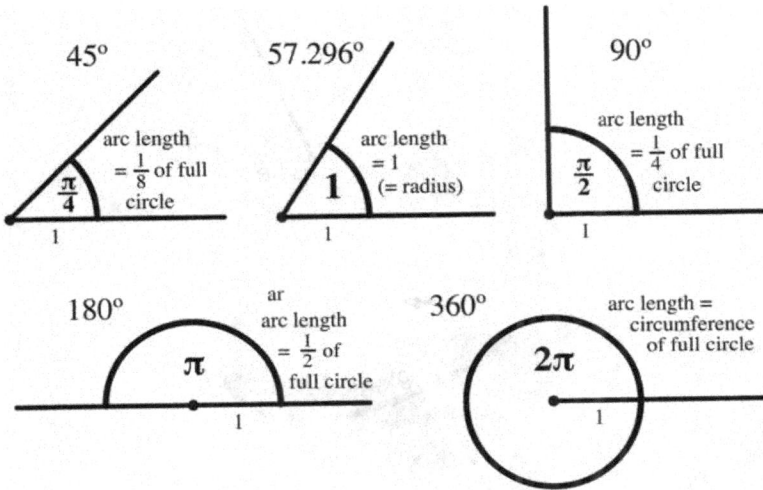

Fig. 4.7. Some common angles given in degrees and in radians (printed in bold letters).

the Earth's radius. For example, the shortest flight path between the cities of Frankfurt and Vancouver has a length of about 8,073 km (5,016 miles). Dividing this distance by the radius of the Earth ($\frac{8073}{6371} \approx 1.267$) provides an angular distance of about 1.267 radians (72.6°). By the way, the east-west distance along the 50° parallel of latitude is about 9,435 km (5,863 miles), considerably longer. The route over Greenland is, therefore, more than 800 miles shorter than the east–west route. The reason for the strange observation that the east–west route usually appears as a shorter path on world maps will be explained in more detail later.

4.6 Axis and Poles

The straight line through the center of a great circle and perpendicular to its plane is called the axis of the great circle (see Figure 4.8). The axis of a great circle passes through the center of the sphere, which, of course, coincides with the center of the great circle. The intersection of the axis with the sphere are two points which are called the *poles* of the great circle. These two poles are antipodal points.

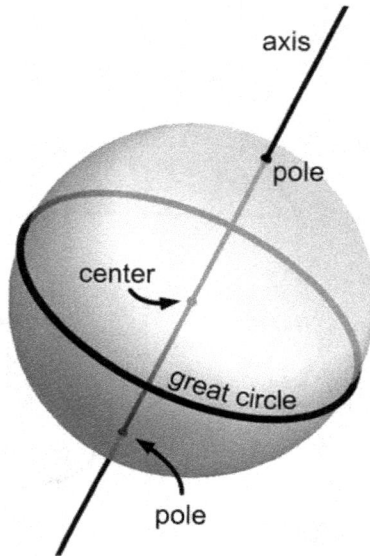

Fig. 4.8.   Axis and poles of a great circle.

All great circles through the pole of a given great circle form right angles with the given circle. Put differently, if the planes of two great circles of the same sphere are perpendicular to each other, each of the circles passes through the poles of the other. Figure 4.9 shows two great circles, passing through the poles of each other and meeting at right angles. The two points of intersection of the two great circles are antipodal points, that is, they both lie on a straight line through the center of the sphere.

In general, all great circles on a sphere are equal. On the Earth, however, the equator is a very special great circle, because its axis coincides with the Earth's axis of rotation and its poles are, of course, the North Pole and the South Pole.

4.7   Regular Digons on the Sphere

While the great circles on a sphere are the best analog of straight lines in plane geometry (as the lines establishing the shortest connection

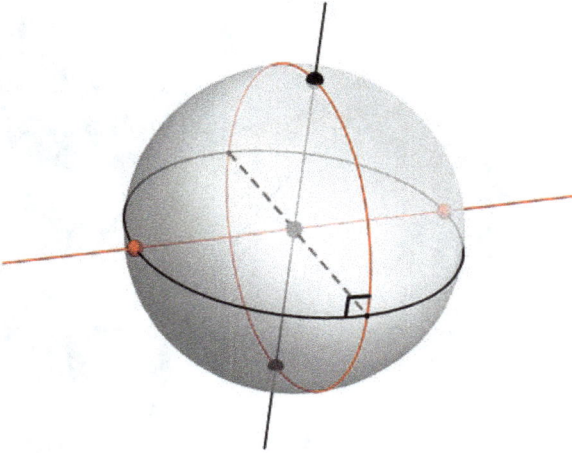

Fig. 4.9. Great circles meeting at right angles.

between two points), they are quite different with respect to another property. In plane geometry, we have the famous axiom that for every given straight line and a point not on it, there is precisely one straight line parallel to the given line through the given point. In the plane, two parallel lines have no intersection. This is quite different from the situation on the sphere.

Any two great circles of the same sphere bisect each other, and there are precisely two points of intersection, which are antipodal points. The resulting shape on the sphere is a sector on the sphere's surface, shown as the shaded area in Figure 4.10. It is a shape determined by two spherical arcs meeting at an angle in the two antipodal vertex points. This geometric figure on the sphere is a *regular digon* — which is a polygon with just two edges and two vertices — and it is also called a *spherical lune*, not to be confused with a lune on a plane which is created by two intersecting circles of unequal size. The word "lune" has its origin in the Latin word *Luna* for the Moon.

We can interpret the two antipodal points of intersection of the two great circles as the poles of another great circle, shown in Figure 4.10. This new great circle intersects the given circles at the points *A* and *B* and cuts the spherical lune into two symmetric halves.

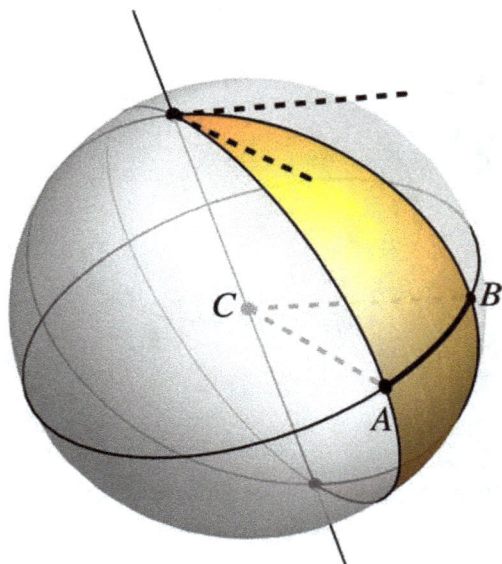

Fig. 4.10.   A regular digon (or spherical lune) on a sphere.

We also notice in Figure 4.10, that the great circle through A and B and the lune's great circles meet at right angles (as do all great circles through the poles of one another). The spherical arc AB cutting through the lune defines a spherical angle (that is, the angle $\angle ACB$). This angle characterizes the lune, because it also describes the angle between the tangents at the two antipodal vertex points, as indicated at the top of the sphere in Figure 4.10. We can also interpret it as the angle between the planes of the two great circles. The angle of any lune must be less than π radians (i.e., less than 180°) since if it were equal to 180°, then the lune would become a great circle. Every spherical lune has a mirror image on the other side of the sphere, formed by the same great circles and subtending the same spherical angle (given by the spherical arc through the antipodal points A' and B'), see Figure 4.11.

The tiling of the sphere by n identical lunes sharing the same antipodal vertex points is called a *hosohedron*, which can be analogous to a regular polyhedron whose edges are not straight lines but

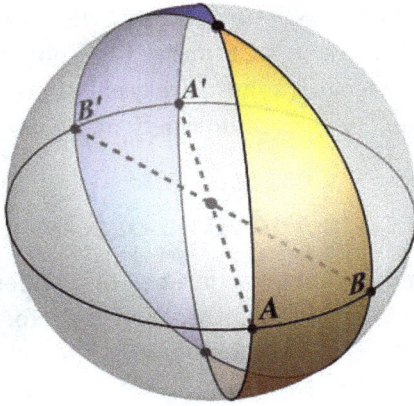

Fig. 4.11.   A spherical lune and its mirror image.

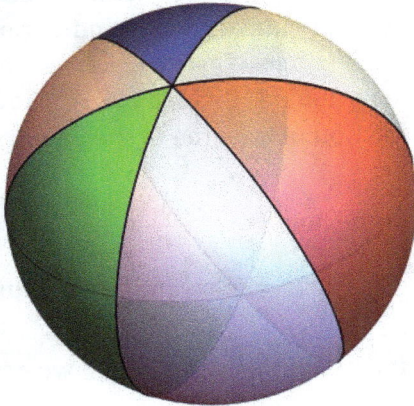

Fig. 4.12.   A beach ball is a hosohedron consisting of six spherical lunes.

spherical arcs and whose faces are parts of a sphere. A hosohedron has two vertices, n edges, and n faces (lunes), and there is a hosohedron for any $n \geq 2$. Each of the spherical lunes of a hosohedron has the same angle of $\frac{2\pi}{n}$ radians ($\frac{360°}{n}$). The beach ball in Figure 4.12 consisting of six lunes is an example of a hosohedron.

In Chapter 6, we will be dealing with polyhedra such as Platonic solids, which have straight edges and planar surfaces (polygons). Each of the polygonal faces of these solids must have at least three edges. As a result, it will turn out that there are only five regular polyhedra (polyhedra whose faces are regular polygons). The edges of polyhedra on the surface of a sphere are great-circle arcs, and the hosohedra form a whole new class of infinitely many regular polyhedra.

4.8 Circumference and Area of a Lune

It is easy to determine the circumference of a spherical lune, since it consists of two halves of great circles, and therefore, it has the same circumference as one great circle, namely $2\pi r$, which holds for any lune, regardless of its angle.

The angle of the lune is the spherical angle between the points A and B as shown in Figure 4.10, which is the angle $\angle ACB$, where C is the center of the sphere. Here, we will measure the angle of the lune in radians. The angle of the lune is a certain fraction of the full angle (2π) of an entire great circle. This fraction is, therefore, given by the angle of the lune in radians, divided by 2π. The area of the lune is given by the same fraction of the area of the whole sphere $(4\pi r^2)$,

$$\text{area of the lune} = 4\pi r^2 \frac{\text{angle of the lune}}{2\pi}$$
$$= 2r^2 \times (\text{angle of the lune}).$$

We will be making use of this result when we compute the area of a spherical triangle.

4.9 Spherical Triangles

Analogous to the definition of a triangle in the plane, we define a spherical triangle by three vertices A, B, and C on the sphere, as shown in Figure 4.13. We want to exclude the case where all three vertices lie on the same great circle, because this would reduce the triangle to a line, just as three collinear points on a plane would not determine a triangle. The sides of the spherical triangle are formed by the spherical arcs that connect any two of these vertices. However, a spherical

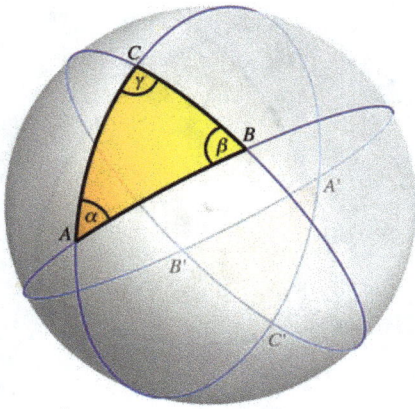

Fig. 4.13.   A spherical triangle and its mirror on the opposite side of the sphere.

arc is uniquely defined only if its endpoints are not antipodal. Therefore, we also want to rule out the case where there is a pair of antipodal points among the vertices. But otherwise, the three points A, B, and C can be chosen arbitrarily on the surface of the sphere and they define a unique triangle, in which each of the three sides is shorter than half of a great circle.

We notice that the three great circles containing the sides of the triangles meet again in the antipodal points A', B', and C', and these antipodal vertices form a "mirror triangle" on the opposite side of the sphere, as shown in Figure 4.13.

While small spherical triangles seem to be analogous to triangles in the Euclidean plane, larger triangles behave very differently. For example, in the plane a triangle can have at most one right angle. On the sphere, all three angles could be right angles, as shown in Figure 4.14. Each of the sides of this "right triangle" is a quarter of a great circle. Also, we can see that each of the three great circles passes through the poles of the other two.

Perhaps one of the places where a spherical triangle departs most radically from a plane triangle is that the sum of the angles of a spherical triangle must be greater than 180° and less than 540°. A plane triangle always has the sum of its angles equal to 180° (π radians). The angle sum of a spherical triangle can be almost three times as much.

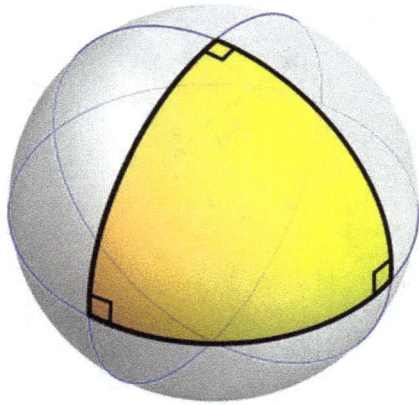

Fig. 4.14.   A spherical triangle with three right angles.

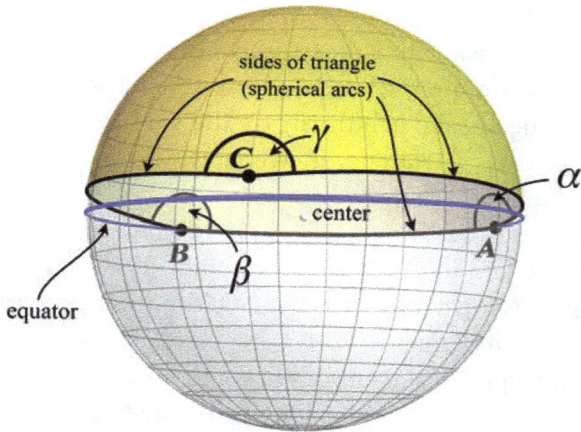

Fig. 4.15.   Triangle *ABC* with angle sum $\alpha + \beta + \gamma$ almost 540° (3π radians).

Consider a triangle, where two vertices are on the equator and one vertex is on the other side of the sphere, just a little bit north of the equator, as shown in Figure 4.15. The area of this triangle occupies almost all of the upper hemisphere, each of the angles is almost 180°. As the point *C* approaches the equator, the angle sum of the triangle tends to 540°. By our definition of a triangle, however, it is forbidden that all points lie exactly on the equator (i.e., on the same great circle).

In the plane, this would correspond to the situation where all vertices are collinear.

4.10 The Area of a Spherical Triangle and Girard's Theorem

There is a truly ingenious trick for computing the area of a spherical triangle. It makes use of the formula obtained earlier for the surface of a lune. In preparation of this technique, we remind the reader that each spherical triangle comes with a mirror triangle on the opposite side of the sphere. The mirror triangle is made from the antipodal points of the vertices of the original triangle, as shown in Figure 4.13. It has the same shape and size, and therefore, occupies the same surface area as the original triangle.

In Figure 4.16, we show how each angle in the triangle defines a spherical lune. For example, the lune with angle alpha is shown in the first image of Figure 4.16. There is a total of six lunes, which we show throughout Figure 4.16 and should convince the reader that the six

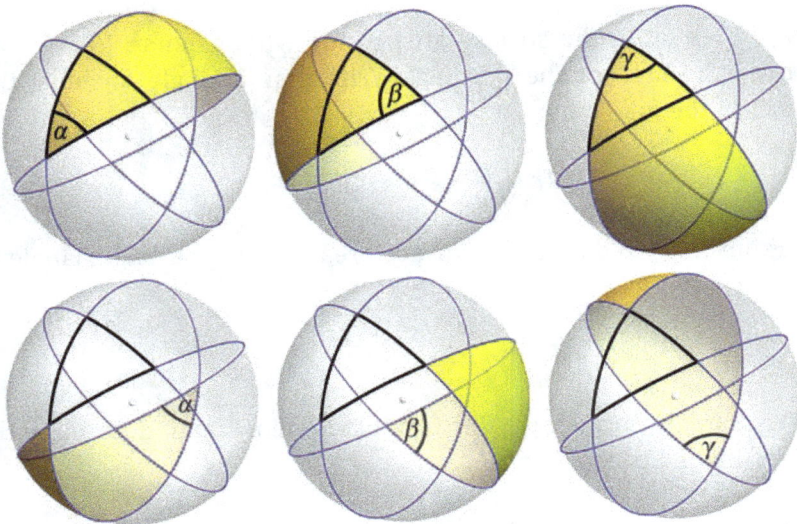

Fig. 4.16.   A spherical triangle and the six associated spherical lunes.

lunes are enough to cover the total surface of the sphere. Every point of the sphere lies within at least one of the lunes. However, both the original triangle and its mirror triangle are covered three times, because three lunes in the top row of Figure 4.16 all contain the original triangle and the remaining three lunes (bottom row of Figure 4.16) all contain the mirror triangle on the back side of the sphere.

If we now add up the surface areas of these six lunes, we obtain the whole surface area of the sphere plus twice the area A of the original triangle and twice the area of the mirror triangle (which has the same area A). It then follows that:

sum of the areas of the six lunes = surface area of sphere plus $4A$.

We also remind the reader that the area of any spherical lune is $2r^2$ times the angle of the lune (in radians). Thus, the first lune in Figure 4.16 has the area $2r^2\alpha$, and the lune below it (first image in the second row) has the same area. Therefore, we obtain the following:

sum of areas of the six lunes = $2r^2 (2\alpha + 2\beta + 2\gamma)$,

and it is equal to the surface area of the whole sphere $(4\pi r^2)$, plus 4 times the area of the triangle. Written as a single formula, this results in the following:

$$4r^2(\alpha + \beta + \gamma) = 4\pi r^2 + 4A.$$

From this we obtain the surface area of the spherical triangle as follows:

$$A = r^2(\alpha + \beta + \gamma - \pi)$$

We can also transform this formula to allow us to determine the angle sum of a triangle as

$$\alpha + \beta + \gamma = \pi + \frac{A}{r^2}.$$

Using radians instead of degrees enables us to stress these relationships in a simple form. This beautiful result is called the *Girard's Theorem*. It is named after its discoverer, the French mathematician Albert Girard (1595–1632) who published this result in 1626, although it was previously discovered by the English mathematician and astronomer Thomas Harriot (1560–1621), in 1603, but who did not publish his findings. Girard's theorem was generalized by Gauss for arbitrary surfaces, and later by Riemann for arbitrary dimensions in the context of Riemannian geometry.

Since the area A of any triangle on the sphere is greater than zero, we see that the sum of the interior angles of a triangle on a sphere is always larger than π radians (the measure of the straight angle corresponding to 180°). The sum of the angles of a spherical triangle is, therefore, always larger than the sum of the angles of a plane triangle (which is always π radians or 180°). The so-called "spherical excess" $\varepsilon = \alpha + \beta + \gamma - \pi$, is the amount by which the sum of the angles exceeds π radians (180°). According to Girard's theorem, the spherical excess is determined by the area of the triangle.

Another interesting way to read Girard's theorem is the following: Assume that we know from some observations that we live on the surface of a sphere. Then we can determine the radius r of the sphere from purely local measurements on the surface. Just take any three points, which do not all lie in the same great circle and connect them with the shortest possible straight lines to form a triangle. Then measure the area A of this triangle and the three interior angles, α, β, and γ, in order to determine the spherical excess ε. Then the radius of the sphere is given by

$$r = \sqrt{\frac{A}{\varepsilon}} \, .$$

It is quite remarkable, that the radius of the sphere can be obtained by simply taking measurements done locally in a small region on the surface of the sphere.

With r being the radius of the sphere, the quantity $\frac{1}{r^2}$ appearing in Girard's formula is called the *Gaussian curvature* of the sphere.

The curvature is also defined for arbitrary curved surfaces in three (and higher) dimensions, and the formula remains valid in this general case, with $\frac{1}{r^2}$ replaced by the curvature of the given surface, and, with some purely mathematical modifications, even in the case where the curvature varies from point to point on the surface.

4.11 Coordinates on the Sphere

In plane geometry, Descartes' idea of using coordinates to describe the location of points has proven to be very fruitful (see Chapter 1). In spherical geometry, the concept of coordinates is of similar importance.

In order to specify a coordinate system on the sphere, we need a coordinate origin, perpendicular coordinate axes, and a unit length. The procedure is essentially the same as in plane geometry. And just as in the flat (plane) geometry, there is some freedom. Any point on the sphere can be chosen as the coordinate origin, because no point is distinguished from any other due to the symmetry of the sphere. Since the great circles on the sphere are analogous to the straight lines in the plane, the coordinate axes are naturally defined by great circles. We can call an arbitrary great circle that contains the origin of the coordinates the "horizontal" axis. The "vertical" axis would then be formed by the great-circle which connects the poles of the horizontal great circle and also contains the coordinate origin, which we can see in Figure 4.17.

In Figure 4.17, the coordinates of a point P on the surface of the sphere would be determined as follows. For the first coordinate we measure the distance along the horizontal axis between the origin and the point P'. The point P' is the intersection of the horizontal axis and the great circle containing the poles and the point P. This great circle is also perpendicular to the horizontal axis. The vertical coordinate is then the distance between the points P' and P. We can measure distances along great circles in degrees. In Figure 4.17, the horizontal coordinate of the point P is the angle OCP', the angle between the origin O and the point P', as seen from the center C of the sphere. The vertical coordinate is the angle $P'CP$ between the points P' and P. Coordinates of points on the other side of the vertical axis

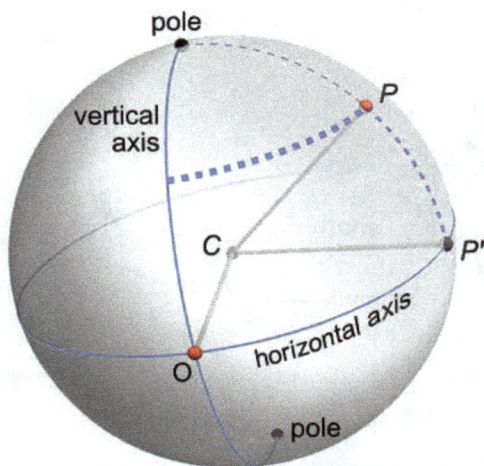

Fig. 4.17. Coordinate system on a sphere.

(the "left side," or "western side"), or below the horizontal axis (on the "Southern Hemisphere") are designated with a minus sign. Therefore, the horizontal coordinate can vary between −180° and +180°, where the angles ±180° both correspond to the antipode of the origin. Likewise, the vertical coordinates can take values between −90° and +90°. The point P in Figure 4.17, for example, has the coordinates (75°, 40°). All points on the equator have vertical coordinate 0°. The poles have the vertical coordinates ±90°, but their horizontal coordinate is undefined. Any two angles within the given range define a unique point on the sphere. The fact, that two coordinates are sufficient to describe arbitrary locations on a sphere means that the sphere is a two-dimensional surface. It is, therefore, sometimes called the 2-sphere, although it lives in three-dimensional space. In Chapter 15, we will get to know the 3-sphere, which is the analogue of the sphere in a four-dimensional space.

4.12 Applications of Spherical Coordinates

Coordinates on a spherical surface are important, for example, in connection with astronomical observations. From the point of view of a

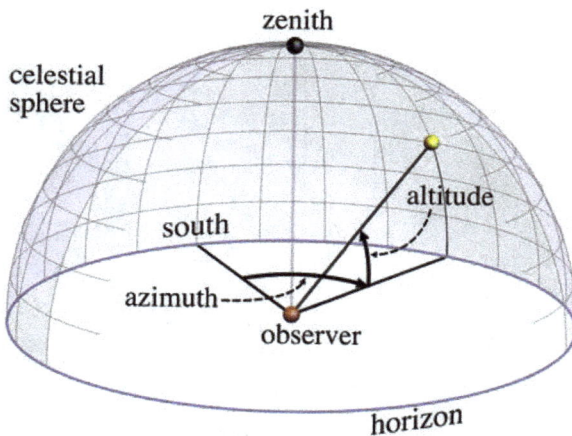

Fig. 4.18.   Celestial sphere and azimuth-altitude coordinates.

human observer, the sky appears like a sphere (the "celestial sphere") that you look at from the inside and from which you can only see the hemisphere that extends just above your position. The horizontal angle on the celestial sphere, measured clockwise from the south in degrees, is often called the *azimuth* (a word of Arabic origin, meaning "the directions"). The altitude of point on the celestial sphere is its vertical angle with the horizon (this angle measured along the great circle passing through the point and the zenith) as can be seen in Figure 4.18.

In geography, coordinates of a point on Earth's surface are its latitude and its longitude. These coordinates enable us to describe positions in a way that is independent of local features of the landscape. It is, therefore, obvious that coordinates became particularly important, as soon as people started to navigate the oceans. And indeed, the geographic coordinate system appeared very early in human history — invented by the Greek scholar Eratosthenes of Cyrene (276 BCE — 194 BCE), who had the truly revolutionary idea to define a global coordinate system for the whole world. Eratosthenes was also the first to calculate the circumference of the Earth, as well as the tilt of the Earth's axis, and he is said to have created the first map of the world. His work has not survived, but its results have been handed down to us by the Greek mathematician Claudius Ptolemy (ca. 100–170 CE), whose book *Geographia* has been most influential in medieval science at the beginning of modern history. It contains a description of the what was then the known world and defines meridians and parallels of latitudes as we use them today.

4.13 Coordinates in Geography

Coordinates on the Earth's surface are called "longitude" and "latitude" and they are defined as spherical coordinates as explained above. On Earth, there are certain points distinguished by physics, namely the North Pole and the South Pole, where the axis of rotation intersects with the Earth's surface. The great circle perpendicular to the Earth's axis of rotation is the equator. It serves as the "horizontal axis" of our coordinate system. In order to define a coordinate origin, we also need a vertical axis, that is, a specified great semi-circle connecting North and South Poles. The choice of this axis, which is called the "prime meridian", is arbitrary but is historically related to the Greenwich Meridian, which passes through the Royal Observatory in Greenwich near London, England. The prime meridian, which is in use since 1984, is the International Reference Meridian, and which also serves as the reference for the global positioning system (GPS). It runs about 102 m to the east of the Greenwich Observatory. These minor local corrections have become necessary to take into account the exact shape of the Earth, which is not exactly spherical, but is deformed into a so-called geoid by Earth rotation and uneven mass distribution.

The Earth with this coordinate system, defined by the equator and the prime meridian, is shown in Figure 4.19 with bold lines. The coordinate origin on the surface of the Earth is where the prime meridian meets the equator, which is in the Atlantic Ocean a somewhat south of Ghana. From the prime meridian, the longitude ("horizontal coordinate") is measured in degrees to the east and degrees to the west, likewise the latitudes are measured from the equator in degrees to the north and to the south. In this coordinate system, the Statue of Liberty in New York is at the longitude of W74°41′21″ and the latitude

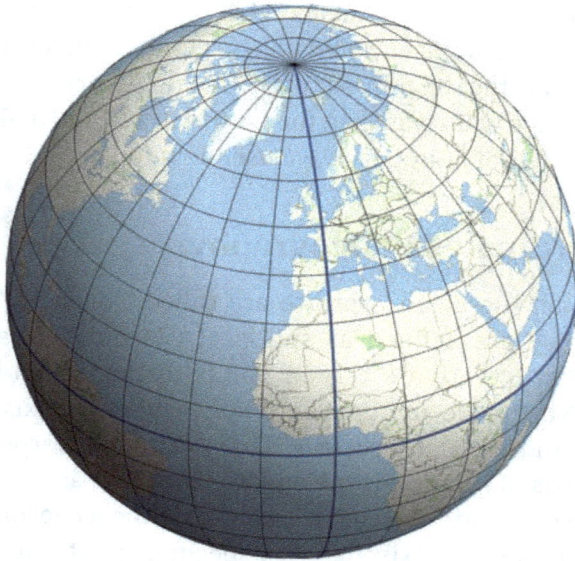

Fig. 4.19. Coordinate system used in Geography, showing the equator and the prime meridian.

of N40°02′40″, which translates into decimal coordinates (−74.044444°, 40.689167°). This means, that from the coordinate origin in the Atlantic, one has to go by 74.044° to the west and then by 40.689° to the north in order to obtain the exact location of the Statue of Liberty.

Figure 4.19 also depicts the lines of constant longitude ("meridians") and the lines of constant latitude ("parallels of latitude"). In this image, the (angular) distance between two parallels of latitude is 10° and the distance between two meridians is 15°. This is exactly the angle by which the Earth rotates within an hour. All parallels of latitude except the equator are "small circles" on the sphere, because their plane does not contain the Earth's center, and, therefore, their radius is smaller than Earth's radius. They are called "parallels", because their planes are parallel to the equatorial plane. Remember, these lines do not play a role in spherical geometry, only great circle lines are considered when studying geometry on the sphere.

4.14 Making Maps

Cartography deals with the problem of representing the Earth on a flat surface, that is, as a region on the plane. The coordinate system of longitudes and latitudes will help us to explain the basic concepts and to understand the features of some of the important principles of map making.

Figure 4.20 shows a sphere with the coordinate lines, along with the meridians as lines with constant longitude and the parallels as lines of constant latitude. Every point on the sphere can be described by these two coordinates, the longitude and the latitude. For example, the point shown in Figure 4.20 has the coordinates 75° longitude and 40° latitude.

A world map can be created by just drawing a plane rectangular grid representing the coordinate lines, as in the background of Figure 4.20. This means that we can simply use the longitude and latitude as coordinates in the two-dimensional plane. The whole surface of the Earth then fits onto a rectangle that extends from −180 to +180 horizontally and from −90 to +90 vertically. This representation of the sphere as a region in the plane is called a *projection*. And this particular projection is called an *equirectangular projection*. A map of the world in equirectangular projection is shown in Figure 4.21.

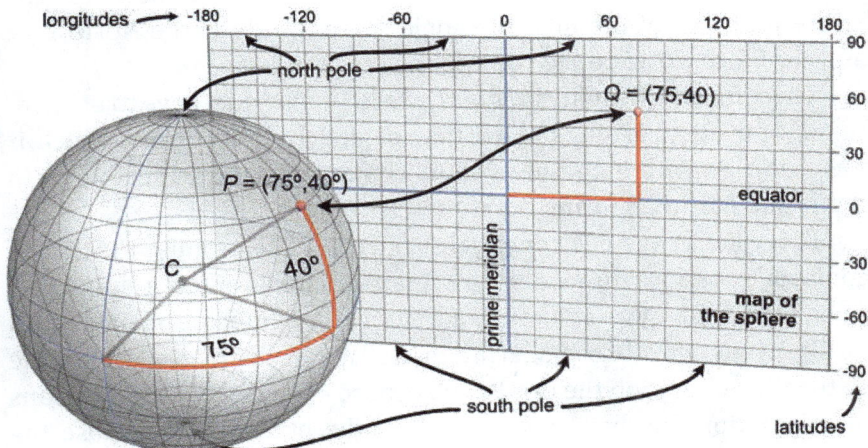

Fig. 4.20. Creating a map of a spherical surface.

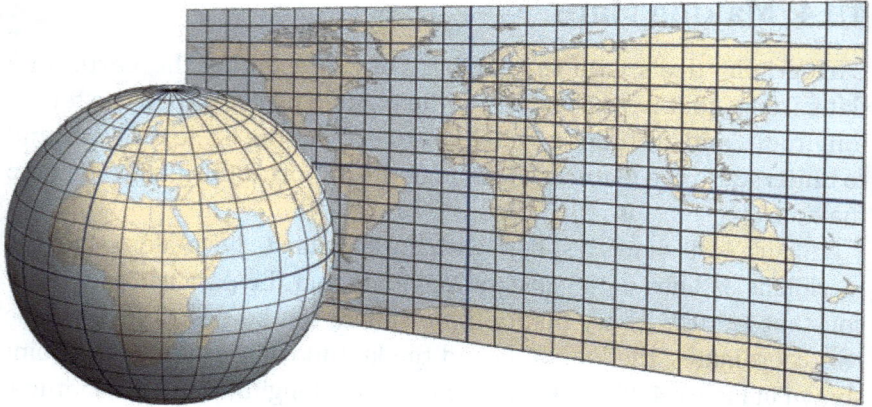

Fig. 4.21.   Map of the world in equirectangular projection.

While this method of map making using an equirectangular projection based on the longitude–latitude coordinate system gives a good impression of the shape of the continents near the equator, it leads to severe distortions as one approach the poles. While distances in north–south direction are reproduced faithfully, distances in west–east direction become larger the closer one gets to the poles. The meridians, while they are actually converging towards the poles on the sphere, they appear as parallel vertical lines on the map. The entire top edge of the map rectangle corresponds to the North Pole and the whole lower edge corresponds to the South Pole.

The left and the right sides of the rectangular map have to be identified, both vertical sides show the so-called *antemeridian*, which is the meridian with longitude ±180°. The antemeridian together with the prime meridian forms a great circle.

Every flat image of a spherical surface is a compromise that has to conform to various disadvantages. A large number of different projections have been developed throughout history. Interest in such maps of the whole world rose sharply in the age of discovery, during the sixteenth century; on the one hand because explorers needed a means of orientation for their travels, and on the other hand because the concept of a spherical and limited world became more and more widespread. Maps, together with technological innovations in navigation enabled the establishment of trade routes and new expeditions that in turn contributed to our geographical knowledge.

4.15 Navigation and the Loxodrome

In 1569, the Flemish cartographer Gerardus Mercator (1512–1594) created a world map (Figure 4.22), which made use of a newly invented projection method. He published his famous map under the name "Nova et Aucta Orbis Terrae Descriptio ad Usum Navigantium Emendate Accommodata" (Latin for "A new and more complete representation of the terrestrial globe properly adapted for use in navigation"). Mercator's projection, which still bears his name, has a particular advantage that should make it the standard for navigation using a compass for the centuries to come.

The particular problem addressed by the Mercator projection is the following. The easiest way to sail a straight course across the ocean is to follow a fixed direction (constant bearing), which can be maintained with the help of Polaris, the north star, or with the help of the ship's compass during daytime. This course, however, typically does not correspond to a straight line on a map. After a long journey, this could result in the actual position deviating considerably from the assumed position. Mercator's map now showed all sailing courses where the bearing of the compass is kept constant as straight lines, which made navigation much easier.

This line of constant bearing is called a *rhumb line* or *loxodrome* and it intersects each meridian at the same angle. A simple example of a rhumb line is a parallel of latitude that one automatically follows when one sails, according to the compass, a course exactly west or exactly east. Another example is shown in Figure 4.23.

Theoretically, people in the 16th century were well aware that the shortest connecting lines on the globe are great circle arcs. In practice, however, it is very difficult to follow such a line because one would have to constantly correct the compass direction. Following a loxodrome (an arc crossing all meridians of longitude at the same angle) with constant bearing is much easier. Mercator's projection, therefore, remained in use until modern times. Figure 4.24 compares a

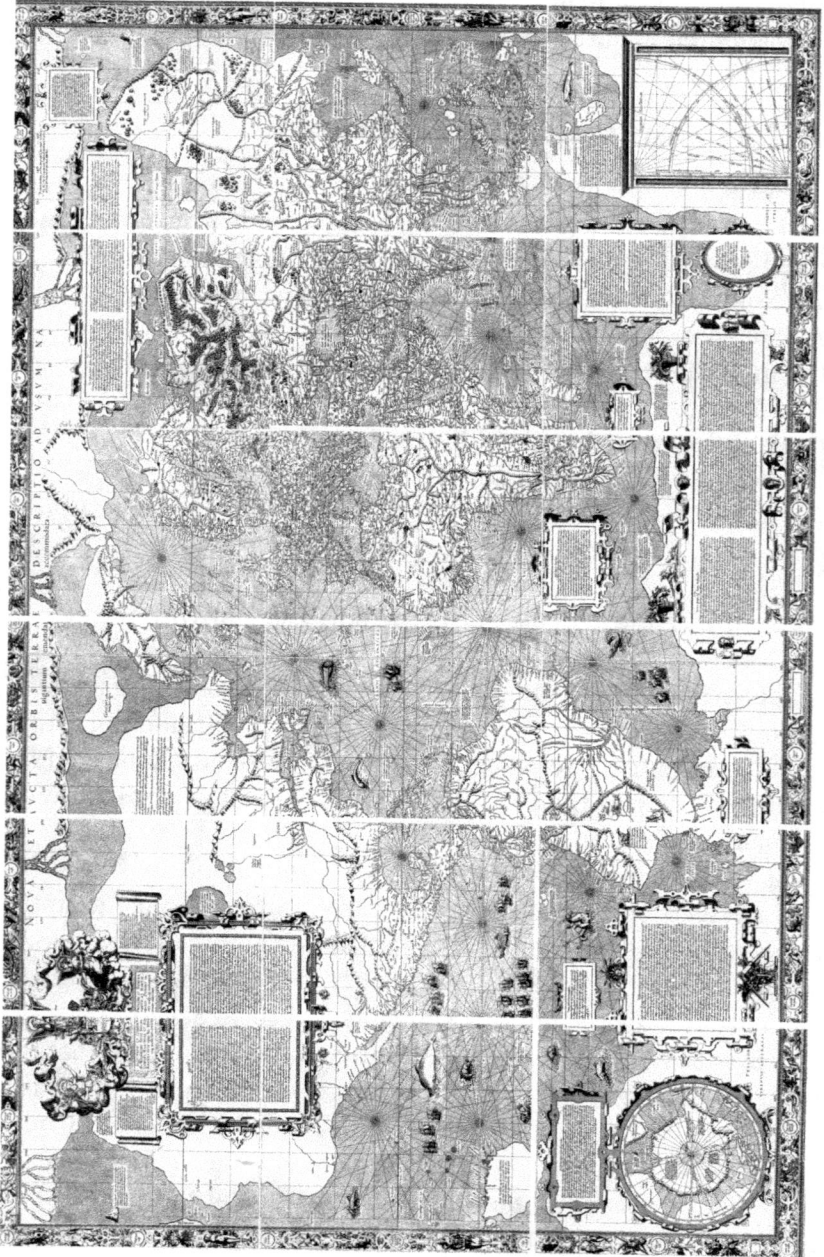

Fig. 4.22.   World map (1569) by Gerardus Mercator.

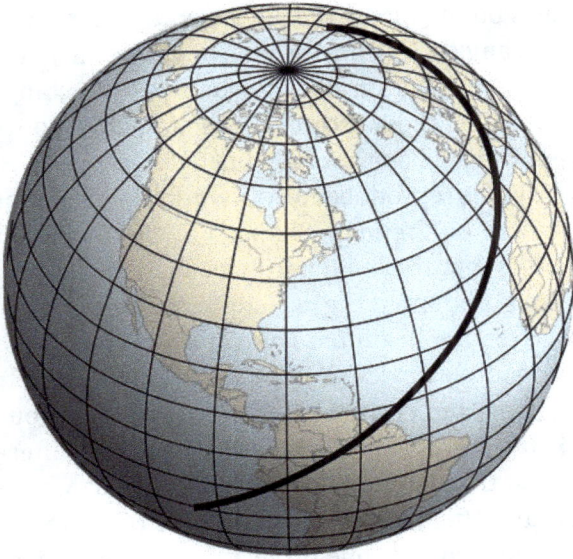

Fig. 4.23.   A rhumb line or loxodrome meets each meridian at the same angle.

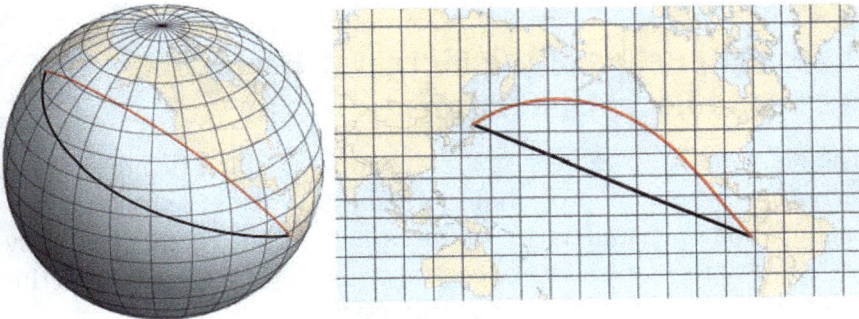

Fig. 4.24.   Loxodrome (black) and great circle arc (red).

loxodrome (black) extending across the Pacific from Peru to Japan with the corresponding great circle arc (red). The great circle arc would be the shortest path, while the loxodrome is considerably longer, but much easier to navigate with a compass. On the Mercator map, the loxodrome appears as a straight line, while the great circle arc is curved.

From today's point of view the big advantage of a Mercator map is that it preserves angles. The image of an angle on the map is the same as the angle on the Earth's surface. Unlike the equirectangular projection discussed earlier, the Mercator projection close to the poles not only stretches distances horizontally but also vertically. Hence, small shapes are faithfully reproduced everywhere. A small square would appear as a square on the map. Shapes will be magnified, if you get closer to the poles, but they will hardly be distorted.

In 2005, Google adopted the variant of the Mercator projection as a standard for Web applications. The Web Mercator projection, now widely used for mobile devices and by map providers on the internet, differs slightly from the classical Mercator projection, but the difference is not visible on the scale of a world map. The difference comes from the different treatment of the shape of the Earth, which is not an exact sphere, but rather resembles a flattened rotational ellipsoid. By the way, one cannot say that one projection is "more correct" than the other. Different maps are simply different models of the Earth's surface, each with its own strengths and weaknesses.

4.16 Properties of the Mercator Projection and Local Distortions

The Mercator projection is an example of a so-called *cylindrical projection*. All cylindrical projections have the meridians as straight vertical lines and the parallels of latitude as straight horizontal lines. They are all variants of the idea of a central cylindrical projection, the principle of which we are going to consider next.

Imagine that the globe is hollow and that there is a bright source of light at the center of the globe and that the surface is in color and transparent. The screen is a cylinder that is tightly wrapped around the equator. The radial light rays emanating from the center then cast an image of the globe's surface onto the cylinder. The cylinder then can be cut, for example, along the anti-meridian (the dashed line at the back of the cylinder in Figure 4.25), and spread out in the plane, which results in the map of the globe in cylindrical projection.

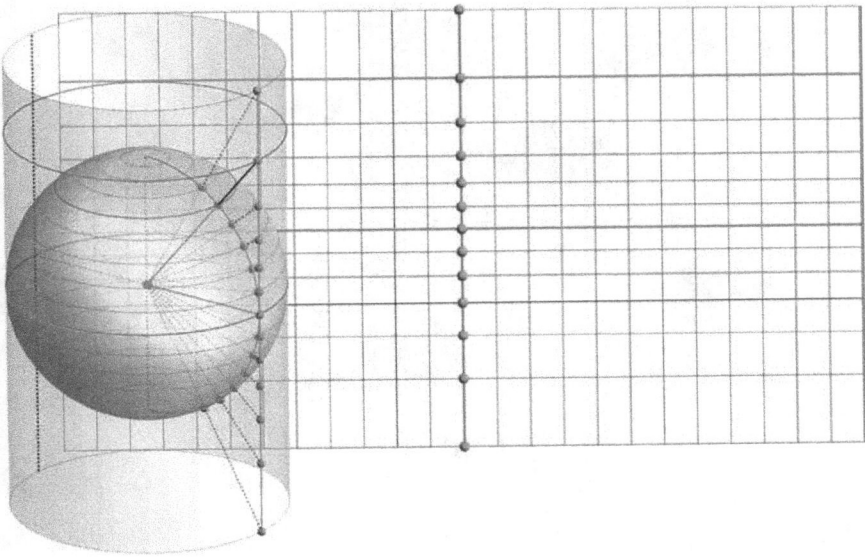

Fig. 4.25.   Central cylindrical projection.

There are some grave disadvantages of this procedure. Figure 4.25 shows a series of equally spaced points on the globe, arranged on a meridian. Their distance on the map, however, increases to the north and to the south. The poles themselves would be mapped to infinity. Thus, the map can only display a certain part of the world. In the example of Figure 4.25, it would be the part between the latitudes of +60° and −60°. The distortions would be extreme in the regions closer to the poles, making it unsuitable for most purposes.

The Mercator projection is a compromise between the central cylindrical projection and the equirectangular projection. Its vertical scale is locally adjusted in such a way that the loxodromes become straight lines, as previously explained. Still, the Mercator map shows strong distortions far to the north and far to the south, as can be seen in Figure 4.26. For example, Greenland appears to be about the same size as Africa, while actually it is less than the size of the Arabic peninsular.

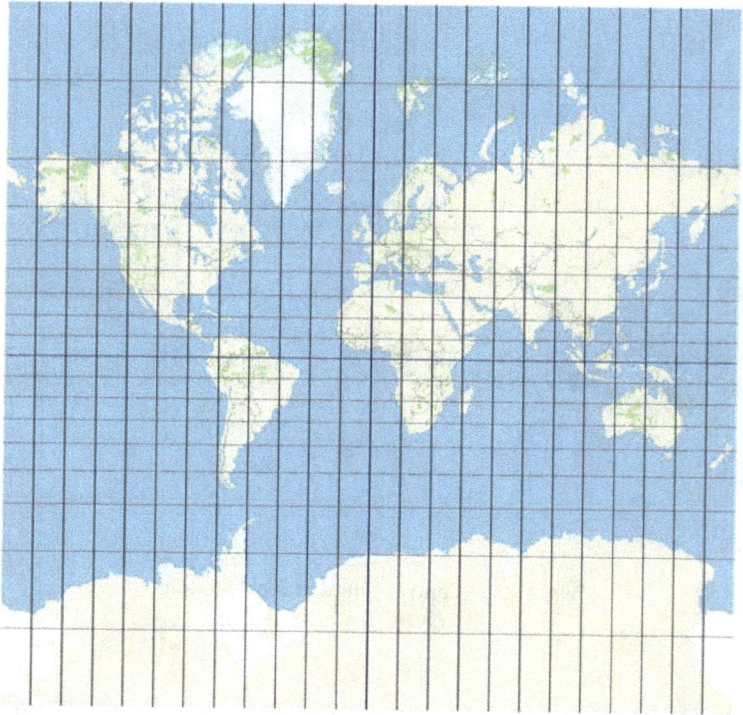

Fig. 4.26.   The Mercator projection shows strong distortions near the poles.

However, the distortion is such that angles are preserved, and hence small figures are mapped to similar figures in the map. Although Iceland appears too large on a Mercator map, its shape is better represented than on an equirectangular map, as can be seen by direct comparison shown in Figure 4.27.

In order to visualize the local distortions introduced by a projection method, the French cartographer Nicolas Auguste Tissot (1824–1897) proposed what is known today as Tissot's method of indicatrices. With this method one just places some small circles on the globe and projects them onto a map, where, in general, they become ellipses. The images of the circles on the map show — for every region of the map — the amount of distortion in shape and size introduced by the projection method.

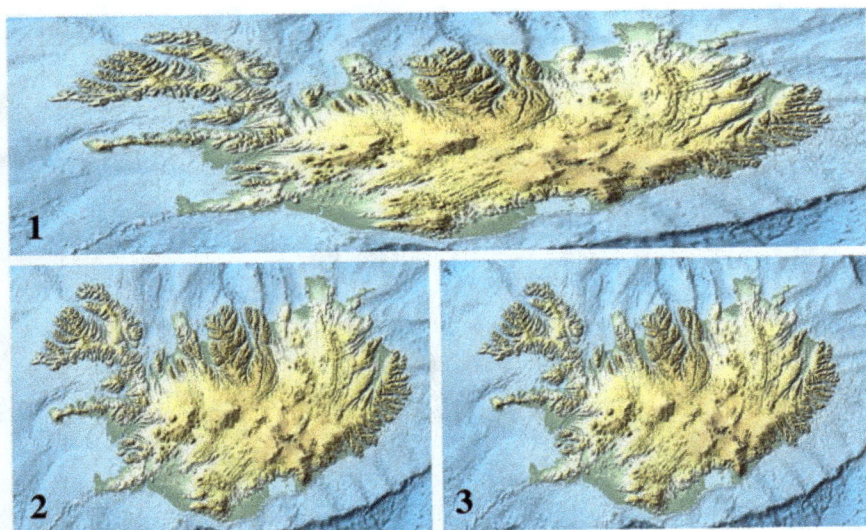

Fig. 4.27. Comparing the equirectangular map (1) and the Mercator map (2) with the true shape of Iceland (3), as seen from a position directly above (orthographic projection).

Figure 4.28 shows the globe with Tissot's indicatrices. We also can see the effect on world maps in equirectangular projection and in Mercator projection.

4.17 Orthographic and Stereographic Projection

The Mercator map of Figure 4.26 extends to latitudes of ±85° latitude and is approximately a square. Theoretically, one could extend it even further towards the north or south, but the poles can never be depicted because they are mapped onto infinity by the Mercator projection. Typically, one chooses some other type of projection in order to display the polar regions and uses these maps as a supplement to other maps in an atlas of the world.

The orthographic projection maps the surface of the sphere to a tangent plane by parallel rays of light that meet the projection plane at right angles. An example is shown in Figure 4.29. The orthographic

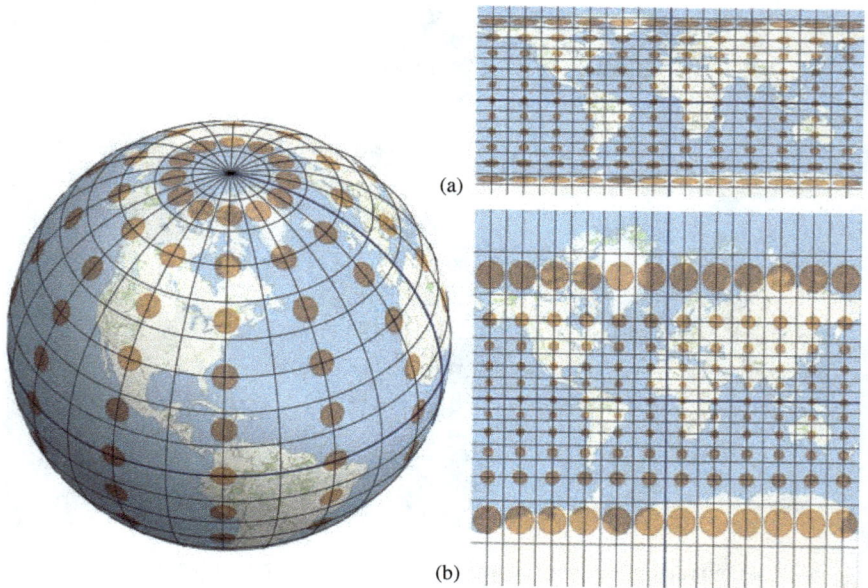

Fig. 4.28. Tissot's indicatrices on a globe are indicators of distortion in maps. The examples show world maps in equirectangular projection (a) and Mercator projection (b).

projection is an example of a parallel projection as discussed in Chapter 3.

The orthographic projection cannot show more than a hemisphere. When the image plane is perpendicular to the Earth's axis, the map would show the polar regions, and, as can be seen in Figure 4.29, the parallels of latitude will appear as circles and the meridians will appear as straight radial lines. Of course, the image plane can be placed anywhere, not just above the poles. The orthographic projection creates a convincing impression of a globe when viewed from a distance. If one adds a shadow, the spatial impression is even more enhanced, as can be seen in Figure 4.30.

The stereographic projection is a perspective projection. The projection center is often chosen as one of the poles. The method is illustrated in Figure 4.31.

From the North Pole we send projection rays through the sphere to an image plane that is tangent to the South Pole. Points from the

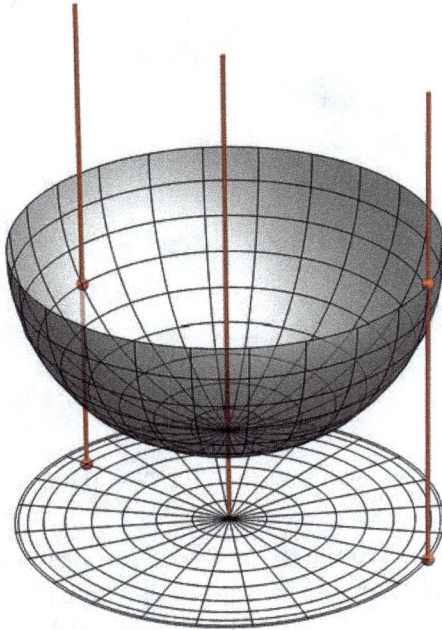

Fig. 4.29.   Orthographic projection of a hemisphere.

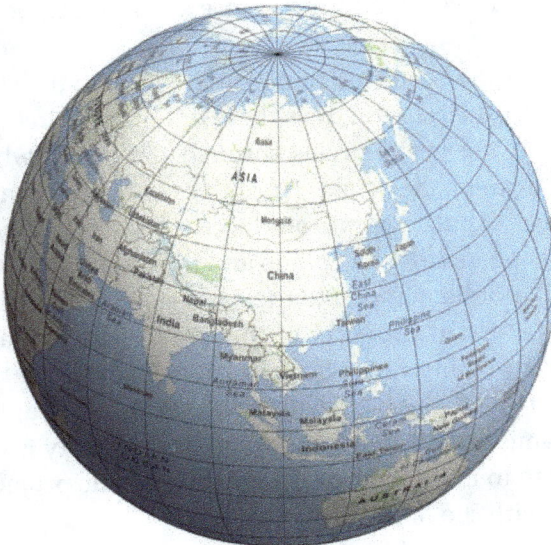

Fig. 4.30.   Orthographic projection with additional shading.

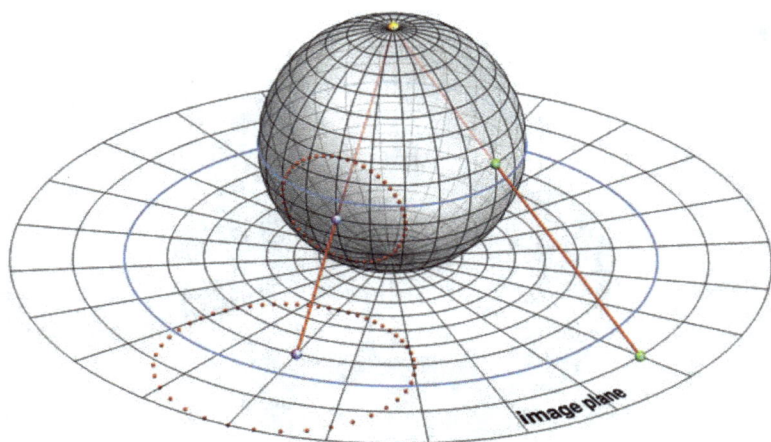

Fig. 4.31.   Stereographic projection of a sphere.

sphere are transferred to the plane along these rays, thereby, creating a map of the sphere. This projection method is called a stereographic projection and it is mathematically interesting because it maps the entire sphere (with the exception of a single point — the North Pole) in a one-to-one fashion onto the two-dimensional plane. Any map of finite size, however, would have to exclude a region around the North Pole.

Figure 4.31 shows another important feature of the stereographic projection: It maps circles on the sphere onto circles in the projection plane — except the great circles through the poles, which become straight lines through the origin. The circles in the plane are, however, distorted. Equally spaced points on the circle will be unevenly distributed on the image-circle in the plane and the midpoint of the circle on the sphere will not be mapped to the center of the circle in the plane.

Another important property of the stereographic map is the following: It preserves angles. This means that when two curves meet on the sphere at a given angle, the curves on the stereographic map will meet at the same angle. A mapping with that property is called a *conformal mapping* in mathematics. The stereographic projection shares this property with the Mercator projection.

While the stereographic map is rather useful for small regions, its practical use for depicting (almost) the whole world in a single map is

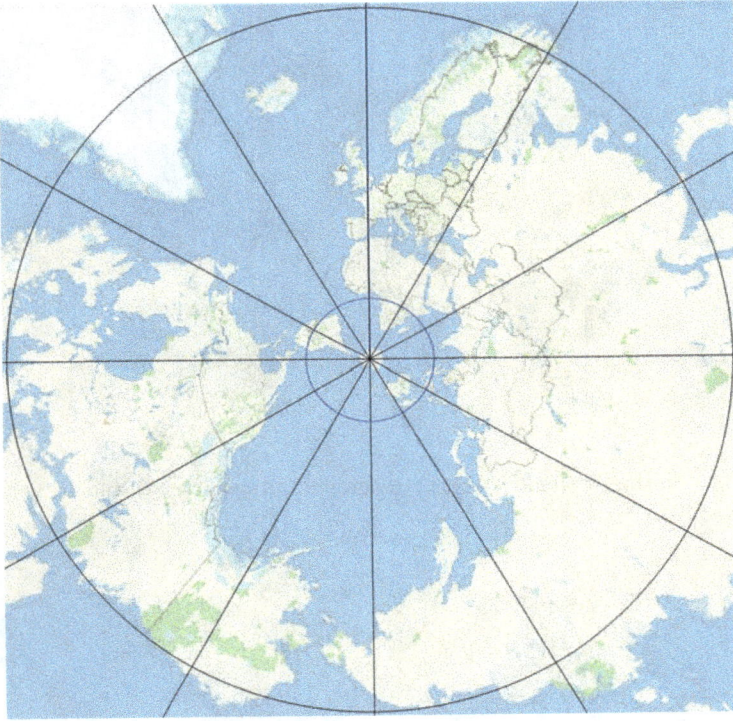

Fig. 4.32. Stereographic projection from the North Pole with the Southern Hemisphere at the center.

very limited, because the map would be huge and because the radial distortions created for the upper hemisphere soon become very extreme, as shown in Figure 4.32. Here, the entire Southern Hemisphere is mapped into the small blue circle indicating the equator. The useful range of a stereographic map is in most cases limited to a hemisphere or less. If one wants to depict the Northern Hemisphere one should choose the South Pole as a center and a tangent plane on the North Pole for the projection surface. This is done in Figure 4.33, where we compare the stereographic projection and the orthogonal projection for the Northern Hemisphere. In both cases, parallels of latitudes are circles, and meridians are radial lines. The amount of distortion in both cases is demonstrated with the help of Tissot indicatrices in Figure 4.34.

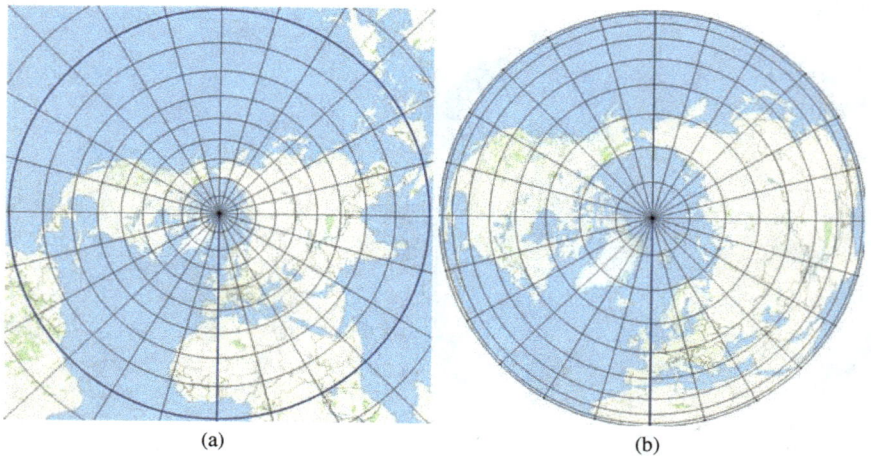

Fig. 4.33. Northern Hemisphere. Stereographic projection (a) and orthographic projection (b).

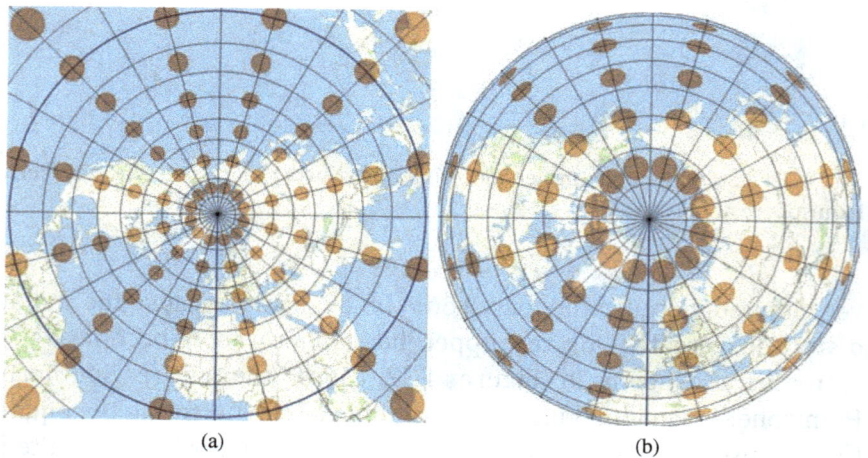

Fig. 4.34. Tissot indicatrices on Southern Hemisphere for the stereographic projection (a) and the orthographic projection (b).

4.18 Other Projections

There are many more methods of turning a spherical surface into a flat map. Various projections only differ in the concrete specification

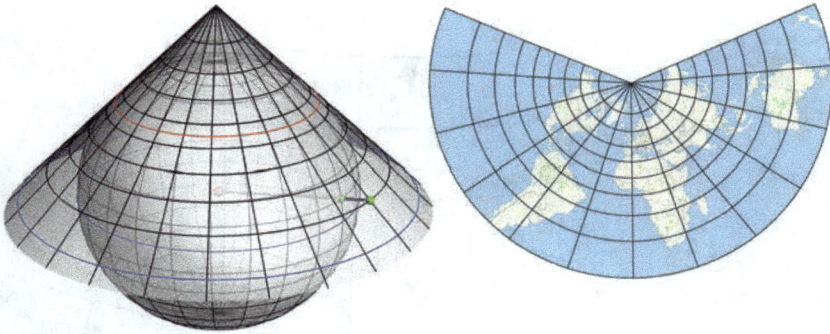

Fig. 4.35.   Conic projection from the center.

of how the two spherical coordinates, latitude and longitude, can be translated into two coordinates x and y of the two-dimensional plane.

A method that is frequently seen for maps of individual countries is conic projection. A cone is placed on the sphere as in Figure 4.35 so that it touches the sphere tangentially along a parallel of latitude. (The variant that the cone intersects the sphere along two circles is also very common.) The center of the projection is the center of the sphere. From there the spherical surface is projected onto the conic screen. It is not possible to map the entire spherical surface onto the cone, so one can only consider an area between a minimum and maximum geographic latitude. Then the cone is cut open and spread out in the plane in order to obtain a somewhat unusually-shaped non-rectangular map.

For maps of a smaller area of the world, and for individual countries, the conic projection is very popular and can often be found in atlases. One can recognize a conic projection by the non-parallel meridians and curved latitudes, as shown in Figure 4.36.

We have now examined some map projections that are commonly used in geography. There are many others that use interesting ideas to get as large a part as possible of the Earth's surface on a flat piece of paper in a shape that is as undistorted and as natural as possible. Once the task is to map the whole world, it becomes difficult to implement and the goal can only be achieved with compromises. That is

Fig. 4.36. Map of the contiguous United States in conic projection.

why there is not only one single, correct method of map display. Each projection has its advantages and disadvantages, and it is up to the cartographer to choose which form of presentation is best suited for what is to be shown.

What you learn about map projections can also be used in other contexts. It is rarely the case that mathematical results are only useful in one area. Understanding these principles, enables useful applications in many areas that at first glance have little to do with each other. For example, wide-angle or panoramic photography also involves projections, since objects that are spherically arranged around the camera have to be projected onto a flat piece of sensor or film.

In this chapter, we had a first look at the wide field of spherical geometry. In a sense, it's just a moderate step from the geometry of the plane to the geometry of three-dimensional space. While spherical geometry can be fascinating as a purely mathematical discipline, it also has some very useful applications. It not only provides the foundation for our ability to navigate on the spherical surface of the Earth, but also helps with observations in the sky. There too, we can observe geometric relationships and dynamic events on a spherical surface, the celestial sphere, only that we look at it from the inside. This will be the topic of Chapter 5.

Chapter 5

The Earth, the Sun and the Moon

5.1 Why the Earth is not Flat

In the previous chapter, we reviewed essentials of spherical geometry and the geography of Earth, especially the role of great circles, navigation paths and map projections, always assuming a spherical Earth. In fact, we consider the Earth as a sphere, albeit, not a perfect sphere but very close to a perfect sphere. What clues do we have that the Earth is actually spherical? Typically, teachers have mentioned it and this was supported when we see images from space. However, when looking out of the window, nothing gives us a clue that we live on the surface of a sphere. So, how over 2,300 years ago did the Greek philosopher and mathematician Aristotle (384–322 BCE) get the idea that the Earth is a sphere? How can we establish Earth as a sphere; or in other words, what are indications that would have us believe that the Earth is a sphere?

Aristotle partly based his assessment of the Earth's shape as a sphere on observations during a lunar eclipse, a phenomenon which only occurs during nights of a Full Moon. Ingeniously, he realized that in this specific case, the Moon moves into the shadow of the Earth, where no direct sunlight can reach the Moon. As the Moon travels through Earth's shadow, observers on the Earth see a dark circle slowly moving across the Moon — the two-dimensional silhouette of

Fig. 5.1.   Partial lunar eclipse (Juli 2018): Earth's shadow on the Moon.

the Earth, as shown in Figure 5.1. Moreover, Aristotle realized that this silhouette is always circular, no matter where in the sky the lunar eclipse occurs. But this can only be the case if Earth is a sphere, because only a sphere casts circular shadows in all directions. If the Earth were a flat circular disk, its shadow would appear elongated as an ellipse when illuminated at an oblique angle. Only if the line drawn from the center of the Sun to the center of the Moon were perpendicular to this disk would its silhouette appear circular. In this case, the lunar eclipse would have to take place exactly at the zenith — which is usually not the case. In fact, lunar eclipses occur in various places in the sky. This means that the angle at which the connecting line between Sun and Moon meets the hypothetical disk would be different each time, and the shape of the silhouette would then also have to be different each time. It is a fact that at any time during any lunar eclipse, we always see a circular silhouette of the Earth, which proves that Earth must be a sphere.

Aristotle had further observations regarding stellar constellations and their appearances over the horizon. For example, the North Star,

Polaris, lies "above" the North Pole. More precisely, it nearly lies on Earth's rotational axis, and since its distance from Earth is more than 50 million times longer than the diameter of Earth's orbit around the Sun, its position relative to Earth practically does not change as the Earth orbits the Sun. Hence, independent of the time of the year, the North Star's positions on the night sky always marks the point where the Earth's rotational axis pierces the firmament. Because the Earth rotates, we see the stars on the night sky revolve around the North Star. This apparent motion can also be explained with a flat Earth model, but there is a decisive difference regarding the change of altitude of Polaris as an observer moves across Earth's surface. If we were to live on a flat Earth, then the altitude of Polaris would be the same everywhere, namely, the angle between the disk representing the Earth and its rotational axis. However, on a spherical Earth, the angle through which an observer sees the North Star, depends on the latitude location. In fact, the altitude of the North Star measured by an observer on the Earth is exactly the geographical latitude of the observer's position, as can be seen in Figure 5.2.

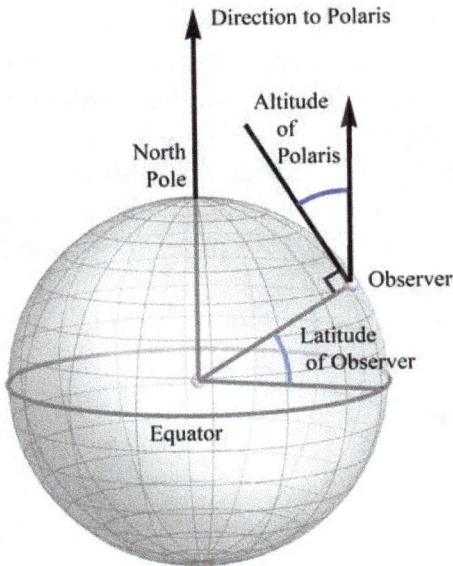

Fig. 5.2. Altitude of Polaris.

The change of altitude of Polaris as one is traveling to the North or South can be verified and, therefore, rules out a flat Earth model. For example, on the geographical North Pole, the North Star would be in the zenith. On the Southern Hemisphere, the North Star lies below the horizon and, thus, cannot be observed at all.

One may realize that a lunar eclipse does not happen every day and atmospheric conditions may not always allow us to determine the shape of Earth's shadow image projected on the Moon with sufficient precision. However, for someone who lives by the sea, the ancient Greeks had one additional argument to support the belief in a spherical Earth. Since Greece consists of a large number of islands, there was a lot of shipping traffic. So, one can imagine the great thinkers in ancient times sitting on the beach and watching the ships come closer. They recognized that when a ship appears on the horizon, one initially sees the sails and then the ship's hull, which we show in Figure 5.3. However, if the Earth were flat, a ship would appear as a tiny spot at the horizon and then increase gradually as one would perceive more details, but with the sails and the ship's hull appearing at the same time.

At this time, it would be quite natural to ask the following question: When we look into the ocean on a clear day, how far away is the horizon? In order to provide an answer to that question, we will start with a sketch of the situation in Figure 5.4, drawing a cross-section indicating the center of the Earth, the position of our eyes, and the point on the horizon at which we would be looking. In the drawing, the line of view is tangent to the circle representing the Earth. Since

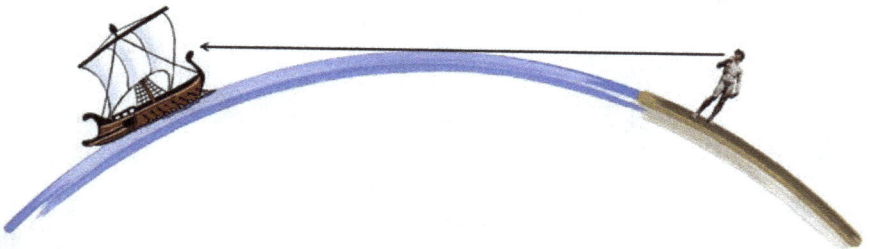

Fig. 5.3.   Watching ships appearing behind the horizon.

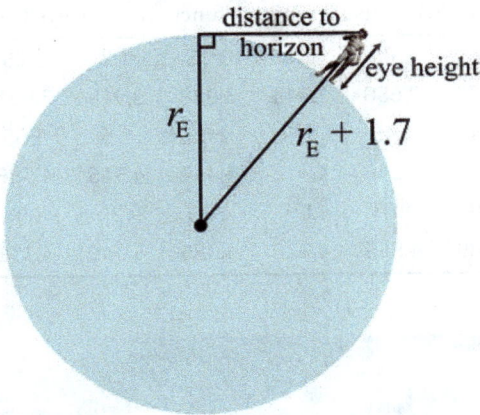

Fig. 5.4.   Visual range of an observer on a sphere — distance to the horizon.

the tangent to a circle is always perpendicular to the radius at the point of tangency, a right triangle is formed. We already know the length of two of its sides. First, the Earth's radius, $r_E = 6,371,000$ m. Note that this is a mean value, since Earth is not truly spherical, but deformed into a so-called geoid, as was mentioned in Chapter 4. The mean value we are using here is defined as the radius of a sphere that has the same volume as the Earth. To obtain the distance from our eyes to Earth's center, we just have to add our eye height (in this case 1.7 m) to the radius of the Earth, giving a value of $r_E + 1.7 = 6,371,001.7$ m.

Applying the Pythagorean theorem, we may calculate the distance from the observer to the horizon, which is $\sqrt{6,371,001.7^2 - 6,371,000^2}$ $\approx 4,654$ m.

In Table 5.1 we have (listed in terms of meters) the visual range at the beach as a function of the eye height above sea level. Here you can find your approximate visual range at the beach.

The average body height in ancient Greece was about 1.65 m, implying that the average eye height was approximately 1.55 m, corresponding to a visual range of about 4.4 km. The same method can be used to determine the distance to the horizon for an observation point at any height. From the observatory of the Burj Khalifa in Dubai, currently the tallest building in the world, at a height of 556 m, the distance to the horizon would be more than 84 km.

Table 5.1. Visual range as function of the eye level.

Eye height	1	1.05	1.1	1.15	1.2	1.25	1.3	1.35
Visual range	3,570	3,658	3,744	3,828	3,910	3,991	4,070	4,147
Eye height	1.4	1.45	1.5	1.55	1.6	1.65	1.7	1.75
Visual range	4,224	4,298	4,372	4,444	4,515	4,585	4,654	4,722
Eye height	1.8	1.85	1.9	1.95	2	2.05	2.1	2.15
Visual range	4,789	4,855	4,920	4,985	5,048	5,111	5,173	5,234

Fig. 5.5. Distant ship photographed from the same position but from different heights above sea level (0.5 m, 2 m, and 4 m). Photo by Ernst Meralla.

The eye height also determines how much one can see of an object that is behind the horizon, as shown in Figure 5.5. The geometry is exactly as shown in Figure 5.3, but here the ship stays at the same distance, only the eye level of the observer above the sea is changed for each photo.

Historically, the famous mathematician Pythagoras of Samos (ca. 572–496 BCE) was most likely the first person who perceived a spherical Earth. Furthermore, he claimed that the Moon was also spherical. However, no records of a proof or any explanation for his hypotheses have been found. As mentioned earlier, Aristotle (384–322 BCE) ingeniously considered the shadow of the Earth during a lunar eclipse to determine its shape. The Greek astronomer and mathematician Aristarch (310–250 BCE) postulated that the Sun is placed in the middle of the known universe, with the Earth orbiting the Sun. He presented the first heliocentric model of the universe. However, his ideas were not accepted and his model fell into oblivion.

It was not until the 15th century when courageous seafarers speculated that Earth might be a sphere. Christopher Columbus (1451–1506) carefully studied the theories and findings of the Greeks from which he inferred the spherical shape of the Earth. Convinced that the Earth is a sphere, but greatly underestimating its radius and thus, the distances on its surface, he made a proposal to the king of Spain, planning to reach India via a new sea route, heading West across the Atlantic instead of sailing around the southern end of Africa. Because of the high custom duties imposed by the Ottoman Empire, the overland routes to India were greatly discouraged. As we all know, he did not reach India, but discovered the American continent (or rather the islands of the Caribbean), so his journey did not do much to prove the spherical shape of the Earth.

The first practical demonstration that Earth is indeed a sphere was achieved by the Portuguese explorer Ferdinand Magellan (1480–1521), who managed (part of) the first circumnavigation of the Earth. Magellan discovered the Strait of Magellan at the southern tip of South America, a natural passage between the Atlantic and Pacific oceans. Reaching the ocean on the other side of the passage, he encountered favorable winds and, therefore, called the ocean *Mar Pacífico*, meaning "peaceful water". Magellan died in a battle after he and his crew arrived at the Philippines. Of his whole crew, 35 men returned to Spain via the Cape of Good Hope. Those men gave the first experimental evidence of a spherical Earth.

Until a general acceptance of the heliocentric model, a few more years went by. Let us now consider the Earth and its path around the Sun from a geometrical point of view.

While orbiting the Sun, the way in which the Earth moves and rotates is responsible for several phenomena, such as the alternation of day and night, the four seasons, ice ages and, in interaction with the Moon, for solar and lunar eclipses.

5.2 Day and Night

Let's take a closer look at the alternation of day and night, phenomena that result from Earth's rotation. The Earth rotates eastwards, hence, if you would be standing exactly on the North Pole, looking at your feet, the Earth would turn counterclockwise around this point.

The rotation of the Earth gives us the impression that Sun, Moon, and stars revolve around the Earth. It is, therefore, quite natural to describe the orbits of celestial objects in a coordinate system centered at the observer's location. For this reason, a very useful concept is the celestial sphere, which we introduced in Chapter 4. For an observer on the Earth, the sky appears like a sphere that we look at from the inside. The position of the observer on Earth's surface represents the center of this celestial sphere. One can imagine the celestial sphere as a spherical projection surface. The position of a star on the celestial sphere is the intersection of the light ray that is emitted from the star to the observer, and the projection surface. In Figure 5.6, we look at the celestial sphere from the outside and indicate the projected position of the Sun at a certain time. If we follow the intersection throughout the day, we obtain the complete path of the Sun on the celestial sphere on that particular day. Actually, the path of any planet or star that is visible from Earth can be drawn on the celestial sphere. This model exactly corresponds to our perception of the sky.

So far, we have not spoken about the radius of the celestial sphere. Although it seems to be rather large, the radius is in fact arbitrary,

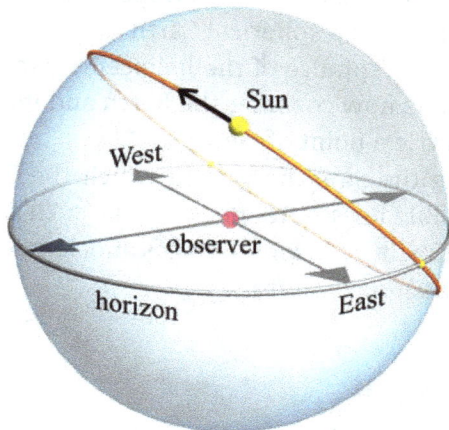

Fig. 5.6.   The celestial sphere and the apparent path of the Sun.

since the coordinates we use on the celestial sphere are independent of the radius of the sphere, just as are the longitudes and latitudes we use to locate a point on Earth's surface.

In Figure 5.6, we should imagine that we are looking at the celestial sphere from a point on Earth's surface. The observer can essentially only see that part of the celestial sphere, which is above the plane tangential to the Earth through the observer's location. The intersection of this plane and the celestial sphere is the horizon, which is here interpreted as a line on the celestial sphere (and not as a line on the Earth's surface indicating the visual range as shown in Section 5.1). We also marked the four directions of the compass in the tangent plane. Aside from mountains, clouds, or other obstacles, the Sun is visible from sunrise to sunset everywhere on Earth. Sunrise and sunset correspond to the intersection points of Sun's path and the horizon. Sunrise always takes place to the east of the observer, the Sun then moves across the sky (as indicated by the arrow in Figure 5.6), reaching its highest point at noon and finally setting in the west. Figure 5.6 shows the typical situation for an observer in the Northern Hemisphere, where the Sun reaches its highest point in the south. If the observer is in the Southern Hemisphere, the Sun's orbit is inclined in such a way that the highest point is in the north.

In Figure 5.7, we have only drawn the part of the Sun's path that is visible from the observer's point of view, that is, we show only the part that lies above the horizon.

As you may have experienced, day length depends on geographical latitude. Therefore, one distinctive feature is still missing in our model. In order to display the latitude, we must introduce the celestial equator. Imagine a plane that goes through the position of the observer and is parallel to the equatorial plane of the Earth. The intersection of this plane with the celestial sphere is called the celestial equator. The celestial equator intersects the horizon in the east point E and west point W. Subsequently, the north point N and the south point S can also be drawn. The highest point on the celestial equator is called the culmination point, and it is denoted by H. The line perpendicular to the equatorial plane is parallel to Earth's axis of rotation. It points towards the celestial North Pole P (the approximate

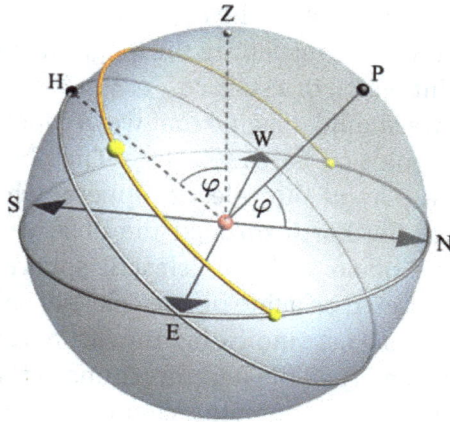

Fig. 5.7. Horizon and celestial equator on the celestial sphere.

Fig. 5.8. Long-time exposure of the night sky.

location of the North Star, Polaris), which is also shown in Figure 5.7. The stars seem to move in circles around the North Star. Figure 5.8 shows a photograph (with a very long exposure time) of the night sky. As a result of their movement, all the stars produce circular lines on the picture, which shows that the entire firmament seems to revolve around the north celestial pole.

The angle φ in Figure 5.7 denotes the geographical latitude, and, as discussed at the beginning of this chapter, it can be defined as the altitude of Polaris, that is, as the angular distance from the celestial North Pole P to the North point N of the horizontal plane. Similarly, the latitude of an arbitrary point on Earth's surface can be defined as the angle between the equatorial plane and the normal line to the tangent plane through this point, that is, as the angular distance between the points Z and H as shown in Figure 5.7. The point Z, which is exactly vertically above the viewer, is called the zenith. We note the following special cases: If $\varphi = 0°$, the observer is located somewhere at the equator, and the celestial North Pole P coincides with north point N. If $\varphi = 90°$, the observer is at Earth's North Pole and the celestial North Pole P coincides with the zenith Z.

In order to describe the position of a celestial object, we need to introduce a coordinate system. The position of the Sun, denoted by S_P in Figure 5.9, can be unambiguously specified by two angles, called the declination, δ, and the hour angle, η.

The declination angle δ measures the angular distance between the celestial equator and a certain celestial body, for example the Sun, as shown in Figure 5.9. The great circle through the points P and S_P

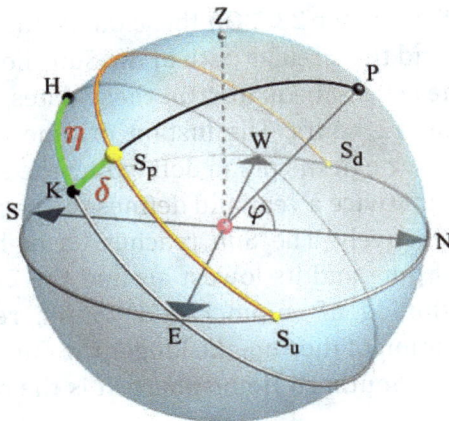

Fig. 5.9.   Declination δ and hour angle η in the celestial sphere.

intersects the celestial equator in the point K. The declination δ is the angle of the great circle arc between the points K and S_p. The hour angle η is the angular distance from the culmination point H to point K along the celestial equator.

Table 5.2 shows the Sun's declination for each day of the year. The declination of the Sun changes continuously as Earth orbits the Sun, because the tilt of Earth's axis causes the Sun to be higher in the sky in the summer months and lower during the winter. This will be explained later in more detail.

The Sun's path on the celestial sphere during a day can be considered approximately as a parallel circle to the celestial equator, and the distance between these two circles is the declination δ. As mentioned before, δ changes continually over the course of the year. Therefore, the path of the Sun on the celestial sphere is actually somewhat spiral. But δ changes very little during a day and hence the deviation from a circle is small. For our purposes, we shall assume that the declination is constant during a day. In this model, the intersection points of the Sun's circle and the horizon represent the sunrise S_u and the sunset S_d. If we calculate the distance measured as an angle along the celestial equator from S_u to S_d, we can determine the length of the day.

Before we go into further detail, consider Table 5.2[1] and note that the declination of the Sun becomes zero twice a year, around March 21st and around September 23rd. On these dates, the Sun crosses the celestial equator and the circular path of the Sun shown in Figure 5.6 coincides with the celestial equator. On these dates, the night is just as long as the day, 12 hours. The instant of time when the center of the Sun passes through the plane defined by the equator, is called an equinox. It occurs twice a year and defines the beginning of spring and autumn, respectively. The Sun reaches its highest declination around the 21st of June and its lowest around the 21st of December, marking the beginning of summer and winter, respectively. The greater the declination of the Sun, the longer the day on the Northern Hemisphere. On the Southern Hemisphere, it is the reverse.

---

[1] https://astronavigationdemystified.com/survival-declination-table/

Table 5.2.   Sun's declination table.

	JAN	FEB	MAR	APR	MAY	JUN	JUL	AUG	SEP	OCT	NOV	DEC
1	−23.1	−17.3	−7.8	4.3	14.9	22.0	23.1	18.1	8.5	−2.9	−14.2	−21.7
2	−22.9	−17.1	−7.4	4.7	15.2	22.1	23.1	17.9	8.1	−3.3	−14.6	−21.8
3	−22.9	−16.8	−7.0	5.1	15.5	22.2	23.0	17.6	7.8	−3.7	−14.8	−22.0
4	−22.8	−16.5	−6.6	5.5	15.8	22.3	22.9	17.4	7.4	−4.1	−15.2	−22.1
5	−22.7	−16.2	−6.3	5.8	16.1	22.5	22.8	17.1	7.0	−4.5	−15.5	−22.3
6	−22.6	−15.8	−5.8	6.2	16.4	22.6	22.7	16.8	6.6	−4.8	−15.8	−22.4
7	−22.4	−15.6	−5.5	6.6	16.6	22.7	22.6	16.6	6.3	−5.3	−16.1	−22.5
8	−22.3	−15.2	−5.1	7.0	16.9	22.8	22.5	16.3	5.9	−5.6	−16.4	−22.6
9	−22.2	−14.8	−4.7	7.3	17.2	22.9	22.5	16.0	5.5	−6.0	−16.6	−22.7
10	−22.1	−14.6	−4.3	7.7	17.4	23.0	22.3	15.7	5.1	−6.4	−16.9	−22.8
11	−21.9	−14.3	−3.9	8.1	17.7	23.0	22.2	15.4	4.8	−6.8	−17.2	−22.9
12	−21.8	−13.9	−3.5	8.5	18.0	23.1	22.1	15.1	4.4	−7.1	−17.5	−23.0
13	−21.6	−13.6	−3.1	8.8	18.2	23.2	21.9	14.8	4.0	−7.5	−17.8	−23.1
14	−21.5	−13.3	−2.8	9.2	18.5	23.2	21.8	14.5	3.6	−7,9	−18.1	−23.1
15	−21.3	−12.9	−2.3	9.5	18.7	23.3	21.6	14.2	3.3	−8.3	−18.3	−23.2
16	−21.1	−12.6	−2.0	9.9	19.0	23.3	21.5	13.9	2.8	−8.6	−18.6	−23.3
17	−20.9	−12.2	−1.6	10.3	19.2	23.3	21.3	13.6	2.5	−9.0	−18.8	−23.3
18	−20.7	−11.8	−1.2	10.6	19.4	23.4	21.1	13.3	2.1	−9.4	−19.1	−23.3
19	−20.5	−11.6	−0.8	10.9	19.6	23.4	20.9	13.0	1.7	−9.7	−19.3	−23.4
20	−20.3	−11.2	−0.4	11.3	19.8	23.5	20.8	12.6	1.3	−10.1	−19.6	−23.4
21	−20.1	−10.8	0.0	11.6	20.1	23.5	20.6	12.3	0.9	−10.5	−19.8	−23.5
22	−19.8	−10.5	0.4	12.0	20.3	23.5	20.4	12.0	0.5	−10.8	−20.0	−23.5
23	−19.6	−10.1	0.8	12.3	20.4	23.5	20.2	11.6	0.1	−11.2	−20.2	−23.4
24	−19.4	−9.7	1.2	12.4	20.6	23.4	20.0	11.3	−0.2	−11.5	−20.4	−23.4
25	−19.2	−9.4	1.6	13.0	20.8	23.4	19.8	10.9	−0.6	−11.8	−20.6	−23.4
26	−18.8	−9.0	2.0	13.3	21.0	23.4	19.6	10.6	−1.0	−12.2	−20.8	−23.4
27	−18.6	−8.6	2.6	13.6	21.2	23.3	19.3	10.3	−1.4	−12.6	−21.0	−23.3
28	−18.4	−8.3	2.8	14.0	21.4	23.3	19.1	9.9	−1.8	−12.9	−21.2	−23.3
29	−18.1	−8.0	3.1	14.3	21.5	23.2	18.9	9.6	−2.1	−13.2	−21.3	−23.2
30	−17.8	—	3.5	14.6	21.6	23.2	18.6	9.3	−2.6	−13.6	−21.5	−23.2
31	−17.6	—	3.8	—	21.8	—	18.4	8.8	—	−13.8	—	−23.1

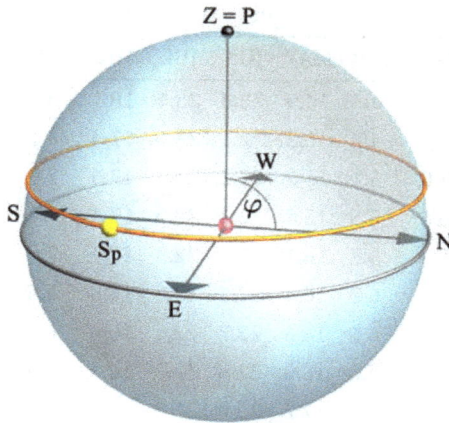

Fig. 5.10.   Celestial sphere at the North Pole, during summer.

In order to get a better understanding of the celestial sphere model, let us consider some specific values for φ. First, suppose that we are at the North Pole, corresponding to a geographical latitude $\varphi = 90°$. Then the points P and Z on the celestial sphere coincide or, equivalently, the celestial equator and the horizon coincide (see Figure 5.10). Between March 21st and September 23rd, the declination of the Sun is always positive. Figure 5.10 shows the path of the Sun on the celestial sphere for this case, a circle above and parallel to the celestial equator. The picture explains why the Sun does not set down during this period. At the North Pole, the Sun rises on March 21st and sets on September 23rd — 6 months of sunlight followed by 6 months of darkness.

By setting $\varphi = 0°$, we are on the equator. Again, some points coincide, as P = N and H = Z, as shown in Figure 5.11. In our model, we assumed that the plane of the Sun's path and the equator's plane are parallel. So regardless of the actual declination δ, at a point on the equator, we can always see half of the Sun's path. Hence, the day length is always 12 hours. Another interesting fact is that on the day of an equinox, the celestial equator and the Sun's path coincide, so, the Sun is at the zenith at midday (high noon).

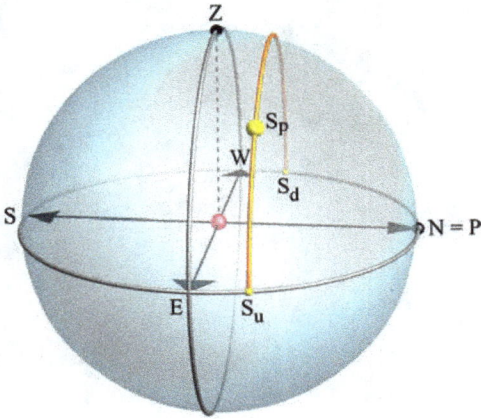

Fig. 5.11. Celestial sphere for an observation point on Earth's equator.

5.3 A Formula for the Length of a Day

For the mathematically ambitious reader, we will derive a formula for the length of a day. For the mathematically ambitious reader, we will derive a formula for the length of a day according to Schuppar (2017).[2] The formula will work for any day of the year and for an arbitrary point on Earth's surface. In order to use the formula, one must know the geographical latitude φ of the place used and the declination δ of the Sun, which for any day of the year can be obtained from Table 5.2. One can then use a calculator to compute the angle:

$$\alpha = 2 \cdot \arccos\left(-\tan\left(\varphi\right) \cdot \tan\left(\delta\right)\right),$$

at which the Sun travels across the sky between sunrise and sunset. The Sun travels exactly 15 degrees in an hour, hence the angle α, divided by 15 gives the day's length in hours.

Our goal is to derive the formula for α, just using some high school trigonometry. Geometrically, the day length corresponds to the angle

---

[2] Schuppar, B. (2017). *Geometrie auf der Kugel. Alltägliche Phänomene rund um Erde und Himmel* [Geometry on the sphere. Everyday phenomena around earth and sky]. Springer.

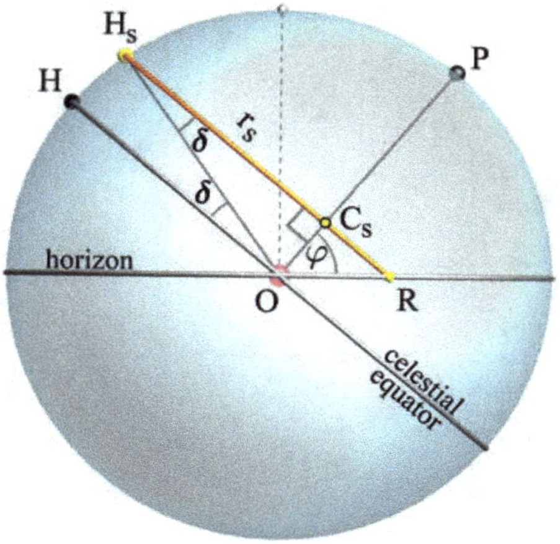

Fig. 5.12.   Vertical cross-section.

that spans the circular sector of the visible Sun's path between the points S_u and S_d. We have already determined the day length for two special locations on Earth: the North Pole and an arbitrary point on the equator. Deriving a general formula for the day length, depending on the geographical latitude φ and the Sun's declination δ, is a bit more involved and we will need two more diagrams: Figure 5.12 shows a vertical cross-section of the sphere shown in Figure 5.7, and Figure 5.13 shows the Sun's orbit, viewed from the top.

For further considerations, we denote the observer by O, and we set the radius of the celestial sphere equal to 1. As mentioned above, the celestial sphere works with every radius. Thus, in Figure 5.12, $OH_S = 1$ (the distance from the observer to the Sun's culmination point H_S). In Figure 5.12, sunrise S_u and sunset S_d coincide and become the point R in the vertical cross-section. We designate the radius of the Sun's circular orbit as r_S, which is represented by the line segment $H_S C_S$ in Figure 5.12 (C_S denotes the center of the Sun's orbit). Furthermore, there are two right triangles $OC_S H_S$ and $RC_S O$. Using trigonometry, we get the following relationships:

$$\sin(\delta) = \frac{OC_S}{OH_S}, \text{ which implies } OC_S = \sin(\delta),$$

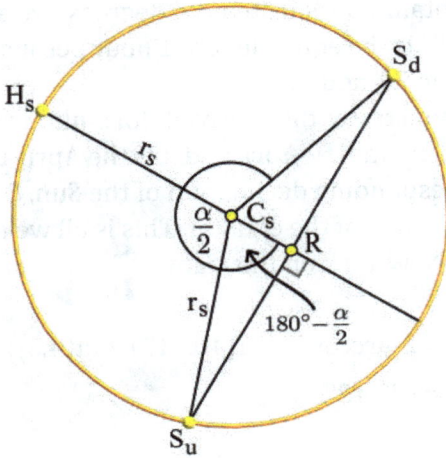

Fig. 5.13.   Top-view of the plane of the Sun's apparent path.

$$\cos(\delta) = \frac{r_s}{OH_s}, \text{ which implies } r_s = \cos(\delta),$$

$$\tan(\varphi) = \frac{C_sR}{OC_s}, \text{ which implies } C_sR = \tan(\varphi) \cdot \sin(\delta).$$

Our goal is to calculate the angle α the Sun travels between sunrise S_u and sunset S_d, which represents the day length. See Figure 5.13, which shows the Sun's path from above.

In Figure 5.13, we consider the right triangle RC_sS_u and express the cosine of the angle at C_s in terms of φ and δ by employing the equations listed above:

$$\cos\left(180° - \frac{\alpha}{2}\right) = \frac{C_sR}{r_s}, \text{ which implies } \cos\left(180° - \frac{\alpha}{2}\right) = \frac{\tan(\varphi) \cdot \sin(\delta)}{\cos(\delta)}.$$

This formula can be further simplified:

$$\cos\left(\frac{\alpha}{2}\right) = -\tan(\varphi) \cdot \tan(\delta),$$

$$\alpha = 2 \cdot \arccos(-\tan(\varphi) \cdot \tan(\delta)).$$

In order to obtain hours instead of degrees, we apply the following conversion: 15° is the equivalent to 1 hour because the Sun travels the angle of 360° in 24 hours.

We can demonstrate the derived formula with an example: Assuming that we are in New York and it is the April 1st, then Table 5.2 provides the corresponding declination of the Sun, $\delta = 4.3°$. New York is situated 40°43′ north of the equator. This is all we need to calculate the day length in New York on this date:

$$\alpha = 2 \cdot \arccos\left(-\tan(40.717) \cdot \tan(4.3)\right),$$
$$\alpha = 187.42°.$$

By dividing this number by 15, we get 12.49 hours or approximately 12 hours and 30 minutes. On the internet, one can find online tools to calculate the precise times of sunrise and sunset for any location on Earth. For example, on the website www.timeanddate.com, for the 1st of April in New York, we find that sunrise occurs at 6:38 and sunset at 19:21, which yields a day length of 12:42. The difference of 12 minutes compared to our calculation can be explained by the various simplifications we made in our model. For example, we treated the Sun as a point, while in reality it appears as a disk on the sky with an angular diameter of about half a degree. Sunset and sunrise are actually defined as the moments when the upper limb of the Sun touches the horizon. Additional corrections come from atmospheric refraction, that lets the Sun appear higher in the sky than it actually is. Hence, the Sun is already visible when it is actually still below the horizon. These effects are taken into account in more accurate computations and because of these effects, daytime is slightly longer than predicted by our simple geometric consideration.

However, the deviation is quite small and the fact that a bit of three-dimensional geometry and trigonometry in the plane were sufficient to derive a formula for the day length depending on the geographical latitude shows the power of mathematics. With the help of Table 5.2 and our formula, we are able to calculate the (approximate) day length for any location on Earth's surface at any time of the year.

There is one further interesting mathematical and historical aspect in connection with observations of the Sun and the planets, which is somehow close to our previous considerations. Nowadays, we are used to time-dependent figures, for example within demographic trends, environmental pollution, etc. The representation of time as an axis from left to right stems from Aristotle and is still used today. However, in 1936, the American mathematician and historian Howard G. Funkhouser (1898–1984) published a note on a tenth century graph. He discovered in a manuscript belonging to the Bayrische Staatsbibliothek in Munich, which is the oldest known example of an attempt to graphically represent values changing with time, in other words the graph of a mathematical function. This graph represents the inclinations of the planetary orbits as functions of time, see Figure 5.14.

We can see the paths of the Sun and the Moon and the planets Venus, Mercury, Mars, Jupiter, and Saturn. In fact, these are the celestial bodies whose movements can be observed by the naked eye. Considering that this graph had been drawn long before Rene Descartes (1596–1650) invented the notion of a coordinate system,

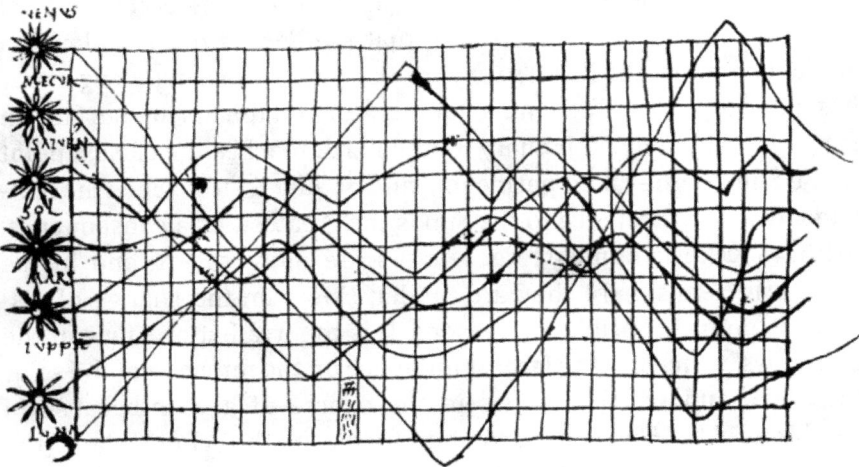

Fig. 5.14.   Oldest graph of a time-dependent function.

Source: Funkhouser, H. G. (1936). A note on a tenth century graph. Osiris, 1, 260–262.

this drawing from the tenth century is very remarkable. It was not until the 17th century, however, that the next time-dependent representation with a time axis can be found in a letter of the Dutch scientist Christiaan Huygens (1629–1695) — about 700 years later!

5.4 Seasons

In the Northern Hemisphere, spring begins (approximately) on March 21st, summer on June 21st, autumn on September 22nd and winter on December 21st. In the Southern Hemisphere, spring starts on September 22rd, summer on December 21st, autumn on March 21st and winter on June 21st. But why are there seasons at all?

Again, we can answer this question with three-dimensional geometry. Zones close to the equator see little annual fluctuations, so there are no four seasons. Instead, one speaks of wet and dry seasons. When the northern regions have a dry season, the southern regions experience a wet season and vice versa. A frequent incorrect response to the question of why there are seasons refers to the distance from the Earth to the Sun. It is true that the distance changes during the year since the Earth's orbit is elliptical. However, the eccentricity of this ellipse is very small, meaning that the Earth's orbit is almost a circle. Therefore, the Earth's varying distance to the Sun during a year has only very little effect on the temperature on the Earth's surface. In fact, approximately on January 3rd the Earth reaches the closest point to the Sun, the so-called perihelion, and on July 5th it reaches the aphelion, where its distance to the Sun is the greatest. As a consequence, in the Southern Hemisphere, winters are more severe and in the Northern Hemisphere they are milder. However, the change of distance is not responsible for seasons. It is the irradiation angle of the Sun's rays, which has the largest influence on the intensity of the solar radiation, and which changes in the course of a year because of Earth's axial tilt. The Earth's rotational axis and the ecliptic plane (the plane of Earth's orbit around the Sun) form an angle of approximately 66.56°, which we show in Figure 5.15. While the Earth orbits the Sun, the orientation of its rotational axis does not change in relation to the

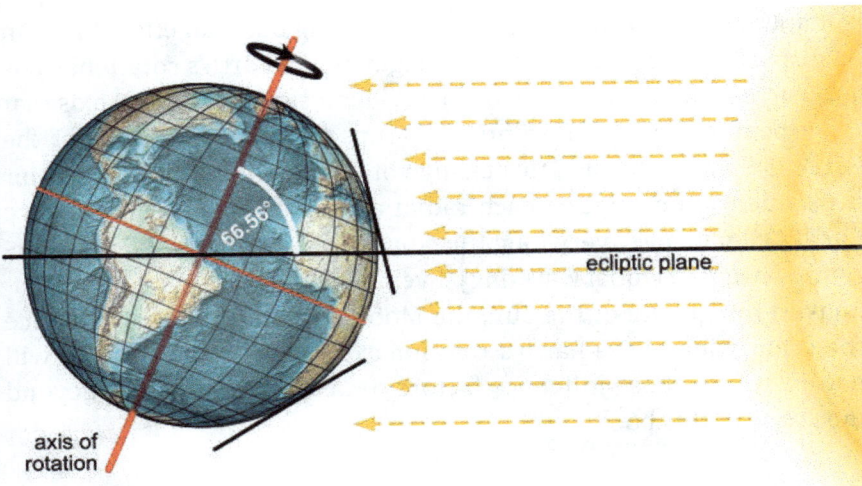

Fig. 5.15. On June 21st, the Northern Hemisphere is tilted towards the Sun.

background of stars, which is a consequence of a law of physics known as conservation of angular momentum. Therefore, from March 21st to September 22nd, the Northern Hemisphere is tilted to the Sun (see Figure 5.15), and from September to March it is the Southern Hemisphere that receives most of the Sun's radiation. In Figure 5.15, we see two tangents to the same geographical latitude except for the hemisphere or the algebraic sign, respectively. If the Northern Hemisphere is tilted to the Sun, the northern tangent is steeper than the southern with respect to the Sun's rays, and therefore, the irradiation angle at the tangent point is bigger than the corresponding one on the Southern Hemisphere. Hence, between March and September, a point on the Northern Hemisphere receives a higher intensity of solar radiation than a point with the same absolute value of geographical latitude on the Southern Hemisphere.

In our examples, we made some simplifications, which, however, could be replaced with by a more elaborate model. Precise calculations of the orbits of celestial bodies are, for instance, required for astronomical observations, but some of the effects we omitted do not play a role on time scales in the order of a human lifetime, such as, the precession of the Earth's rotational axis. Precession is a change of orientation of the rotational axis of a rotating body, an effect that can be observed with a spinning top. Imagining the Earth's rotational axis as an arrow, the tip of this arrow describes a circle around an axis that is perpendicular to the ecliptic and runs through the center of the Earth. This motion leads to a change of the Earth's tilt and in further consequence a change of each radiation angle, which will influence Earth's climate over very long timespans (but cannot explain the currently observed global warming). Events like tidal interactions, earthquakes, the gravity of the Sun, the Moon and other bodies influence the orientation of the Earth's rotation axis as well. However, we will not delve into those influence factors because this would go beyond the scope of this book.

5.5 Phases of the Moon

Considering the night sky, many people are fascinated by the Moon. Some of them even credit the moon is having an impact on human behavior. Since the time of the ancient Greeks (and perhaps even earlier) the Moon has been associated with insanity, suicide, murder, violence and illegal activity. Aristotle was convinced that the brain, mostly consisting of water, is influenced by the Moon in a way similar to the tides. Nowadays, the lunar effect states an increase of traffic accidents, admissions to psychiatric hospitals, murder and births during a Full Moon. Although scientists have debunked all of these myths, people still fall for theories claiming that the lunar cycle correlates with psychological changes in human behavior on Earth.

While most of the claimed effects refer to phases of the Full Moon, the Moon is, of course, there all the time, only its appearance changes. We all know pictures of the different Moon phases as shown in Figure 5.16.

So, why does the Moon change its appearance during a lunar cycle? At any time, half of the Moon's surface is lit by the Sun (on this hemisphere it is "day" on the Moon), except during a lunar eclipse. On the other half, it is "night". Therefore, only half of the Moon appears bright. Depending on the position of the Moon on its orbit around the

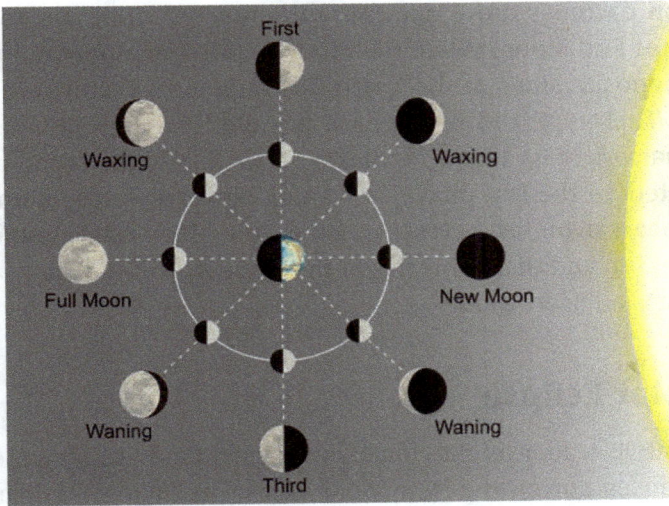

Fig. 5.16. Moon phases.

Earth, we see different portions of this bright half. Astronomers divided the journey of the Moon around the Earth into different phases, which are shown in Figure 5.16.

Let us start with the phase of New Moon: In this constellation, the Sun and the Moon are on the same side of the Earth, hence the Moon is not visible because it is hidden in the Sun's glare. As the Moon travels further along its orbit, in a counterclockwise direction in Figure 5.16, we see a bright crescent. This phase is called the waxing crescent. Many people believe that the shadow on the Moon arises because the Earth shields the Moon from sunlight. However, this is only the case during a lunar eclipse. Throughout the different phases of a lunar cycle, we simply see different amounts of the bright half of the Moon, as is illustrated in Figure 5.16. When the Moon has completed one quarter of its orbit around the Earth, we see a Half-Moon (see Figure 5.16). This phase of the lunar cycle is called "first quarter", referring to the fraction of the Moon's orbit that has been covered since New Moon, not to the part of the Moon that appears illuminated, as many people believe. During the so called waxing gibbous phase, the Moon becomes more and more illuminated until it reaches the position of Full Moon. When the Moon is full, the Moon and the Sun are on opposite sides of the Earth. Subsequently, the lighted area is waning. Therefore, the next phase is called waning gibbous. After three quarters of its way around the Earth, half of the Moon appears illuminated. In the last phase — waning crescent — the illuminated area appears again as a crescent. When the Moon has completed its orbit, the Sun and the Moon are on the same side of the Earth and the cycle starts all over again.

5.6 Solar Eclipse

The attentive reader might wonder why we do not always experience solar eclipses during the New Moon phase, as Figure 5.16 suggests. Thus far, we did not mention one important feature of Moon's orbit that is not visible in Figure 5.16, namely, that the Moon's orbit is tilted by an angle of 5.146° with respect to the ecliptic (the plane defined by the orbit of the Earth around the Sun), and the orientation of the lunar orbit with respect to the ecliptic remains essentially unchanged as the entire Earth–Moon system moves around the Sun, as shown in Figure 5.17. Therefore, its shadow usually misses the Earth and a solar eclipse does not occur during every New Moon phase.

But under which circumstances does a solar eclipse occur? As we have already noticed, the Sun, the Moon, and the Earth must be aligned, with the Moon in the middle, hence, it must be the New Moon

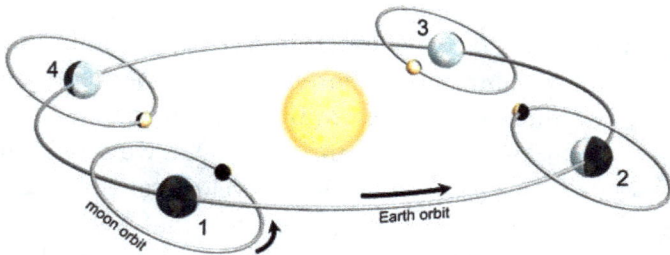

Fig. 5.17. Schematic representation showing the inclination of Moon's orbit with respect to the ecliptic (the plane defined by Earth's orbit around the Sun).

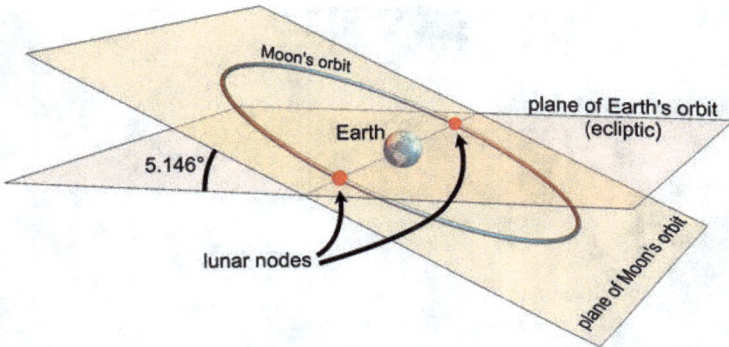

Fig. 5.18.   The Moon's orbit and the two lunar nodes (in red).

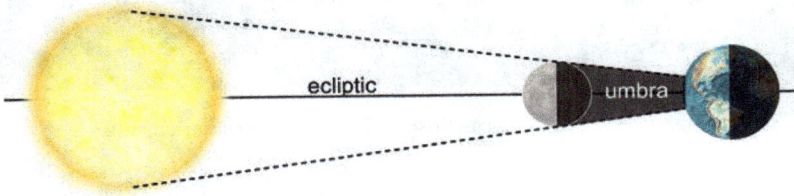

Fig. 5.19.   Schematic representation of the umbra of the Moon (not to scale).

phase. Figure 5.17 shows four possible positions of the Earth and New Moon. Because of the inclination of the Moon's orbit, a solar eclipse can only occur in positions 1 and 3, while in positions 2 and 4 the Moon is either too high above the ecliptic or too far below. For an eclipse, the Moon has to be close to one of two special points along its orbit, the so-called lunar nodes. These are the points where the Moon's orbit crosses the ecliptic, and they are shown in Figure 5.18.

When the Moon crosses the ecliptic exactly at New Moon, the Sun, the Moon and the Earth are in a straight line, see Figure 5.19. In this case, a portion of the Earth is engulfed in a shadow cast by the Moon. Figure 5.19 shows the innermost and darkest part of this shadow, the so-called umbra. Depending on the location on the Earth's surface, the Moon then partially or totally blocks the sunlight, resulting in a partial or total solar eclipse. For observers in the umbra, the light of the Sun

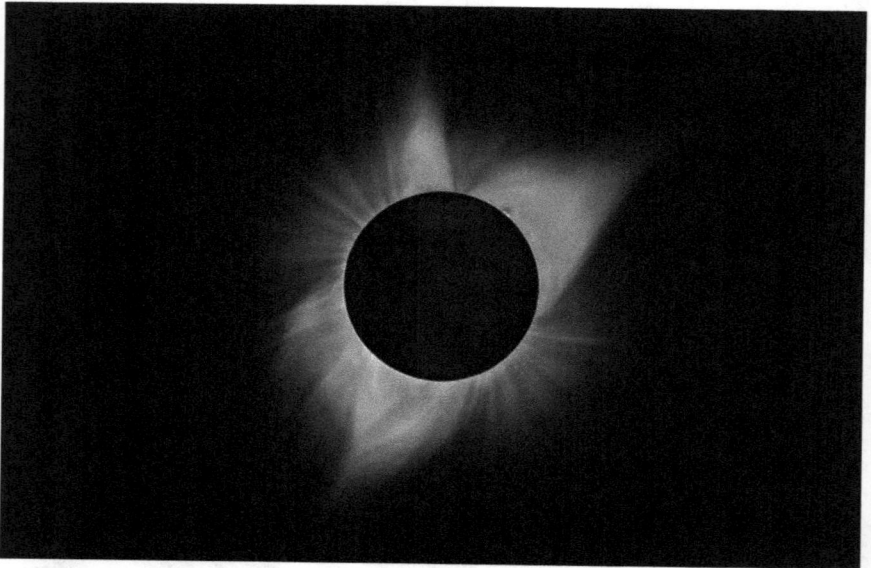

Fig. 5.20.   Total solar eclipse of 2017 and the solar corona.

is completely blocked by the Moon and they experience a (total) solar eclipse. The shape of the shadow is a cone where the Earth truncates its apex and the Moon cuts off a bit of its bottom.

On August 21st, 2017 the umbra of the Moon took a path across the USA from the Pacific to the Atlantic coast. The total solar eclipse was visible only within a narrow path of about 110 km in diameter. Figure 5.20 shows a high-dynamic-range image of the view as seen in the State of Wyoming, United States. With the naked eye, the solar corona can only be seen during a total eclipse — it is a rare and mesmerizing sight. It consists of thin plasma that is very hot (1–3 million degrees Celsius) and is shaped by the twisted magnetic field of the Sun.

5.7 Size of the Moon's Shadow

Having clarified the geometric conditions for a solar eclipse, we shall now go a bit further and calculate the size of the Moon's shadow on

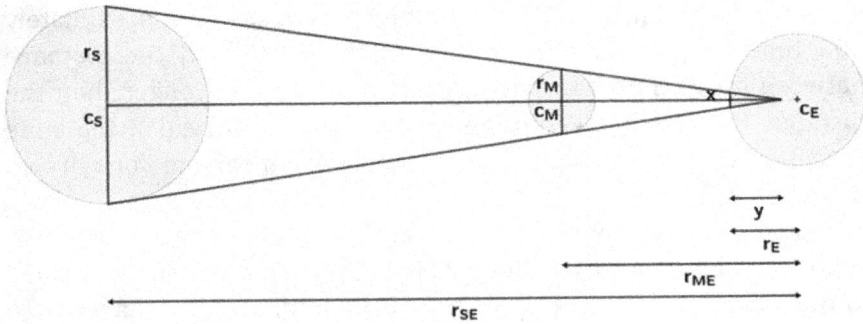

Fig. 5.21. Geometry of a solar eclipse, not drawn to scale.

the Earth's surface. The set of all lines that are tangent to the Sun and the Moon form a cone. A cross-section of this cone is shown in Figure 5.21. We denote the radius of the shadow at Earth's surface by x and the distance from Earth's surface to the vertex of the cone by y. The radius of the Sun r_S is to the distance from the center of the Sun to the vertex, $r_{SE} - r_E + y$, as the radius of the Moon, r_M, is to the distance from the center of the Moon to the vertex, $r_{ME} - r_E + y$. Using the relationship from two parallel lines cut by transversals (intercept theorem), we get

$$\frac{r_S}{r_{SE} - r_E + y} = \frac{r_M}{r_{ME} - r_E + y}.$$

Except for y, we are aware of all values. The average distance between the Earth and the Sun is $r_{SE} = 149{,}600{,}000$ km and the average distance between the Moon and the Earth is $r_{ME} = 384{,}000$ km. The radii of the involved celestial bodies have the following average values: $r_S = 696{,}340$ km, $r_E = 1{,}737$ km, and $r_E = 6{,}371$ km. We consider average distances and radii because these bodies move along ellipses and they are not perfect spheres. Inserting these values into the equation and solving for y, we obtain $y = 23{,}433.22$ km. Thus, the value of y is, in fact, larger than the diameter of the Earth, other than the diagram shown in Figure 5.21. However, this does not matter, since it is just a schematic drawing that helps us to set up the equations needed to determine y and x. In fact, the radius of the Sun is approximately 100 times bigger than the radius of the Earth and the distance between the Sun and the Moon is about 400 times longer than the distance between the Moon and the Earth. In particular, the cone representing the umbra is significantly narrower than the one drawn in Figure 5.21.

Knowing y, we apply the intercept theorem a second time to calculate x. The radius of the Moon r_M is to the distance from the center of the Moon to the vertex $r_{ME} - r_E + y$ as the radius of the shadow x to y. This leads to the following equation:

$$\frac{r_M}{r_{ME} - r_E + y} = \frac{x}{y}.$$

From this we obtain $x \approx 109$ km. Hence, the diameter of the circular shadow cast on the surface of the Earth is approximately 218 km.

5.8 Annular Solar Eclipse

As mentioned before, the celestial bodies move along elliptical paths, and therefore, their relative distances change continually and so does the shadowing. The odds are that, although the Moon is in the New Moon position and in one node point, it need not cover the whole Sun. In this case, we speak of an annular solar eclipse, during which the

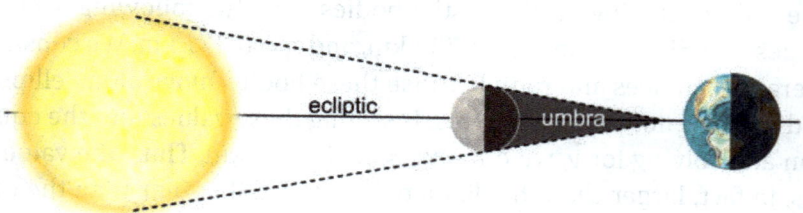

Fig. 5.22.   Annular solar eclipse (not to scale).

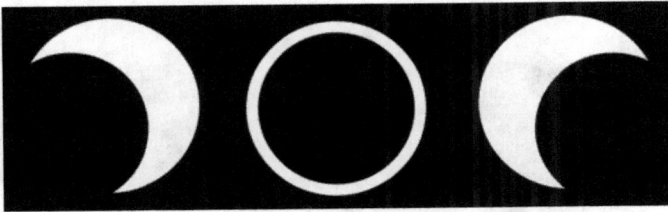

Fig. 5.23.   Appearance of a partial and an annular solar eclipse as seen from Earth.

distance between the Sun and the Moon is smaller in comparison to the full solar eclipse, as shown in Figure 5.22.

In the culmination phase of an annular eclipse, the Sun appears as a very bright ring, or annulus, around the dark disk of the Moon (see Figure 5.23). Although the Moon covers a large amount of the area of the Sun's disk, daylight remains and there are no stars or planets visible. Due to the Earth's rotation and the constant movement of the involved celestial bodies, annular eclipses start and end with partial eclipses.

5.9 Frequency of Eclipses

So far, we know, which events have to take place for a solar eclipse, but we do not know, how often such an eclipse occurs. In order to determine the frequency of recurrence of solar eclipses, we need two more numbers: the orbital period of Moon, that is the time that the Moon takes from one lunar node to return to the initial position, in other words to complete one orbit around the Earth, and the duration from one New Moon phase to the next. One might think that these time spans must be equal, but this is not true. Since the Earth moves along its orbit around the Sun, while the Moon is orbiting the Earth, the Moon has to travel a bit more than one orbit to reach the New Moon position again, which is shown in Figure 5.24. This duration is called synodic period or synodic month.

Generally speaking, the synodic period is the period of time that it takes the Moon to orbit the Earth with respect to the line joining the center of the Sun and the center of the Earth. The draconic period, on

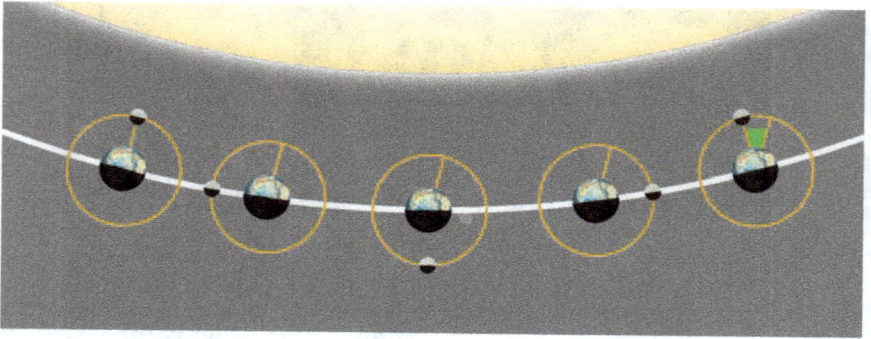

Fig. 5.24.   From one New Moon to the next.

the other hand, is the time it takes one object between two passages of its ascending node. This is the point of its orbit where it crosses the ecliptic from the southern to the Northern Hemisphere. The notion draconic period is closely related to solar eclipses, it stems from the imagination that a dragon scars the Sun during a solar eclipse. In short, the synodic period lasts 29 days 12 hours 44 minutes and the draconic period encompasses 27 days 5 hours 5 minutes.

The calculation of the frequency of solar eclipses brings us back to elementary arithmetic and number theory. To calculate the time between two total solar eclipses, we have to find the least common multiple of the synodic period and the draconic period. In fact, 223 New Moon cycles correspond to 6,585.31 days and 242 node cycles amount to 6,585.26 days. Although these two numbers are not exactly equal, the difference is very small. Thus, approximately every 6,585 days or 18 years, the relative geometry of the Sun, the Earth, and the Moon repeats itself. Already in ancient times, people were aware of this period of 18 years. The English astronomer Sir Edmond Halley (1656–1741) called it Saros cycle, referring to an ancient word used in the Babylonian sexagesimal system (base 60). Since the Saros period of 223 synodic months represents only an almost least common multiple of the synodic and the draconic period, an exact solution is impossible because both round-trip times are incommensurable, meaning that there are no two integers a and b, such that the proportion of the synodic period to the draconic period can be written in the

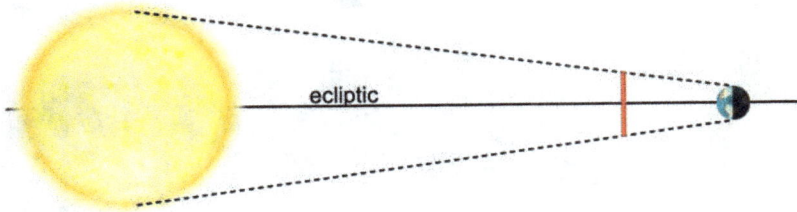

Fig. 5.25. "Elbow-room" for a solar eclipse.

form $a:b$. The French mathematician Henri Poincaré (1854–1912) was the first to discover this phenomenon of round-trip times of planets while considering differential equations. The orbits of planet systems exhibit more stability when having irrational round-trip times. A Saros cycle (just) holds for approximately 75 eclipses or 1,500 years. There are several concurrent Saros cycles, because eclipses do not only occur when the Moon is exactly at a lunar node. There is certain "elbow-room" for the ecliptic latitude of the Moon, represented by the red segment in Figure 5.25.

When the Moon passes between the Earth and the Sun, it only generates a solar eclipse somewhere on the Earth, if it crosses within the red segment. Since the red segment is considerably larger than the diameter of the Moon, there are, therefore, many simultaneous Saros cycles. Every year there are at least four eclipses, including partial shade events and lunar eclipses.

5.10 Lunar Eclipses

Having investigated solar eclipses in some detail, we will now turn our attention to lunar eclipses. Lunar eclipses proceed analogously to solar eclipses, but the roles of the Earth and the Moon are interchanged. The Moon moves through the shadow cast by the Earth, as shown in Figure 5.26.

A lunar eclipse happens when the Moon enters the deepest shadow, the Earth's umbra. In order for a lunar eclipse to occur, there has to be a Full Moon. However, just as with solar eclipses, lunar

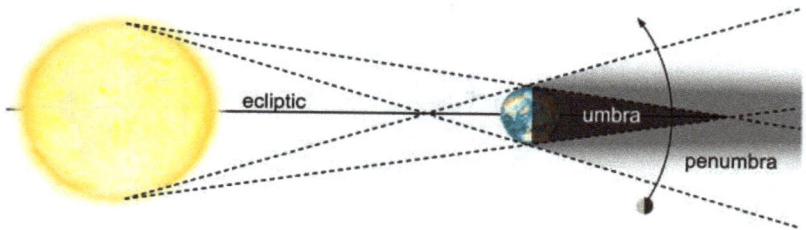

Fig. 5.26.   The Earth's umbra.

Fig. 5.27.   The reddish appearance of the Moon during a total lunar eclipse ("Blood Moon", July 2018).

eclipses do not occur during every revolution of the Moon around the Earth. Because of the inclination of the lunar orbit with respect to the ecliptic, the Moon usually passes by the Earth's umbra. However, the Moon does not need to be exactly at the node point to generate a total lunar eclipse because the Earth casts a much larger shadow on the Moon during a lunar eclipse than the shadow that the Moon casts on the Earth during a solar eclipse.

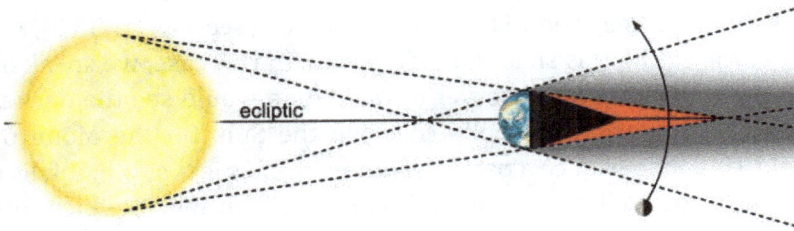

Fig. 5.28. Earth's umbra — why the Moon appears red.

Having previously observed a lunar eclipse, you may have noticed that even during a full eclipse, the Moon is not invisible. Instead, it appears in as a reddish color, as we can see in Figure 5.27. In fact, it is not true that the Earth blocks all of the sunlight. On their way through space, the sunrays hit the Earth's atmosphere. While the blue light is scattered in the atmosphere, the red light passing through the atmosphere gets refracted around the Earth and hits the surface of the Moon, thus, resulting in a red Moon.

Due to the refraction caused by the atmosphere, the totally dark part of the Earth's umbra cone only extends to approximately 40 times the Earth's radius, which is about 250,000 km. The average distance from the Earth to the Moon is about 375,000 km. Figure 5.28 is a refinement of Figure 5.26, indicating both the inner and completely dark part of Earth's umbra and the outer part, in which there is still some red light bent around the Earth by refraction in the atmosphere. The dustier the atmosphere is, the more reddish the Moon appears, because blue light is strongly absorbed by dust. Thus, the more dust particles in the atmosphere caused, for example, by a large volcanic eruption, the more impressive the so-called Blood Moon appears.

Moreover, lunar eclipses are more easily observable than solar eclipses because one can watch this phenomenon from every point of the Earth, where it is night, and the duration lasts up to approximately 100 minutes, if the Moon centrally crosses Earth's umbra. Hence, it is assumed that people in ancient times were more aware of lunar eclipses than of solar eclipses. The Saros cycle was probably discovered in relation to lunar eclipses rather than solar eclipses.

As long as the Moon is in the penumbra (see Figure 5.26), only part of the sunlight is shaded by the Earth. In this case, we speak of a penumbral eclipse. An observer on the Moon would see the dark silhouette of the Earth partially covering the Sun. On the Moon one would, thus, experience a partial solar eclipse. As less sunlight falls on the Moon, the Full Moon shines less brightly on Earth, which often goes unnoticed, because the brightness of the Moon does not decrease very dramatically, and therefore, a penumbral eclipse is often mistaken for a regular Full Moon. Typically, they are observable to the naked eye, when at least about 70% of the Moon's diameter immerses in the penumbral area. A total penumbral lunar eclipse represents a very rare type of lunar eclipses, where the entire Moon passes through the penumbral area and there is just a little margin. In order to get an idea of the lunar paths and the related types of eclipses, consider Figure 5.29.

A lunar eclipse will certainly be noticed when the Moon enters the deepest shadow, the umbra of the Earth. Then the circular shadow of the Earth slides over the Full Moon. In this case, an observer on the Moon would experience a total solar eclipse, because the Earth's silhouette would completely cover the Sun.

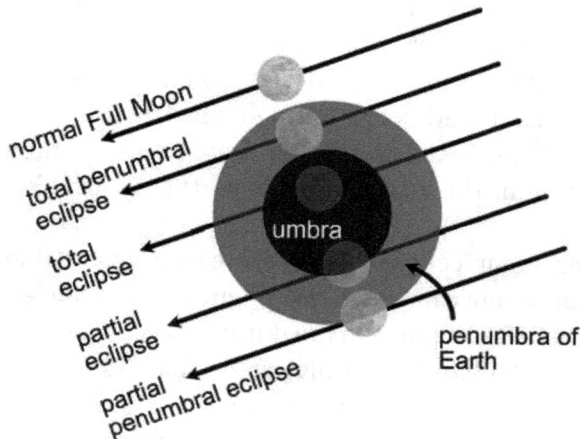

Fig. 5.29. Lunar eclipse paths.

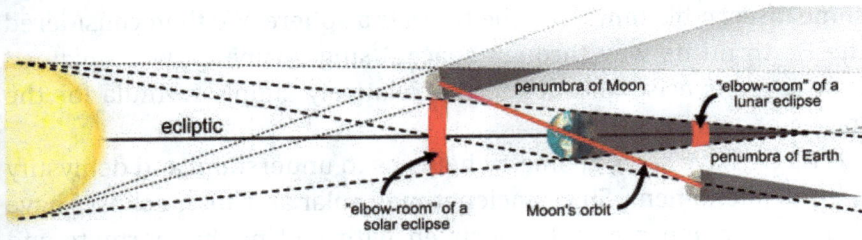

Fig. 5.30. Difference of the elbow-rooms for solar eclipses and lunar eclipses.

We shall consider one more interesting fact about eclipses: As you may know from experience, for a certain point on Earth's surface, there are always more lunar eclipses observable than solar eclipses. However, from a global perspective, counting partial phases, there occur slightly more solar eclipses than lunar eclipses. These statements can be explained geometrically. First, a lunar eclipse is observable from every point on the Earth, where it is night, covering a much bigger area than the shadowing of a solar eclipse, which has a diameter of approximately 200 km. However, as shown in Figure 5.30, the region of ecliptic latitudes of the Moon, for which a partial or total eclipse occurs somewhere on the Earth (the "elbow-room" of the eclipse), is narrower for lunar eclipses than for solar eclipses. This figure represents a vertical cross-section where the Moon's orbit is seen from the side. A solar eclipse happens whenever the Earth enters the penumbra of the Moon, whereas a lunar eclipse happens whenever the Moon enters the umbra of the Earth. So, the elbow-room for a solar eclipse is slightly larger than the elbow-room for a lunar eclipse, as shown in Figure 5.30. This is the reason for the fact that there are slightly more solar eclipses than lunar eclipses.

5.11 Summary

Space is full of interesting phenomena. In this chapter, we examined some aspects of space with respect to three-dimensional geometry. Typically, we perceive the Earth as a flat plate, although it is (almost) a sphere. However, at the beginning of this chapter, we presented some discernible hints that the Earth is a sphere. We then considered the Earth on its way through space. Using geometry, we explained how seasons arise and deduced a relatively simple formula for the length of a day.

As we have seen, geometry helps us to understand and demystify natural phenomena. Since ancient times, solar and lunar eclipses have caused fear and are still associated with bad myths, portents and superstitions, which can be traced back to a lack of knowledge. Humans quickly declare phenomena that cannot be explained as supernatural. For example, in ancient China, solar eclipses were thought to be caused by a terrifying dragon eating the Sun. The designation draconic period has its origin in this story. This is still the case today, for example, in the appearance in 2017 of the first known interstellar object passing through the Solar system, which was called Oumuamua, immediately led to speculations about an invasion of aliens. Supposedly, Thales of Miletus (ca. 624 — ca. 546 BCE) was the first person to exactly predict a solar eclipse. In more detail, it was the eclipse of May 28, 585 BCE. This eclipse occurred during the battle between the Medes and the Lydians. Frozen with fear, both sides put down their weapons and declared peace. This would imply that such astronomical phenomena can also be considered to have had positive results. A motivated reader will now have the tools to pursue further investigation into the boundless phenomena that space, that is three-dimensional space, offers us.

Chapter 6

Simple Solid Shapes

In our everyday world, we often encounter solid shapes which exhibit regularity or symmetry, a sample of which we show in Figure 6.1. The symmetry of shapes has typically generated some fascination. The most basic symmetric shapes are usually classified as cubes, prisms, pyramids, cylinders, cones, or spheres. Industrial design often uses these solid shapes as basic elements. Many buildings have the form of a rectangular solid, but there are also buildings today in the shape of cylinders, pyramids and other unusual solids.

The body with the greatest symmetry is, of course, the sphere, which we considered in Chapter 5. Here, we will consider shapes with polygonal sides, as in Figure 6.2.

6.1 Pyramids

The Egyptian pyramids are impressive proof of the fascination that ancient people had for distinctive geometric shapes. In particular, the great pyramid of Giza, the tomb of Cheops, the second pharaoh of the fourth dynasty, has intrigued generations of researchers as well as amateurs. The square at the base of the pyramid of Giza has a side length of exactly 440 royal cubits[1] (230.36 m) and an original height

---

[1] The royal cubit is an ancient Egyptian unit of length and corresponds to approximately 52.35 cm.

Fig. 6.1.   Toilet paper roll (cylinder), traffic cone, nut and bolt (hexagonal prism).

Fig. 6.2.   Icosahedron as a popular climbing frame for children or as a decorative object.

of 280 cubits (146.59 m). Unfortunately, its original shape has been destroyed to some extent by stone robbery and by some erosion. But a pyramid with these proportions has several remarkable properties.

1. The shaded right triangle in Figure 6.3 is a Golden right triangle to high accuracy, that is, it is a right triangle, where the ratio of the hypotenuse to the shorter leg is the Golden ratio.

2. The square of the height length is approximately equal to the area of every face of the pyramid (see Figure 6.4).

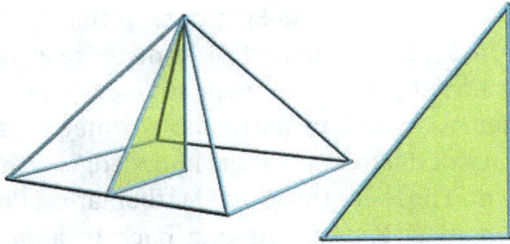

Fig. 6.3.   The Golden triangle in the Cheops pyramid.

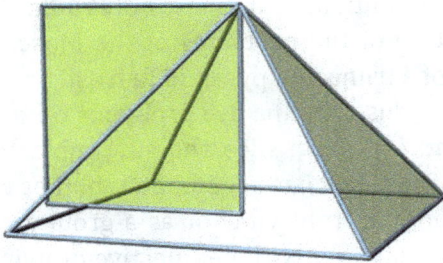

Fig. 6.4.   The two shaded areas are equal.

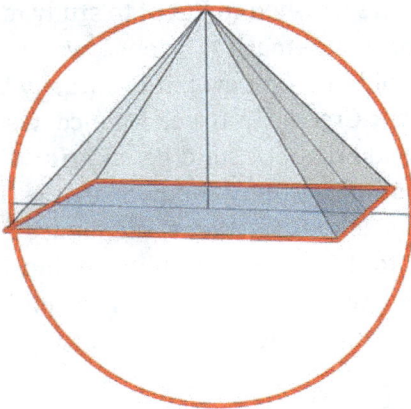

Fig. 6.5.   The square at the base and the circle have the same circumference.

3. The perimeter of the base square of the pyramid is equal to the circumference of a circle whose radius is equal to the height of the pyramid (see Figure 6.5).

There is no evidence that the Egyptians at that time (about 2,600 BCE) had any in-depth mathematical knowledge concerning these and other geometric figures. The main sources of our knowledge of Egyptian mathematics, and in particular geometry, are two papyri that give evidence of the mathematical knowledge of the later Middle Kingdom of Egypt. These are the Rhind Mathematical Papyrus and the Moscow Mathematical Papyrus (dating back to about 1,700 BCE). These texts are collections of practical mathematical problems along with solutions, and which contained no abstract results such as theorems and proofs. Essentially, they show only rudimentary knowledge of solid shapes. One of the problems of the Moscow Papyrus deals with the volume of a truncated pyramid (a frustrum) in a special case. And the Rhind Papyrus contains five problems on the volume of rectangular and cylindrical grain silos and six problems on the slope of pyramids. Descriptive sketches of geometric bodies were occasionally augmented, but only in front view or as a ground plan, because the Egyptians were not familiar with perspective drawing.

Egyptian mathematics eventually formed the foundation for Greek mathematics. Important Greek scholars such as Thales, Pythagoras and others traveled to Egypt to study mathematics. But as in Egyptian culture, mathematical science was used exclusively for solving practical problems, however, it eventually became a theoretical science in ancient Greece. In Greek science, we find a rich source of mathematical wisdom about solid geometric shapes. In particular, in Plato's school of philosophy, known as the Platonic Academy (depicted in Rafael's School of Athens, see Figure 3.20), mathematics was treasured as a tool for acquiring philosophical knowledge in the world of ideas, and was developed as a purely deductive science devoid of immediate applications.

6.2 Platonic Solids

Plato's dialogue, *Timaeus*, written ca. 360 BCE, was named after Timaeus of Locri, who, in a long monologue, explains the purpose and the properties of the Universe. His explanation involved the four elements (fire, air, water, and earth), and the constituent particles of

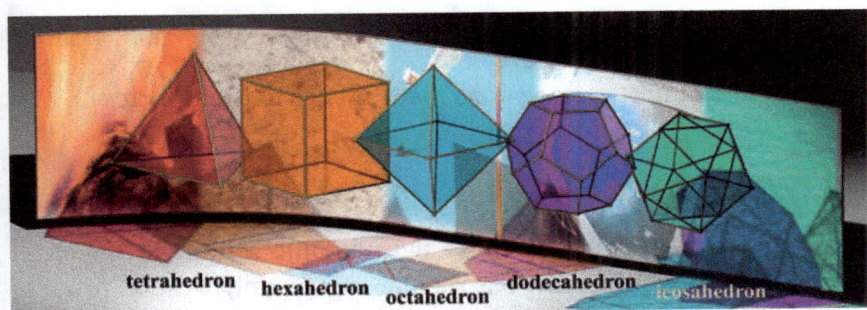

Fig. 6.6.   The Platonic solids.

these elements, that cannot be seen because of their small size. According to Timaeus, they would have the shapes of the most beautiful of all geometric solids — associated to the elements according to their "lightness and power of penetration" (see Figure 6.6): tetrahedron (fire), octahedron (air), icosahedron (water), and cube (earth). And there is a fifth figure, the dodecahedron (consisting of 12 pentagonal faces), which would be a model for the 12-fold division of the Zodiac.

Because of their role in Plato's writings, these five solids have become known as "Platonic solids." Although some of these solids might have been previously known to Pythagoras 150 years earlier, Theaetetus of Athens (ca. 417–369 BCE), a contemporary of Plato, is usually credited with the discovery of the octahedron and the icosahedron. In modern mathematical language, they are called regular convex polyhedra. A body is called "convex", if it has no holes, notches or indents. More precisely, a body is convex, if any line segment joining two points of the body lies on or within the body. Thus, a polyhedron is a shape with flat faces and straight edges. In classical Greek, "poly" means "many" and "hedron" means "face", so it is a shape with many faces, each face being a polygon (such as, triangle, quadrilateral, pentagon, etc.). It is called a *regular* polyhedron, if it meets the following criteria:

- all its polygonal faces that are identical in shape and size (i.e., congruent),

- the faces are regular polygons, which means that their angles are equal and side-lengths are equal,
- the same number of polygons meet at each corner (vertex).

6.3 What Can Be Learned from Folding Vertices

Surprisingly, there are only these five Platonic solids that meet all these criteria, and are known as regular convex polyhedra. A proof of this fact is probably due to Theaetetus and is described in the 13th book of Euclid's *Elements* (ca. 300 BCE). Following is a presentation of this proof.

The vertices of the Platonic solids are "spatial corners", where at least three of its faces meet in a point. These are known as polyhedral angles. We have already established that each face of a Platonic solid is a regular polygon. The most familiar regular polygon is a square. In order to form a spatial corner from three squares, we place these congruent squares on a flat surface and then fold them up, as shown in Figure 6.7. This will give us one of the corners of a cube.

The most important aspect of this is the following: This only works with exactly three squares. Two squares can meet on one edge, but two squares are insufficient to form a spatial vertex, or polyhedral angle. On the other hand, were we to use four squares they would lie flat on the plane with no gap in between. One cannot fold them up to create a spatial corner. But this already shows that it is only possible

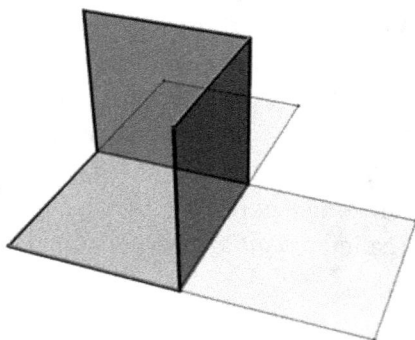

Fig. 6.7.   Folding a vertex from three squares.

to form a solid body using square sides so that exactly three squares would meet at each of these vertices. And this solid body is the familiar cube. This essentially shows that the cube is the only Platonic solid whose faces are squares.

The simplest regular polygon is an equilateral triangle. All interior angles of an equilateral triangle are 60°. One needs at least three triangles to form a vertex in space, but it is also possible to form a vertex with four or five triangles meeting at a point, since the total angle sum at the vertex will be less than 360°, as shown in Figure 6.8. When six equilateral triangles meet at a point, as shown in Figure 6.8(d), the interior angles of these triangles have a sum of 360° and the angles simply lie in a plane, and as a result they cannot form a three-dimensional angle. This is analogous to the four squares (above) that lie in a plane. This shows that the only Platonic solids whose faces are equilateral triangles are the tetrahedron, where three triangles meet at each vertex, as shown in Figure 6.8(a), the octahedron where four triangles at each vertex, as shown in Figure 6.8(b), and the icosahedron, where five triangles at each vertex, as shown in Figure 6.8(c).

A regular pentagon is a polygon with five sides of equal length and five identical angles, each of which is 108°. If we place three pentagons in a plane such that they meet at a common vertex, there still

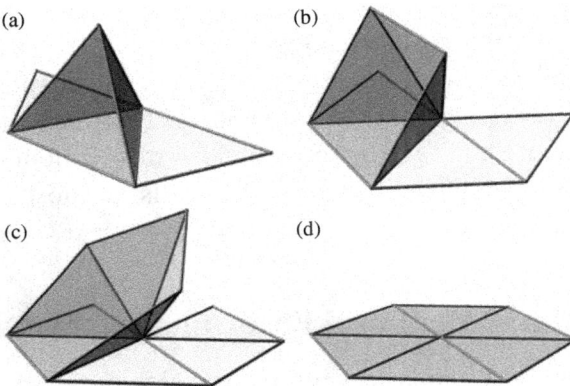

Fig. 6.8. Spatial vertices can be formed by (a) 3 equilateral triangles, (b) 4 equilateral triangles, (c) 5 equilateral triangles, or (d) 6 equilateral triangles that lie flat on the plane.

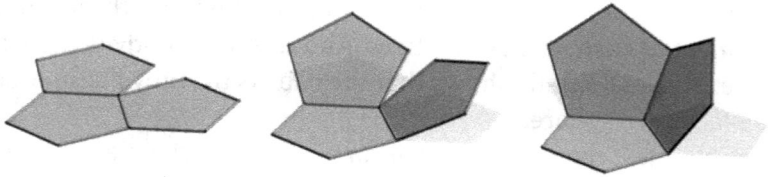

Fig. 6.9.   Folding three pentagons to form a spatial vertex.

remains a small gap of $360° - (3 \times 108°) = 36°$. Thus, we can fold them up to form a spatial corner (vertex), as shown in Figure 6.9.

There is, of course, no space to add another pentagon; therefore, the only way to form a vertex with congruent regular pentagons is by using three pentagons, as shown in Figure 6.9. Thus, we can conclude that the only Platonic solid made of identical regular pentagons is the dodecahedron, with each vertex formed by exactly three pentagons.

It is not possible to create a corner (vertex) using hexagons, or other regular polygons with more than five sides. Each of the interior angles of a regular hexagon is 120°. If you place three regular hexagons (the minimum number required to create a spatial corner) so that they meet at one vertex, the sum of the interior angles at that vertex would be 360°, which then has the hexagons lying flat in the plane surface without a gap. This is analogous to the way the six equilateral triangles lay in a plane as mentioned earlier. It is, therefore, not possible to create a regular convex polyhedron using only regular hexagons. And regular polygons with more sides (heptagons, octagons, etc.) have even larger interior angles, so there is no possible way to place three of them together at a single vertex without overlap. We have therefore established that there can exist at most five Platonic solids.

6.4 Angular Defect and Euler Characteristic

Our previous discussions lead us to consider the sum of the face angles at a vertex of a Platonic solid. As we have seen, these sums are always less than 360°. For example, when three squares meet at each vertex of a cube, the sum of the face angles is 270°. The 90° which are missing from a full 360°, are called the *angular defect* of the vertex. In Figures 6.8(a)–6.8(c), the angular defect is the gap that enabled us to fold the polygons up to spatial vertices. For the dodecahedron, the angular defect at each vertex is 36°. The French mathematician René Descartes (1596–1650) had the idea to find the sum of the angular defects for all vertices of a polyhedron. For example, a cube has eight vertices, each having an angular defect of 90°; hence, the total angular defect of the cube is $8 \times 90° = 720°$. The dodecahedron has 20 vertices and an angular defect of 36°, hence the total angular defect of the dodecahedron is $20 \times 36° = 720°$, again. A coincidence? Take a look at Table 6.1. It reveals the amazing fact that all five Platonic solids have the same total angular defect of 720°.

In fact, one can show that all convex polyhedra (whether regular or not) have the property that the total angular defect is 720°, twice the full angle of 360°. It even generalizes to all polyhedra that are topologically equivalent to a sphere. This means, that a polyhedron can be continuously deformed (just bending, stretching, or shrinking without tearing it apart) into a sphere. This idea of turning polyhedra into a sphere by continuous deformation (e.g., by inflation) will be applied in the next chapter to the classification of balls used in sports.

There is another interesting observation in Table 6.1. For every Platonic solid, we can verify that the number of vertices (v) plus the number of faces (f) minus the number of edges (e) equals two. This well-known formula, $v + f - e = 2$, was originally discovered in 1758 by the Swiss mathematician Leonhard Euler (1707–1783), and is known as "Euler's Polyhedral Formula." As is the case with the formula for the total angular defect, the polyhedral formula also extends to all polyhedra that are topologically equivalent to a sphere. The number obtained by ($v + f - e$) is called the *Euler characteristic* of a polyhedron. Other non-convex bodies (e.g., solids with holes) might have other Euler characteristics. There is a significant mathematical connection between the Euler characteristic and the total angular defect, which is given by the product of the Euler characteristic and a full angle of 360°. In the case of a body that is topologically equivalent to a sphere (Euler characteristic of 2), the total defect is, therefore, 720°.

Table 6.1. Properties of regular convex polyhedral (Platonic solids).

Polyhedron	# Vertices v	# Faces f	# Edges e	Angular defect	Total angular defect	v + f − e
Tetrahedron	4	4 ("tetra-")	6	180°	720°	2
Hexahedron	8	6 ("hexa-")	12	90°	720°	2
Octahedron	6	8 ("octa-")	12	120°	720°	2
Dodecahedron	20	12 ("dodeca-")	30	36°	720°	2
Icosahedron	12	20 ("icosa-")	30	60°	720°	2

6.5 Existence of Platonic Solids and the Golden Ratio

We have shown that there are at most five Platonic solids. For mathematicians, it is also important to understand why they all actually exist. It becomes obvious that this existence question is a real problem, if one begins to assemble, say, an icosahedron from 20 equilateral triangles. One can do this by starting with a vertex, as shown in Figure 6.10, and then adding more triangles step by step. During this process, one might wonder whether everything will eventually fit together properly. Will the pieces meet on the other side at exactly the right place? Finally, it is somewhat surprising when the last piece fits exactly into the remaining space, without having to squeeze or stretch or distort it in the slightest. Was it just a weird coincidence or was it mathematically predetermined? And the question which then arises is: Why does it work just as well with the dodecahedron?

In Book 13 of his *Elements*, Euclid shows exact constructions for all Platonic solids, thereby, proving their existence in an exact, mathematical sense. There is an ingenious fashion in which one can construct an icosahedron. This construction provides an unexpected mathematical connection with *golden rectangles*, which are rectangles whose sides are in the proportion of the Golden ratio ϕ, which is $\frac{l}{w} = \frac{l+w}{l}$, where w represents the rectangle's width and l represents the rectangle's length. If the shorter side, w, is one unit of length, then the longer side is precisely:

$$\phi = \frac{1+\sqrt{5}}{2} \approx 1.618\cdots$$

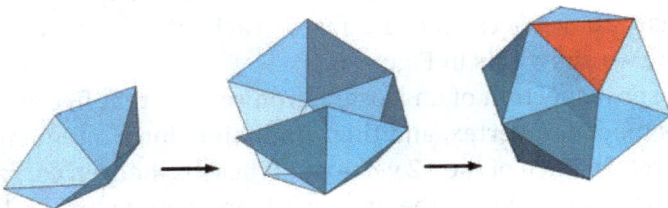

Fig. 6.10. Assembling an icosahedron from triangles — the last piece fits in exactly.

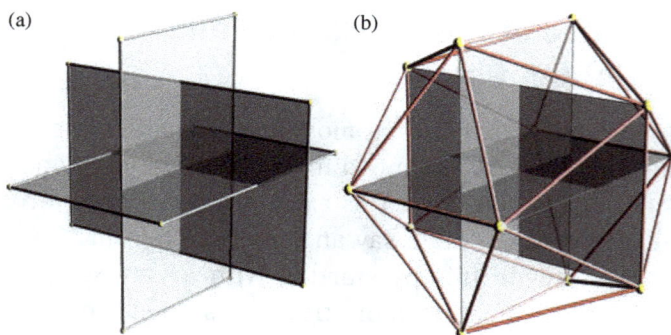

Fig. 6.11.   Creating an icosahedron from Golden rectangles.

To do the construction, one arranges three congruent Golden rectangles orthogonal (perpendicular) to each other, in three-dimensional space, as shown in Figure 6.11(a). Keep in mind that the three rectangles have a total of 12 vertices, which in turn will become the vertices of the completed icosahedron. It can now be shown that in the resulting three-dimensional figure, all neighboring vertices are the same distance from each other. If we connect them with straight lines (as shown in Figure 6.11(b)), then all these lines have the same length. This gives us a body that is determined by equilateral triangles and has exactly 12 vertices — which is the resulting icosahedron.

6.6 Duality

The next logical step is the construction of another Platonic solid, making use of the complete symmetry of an icosahedron with its 20 faces, each of which is an equilateral triangle. If we connect the centers of each of these triangles the result would be a dodecahedron, which has 20 vertices and 12 faces, each of which is a regular Pentagon, we show this in Figure 6.12.

A closer inspection of an icosahedron reveals that five equilateral triangles meet at a vertex, and their midpoints form a regular pentagon. Therefore, each of the 12 vertices of the icosahedron corresponds to a pentagonal face of the inscribed body in Figure 6.12. This inscribed body whose 12 faces are regular pentagons is, by definition,

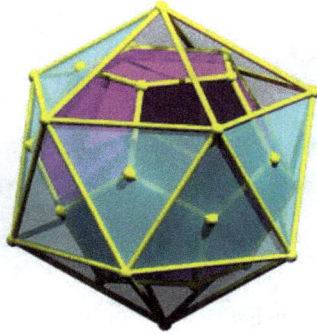

Fig. 6.12.   The dodecahedron inside the icosahedron.

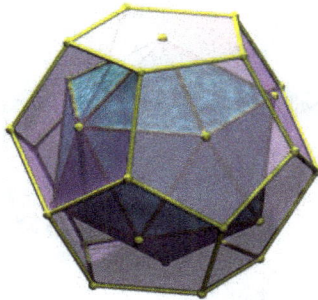

Fig. 6.13.   The icosahedron inside a dodecahedron.

a dodecahedron. This establishes the existence of a dodecahedron based on the proven existence of the icosahedron. The construction also shows that the number of vertices of the dodecahedron equals the number of faces of an icosahedron (20), and vice versa: Analogously, the number of faces of a dodecahedron equals the number of vertices of the icosahedron (12), which can also be seen in Table 6.1.

We can see from the above description, that the dodecahedron and the icosahedron are closely related, which is a concept called *duality*. It works in both directions, as we can see since the body inscribed in a dodecahedron by joining the midpoints of the pentagonal faces is an icosahedron, as shown in Figure 6.13.

Inspecting Table 6.1 again, we see that a similar relationship seems to hold true between the hexahedron (cube) and the

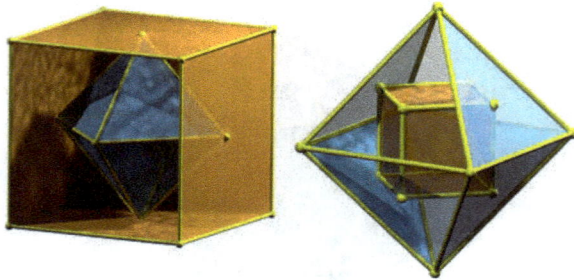

Fig. 6.14.   Duality between hexahedron and octahedron.

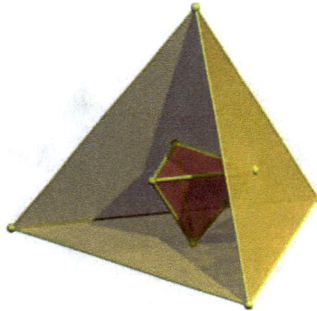

Fig. 6.15.   Self-duality of the tetrahedron.

octahedron. Indeed, a cube has six faces and eight vertices, and the octahedron has six vertices and eight faces. Therefore, we can obtain one solid from the other by using the midpoints of the faces of one solid as the vertices of the other, as shown in Figure 6.14.

This leaves the tetrahedron without a partner Platonic solid. What would we get by connecting the midpoints of the faces of the tetrahedron to establish the vertices of a new solid? Figure 6.15 shows that we get another tetrahedron (just upside down and a little bit smaller). We, therefore, consider the tetrahedron self-dual.

Suppose we let the inside dual Platonic solid grow until the edges intersect. We then obtain beautiful, star-shaped (non-convex) polyhedra, as shown in Figure 6.16. Furthermore, we notice that the edges of the dual polyhedra bisect each other. An ambitious reader might look for other curious relationships among these solid figures.

Fig. 6.16. Composite of a Platonic solid and its dual.

6.7 Kepler-Poinsot Solids

As we continue with our journey through the Platonic solids, we will consider the vertices of an icosahedron, and notice that we can draw several regular pentagons inside the icosahedron. A regular pentagon can be formed by joining vertices of the icosahedron as shown in Figure 6.17(a). This pentagon is a surface cutting diagonally through the icosahedron. If we draw analogous congruent pentagons for all groups of five vertices of the icosahedron, a total of 12 congruent regular pentagons will result. These pentagons form intersections, which we choose to ignore (that is, we do not consider the lines of intersection as edges of a new body). Thus, we obtain a shape with 12 self-intersecting pentagonal faces, with five pentagons meeting at each of its 12 vertices. The number of edges equals the number of edges of the icosahedron (30). While it is not a Platonic solid, it could be considered a regular non-convex dodecahedron. An image of this solid, which is called *great dodecahedron* is shown in Figure 6.17(b).

The great dodecahedron is a three-dimensional analogue of the pentagram, which is a five-pointed star formed by the diagonals of a regular pentagon (see Figure 6.18). It is a two-dimensional figure with five equal sides and five vertices, which also forms a regular pentagon at its center, and whose lines intersect each other in the Golden ratio. Typically, one does not introduce the intersections of the diagonals as new vertices (as we indicated in Figure 6.18, where the diagonals are drawn as if they would avoid each other in the third dimension). In the three-dimensional great dodecahedron, the

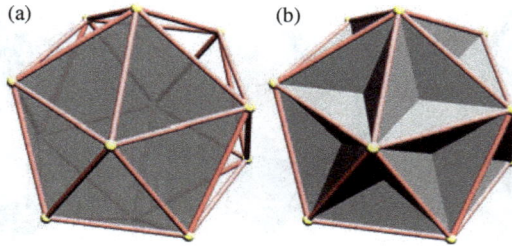

Fig. 6.17.   Pentagon inside an icosahedron (a) and the great dodecahedron (b).

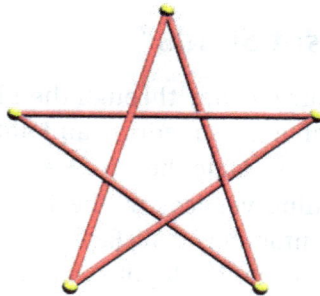

Fig. 6.18.   Pentagram formed by the diagonals of a regular pentagon.

pentagonal faces also enclose an empty space in the middle of the figure — this is a regular (Platonic) dodecahedron.

Because a great dodecahedron has 12 faces, 12 vertices, and 30 edges, it does not fulfill Euler's polyhedral formula. Its Euler characteristic is $v + f - e = -6$. The great dodecahedron is not topologically equivalent to a sphere, it cannot be continuously deformed into a sphere without destroying the pentagonal faces.

The Great Dodecahedron belongs to a type of solids known as *Kepler–Poinsot polyhedra*, all being related to the two-dimensional pentagram. Without going into detail beyond the scope of this book, we show images of these evolving beautiful shapes in Figure 6.19.

The great icosahedron has 12 vertices (arranged like the vertices of an icosahedron), 20 faces (equilateral triangles), and 30 edges. The vertices are marked in Figure 6.19, the intersections of the edges do not count as new vertices (they are sometimes called "false vertices").

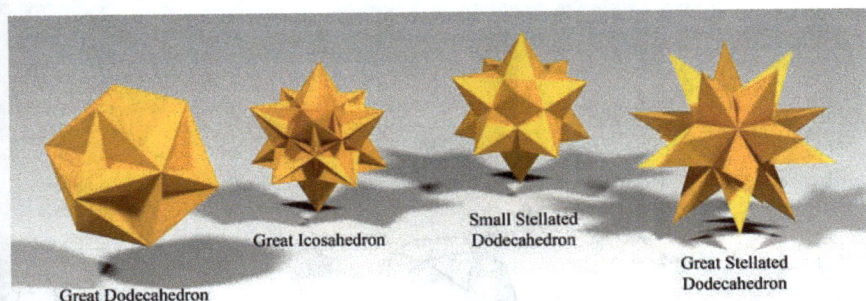

Fig. 6.19.   The four Kepler-Poinsot solids.

The small stellated dodecahedron, like the great dodecahedron has 12 vertices, and 30 edges. Here the 12 faces have the shape of a regular pentagram (where the faces of the great dodecahedron are pentagons). Finally, the great stellated dodecahedron has 20 vertices (arranged as the vertices of a dodecahedron) and 12 faces (pentagrams) formed by a total of 30 edges. The small stellated dodecahedron, as its dual, the great dodecahedron has an Euler characteristic of −6, while the great icosahedron and its dual, the great stellated dodecahedron, both have an Euler characteristic of 2. As a help for visualization, one can imagine that the shape of a small stellated dodecahedron can be created by adding five-sided pyramids to the faces of a dodecahedron, and the great stellated dodecahedron can be created by adding three-sided pyramids to the faces of an icosahedron.

6.8 Archimedean Solids

Archimedean solids are convex polyhedra whose faces are regular polygons. The vertices are all congruent, and all edges have the same length. What makes them different from Platonic solids, is that their faces need not be identical. While Platonic solids are composed of only one type of polygons, Archimedean solids consist of two or three different types of regular polygon faces. Excluding the so-called *prisms* and *antiprisms* discussed below, there are 13 Archimedean solids. These are all shown in Figure 6.20.

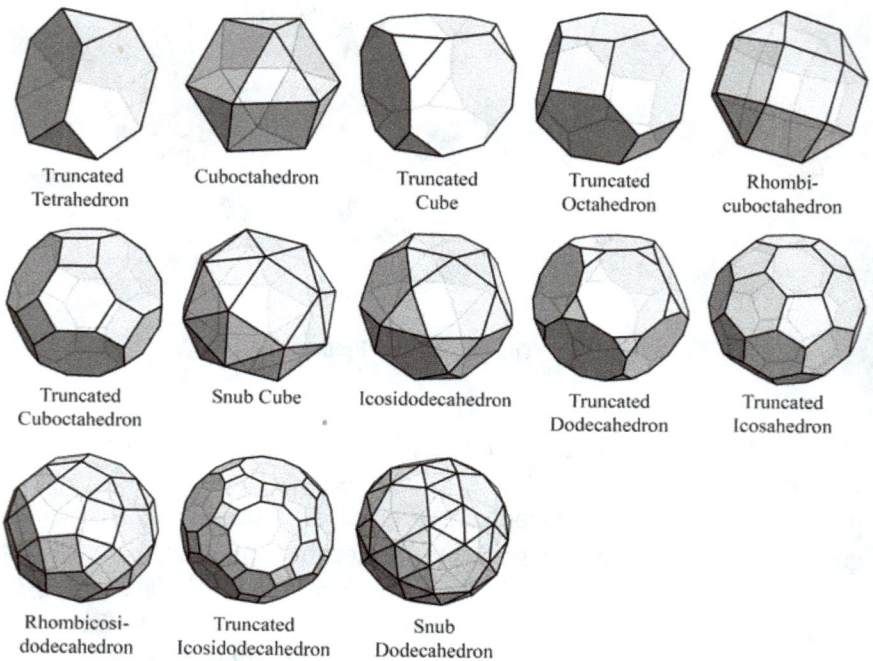

| Truncated Tetrahedron | Cuboctahedron | Truncated Cube | Truncated Octahedron | Rhombi-cuboctahedron |

| Truncated Cuboctahedron | Snub Cube | Icosidodecahedron | Truncated Dodecahedron | Truncated Icosahedron |

| Rhombicosi-dodecahedron | Truncated Icosidodecahedron | Snub Dodecahedron |

Fig. 6.20.   The 13 Archimedean solids.

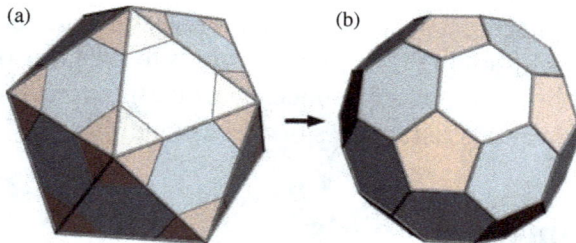

Fig. 6.21.   Truncating (sanding down) the vertices of an icosahedron.

Archimedean solids are obtained, for example, by cutting off the vertices of Platonic solids. In this process, one has to make sure that one cuts off the vertices far enough from the vertex point so that the resulting faces become visible regular polygons. For example, consider the icosahedron in Figure 6.21(a). We have marked the pentagon at each vertex. We want to truncate these vertices, so that the gray hexagonal areas in between become regular hexagons. The result would be a solid called the truncated icosahedron, whose resemblance with a soccer ball will be a topic in Chapter 7.

In the manner in which the truncated icosahedron was created, it is easy to infer its properties. Each of the 12 vertices of the icosahedron produces five vertices to the truncated icosahedron, which, therefore, has $12 \times 5 = 60$ vertices. It has 32 faces, which results from the sum of the vertices and faces of the icosahedron. The number of edges is easily obtained from Euler's polyhedral formula, namely, $v + f - e = 2$. Inserting $v = 60$ and $f = 32$, we obtain $60 + 32 - e = 2$, therefore, $e = 90$. Thus, the truncated icosahedron has 90 edges!

What happens if we grind down the pentagonal areas of the truncated icosahedron even further? Then the pentagons get bigger and bigger and the originally hexagonal surfaces between them get smaller and smaller and become triangles at the moment the vertices of the pentagons meet each other, as we can see in Figure 6.22(a). We then get an Archimedean body, which is composed of 12 regular pentagons and 20 equilateral triangles, the so-called icosidodecahedron, which is then clearly pictured in Figure 6.22(b). Since two of the vertices of the truncated icosahedron have become a single vertex of the icosidodecahedron, it has only 30 vertices, which is half as many as the truncated icosahedron. Applying Euler's polyhedral formula, we find that there are then exactly 60 edges. We can also calculate the number of edges as follows: The 12 pentagons and 20 triangles

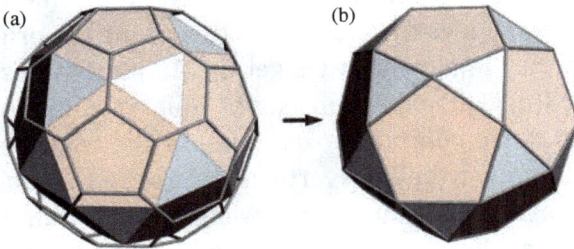

Fig. 6.22.   Truncated icosahedron (a) and icosidodecahedron (b).

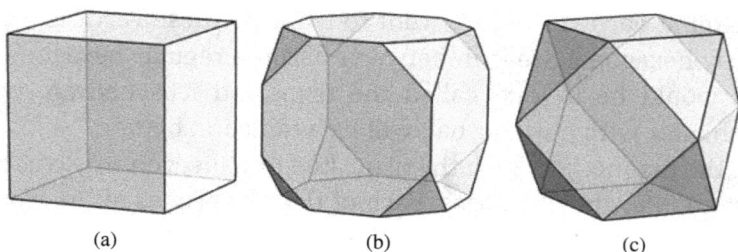

Fig. 6.23. (a) Cube, (b) truncated cube and (c) cuboctahedron.

have a total of $(12 \times 5) + (20 \times 3) = 120$ sides. However, two faces come together with their sides along one edge, so we essentially counted the number of sides twice and, therefore, the solid only has 60 edges.

A motivated reader might wish to inspect the other Archimedean solids. Which ones can be obtained from Platonic solids through a similar procedure? What kind of solid will we get, if we truncate a tetrahedron or a dodecahedron? For example, let's take a closer look at the two bodies that one gets when one cuts the corners of a cube (see Figure 6.23).

When you start flattening by grinding down the corners of a cube, triangles are created, which replace the original vertices. The remaining cube's faces consequently become octagons. When the corners have been completely removed so that the triangles and octagons have the same edge length, we get an Archimedean body, which is called the *truncated cube*. It is the solid shown in Figure 6.23(b). By further grinding down the triangular surfaces become even larger, and their vertices move towards each other along the original cube's edges. When they finally meet, we get the Archimedean solid shown in Figure 6.23(c). This body consists of squares and equilateral triangles and is called *cuboctahedron*. Because the triangles were created by grinding down the eight vertices of the cube, the cuboctahedron has eight triangles. The squares are what remains from the original faces of the cube, hence the cuboctahedron must have six squares, the cuboctahedron thus has a total of $f = 14$ faces. It is now easy to determine the number of vertices. The six squares contribute

$6 \times 4 = 24$ vertices, and the eight triangles contribute $8 \times 3 = 24$ vertices. This would give a total of 48 vertices. However, each vertex of the cuboctahedron (of which all are identical) is the meeting point of precisely four faces. Hence, by counting the vertices of all faces, we have counted each vertex four times. The number of vertices of the cuboctahedron is therefore, $v = \frac{48}{4} = 12$. By Euler's polyhedral formula, $v + f - e = 12 + 14 - e = 2$, we find that $e = 24$.

Finally, we would like to note that there are two other types of solids that meet the conditions for Archimedean solids, although they are generally not considered to belong to this group. These are the n-sided prisms and n-sided anti-prisms (where n is any natural number greater than 2). Figure 6.24(a) shows prisms and anti-prisms (Figure 6.24(b)). They consist of two types of regular polygons and have congruent vertices. The three-sided antiprism is just an octahedron (see Figure 6.24), and "four-sided prism" is just another name for a cube.

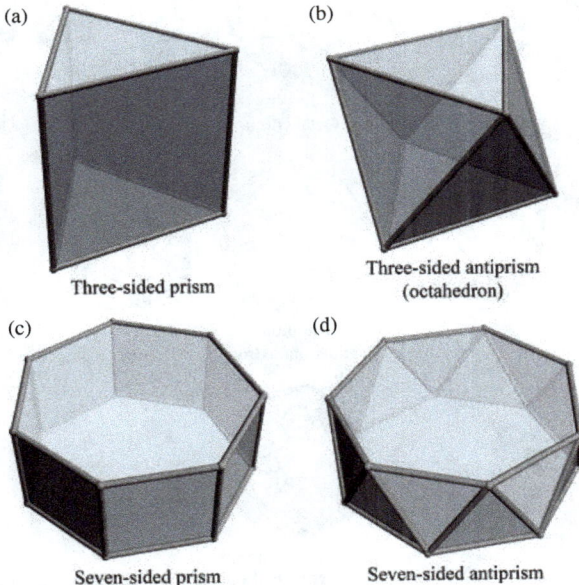

(a)

Three-sided prism

(b)

Three-sided antiprism
(octahedron)

(c)

Seven-sided prism

(d)

Seven-sided antiprism

Fig. 6.24.   Some prisms and anti-prisms.

6.9 Duals of Archimedean Solids

Any polyhedron has a dual polyhedron. The duals of the Archimedean solids are sometimes called *Catalan solids*. Catalan solids consist of identical faces that are, however, not regular polygons. Furthermore, their vertices are not all congruent. For example, the rhombic dodecahedron has congruent rhombic faces, but it has eight vertices of one type, where three rhombi meet and six vertices of another type, where four rhombi meet in a point. Figure 6.25 gives an overview of the thirteen dual Archimedean solids, listed in the same order as those shown in Figure 6.20.

The dual polyhedron has a vertex for each face of the original polyhedron. In other words, every vertex of the original polyhedron there corresponds a face of the dual polyhedron. The dual of the cuboctahedron (see Figure 6.23) is therefore, a solid with 24 edges,

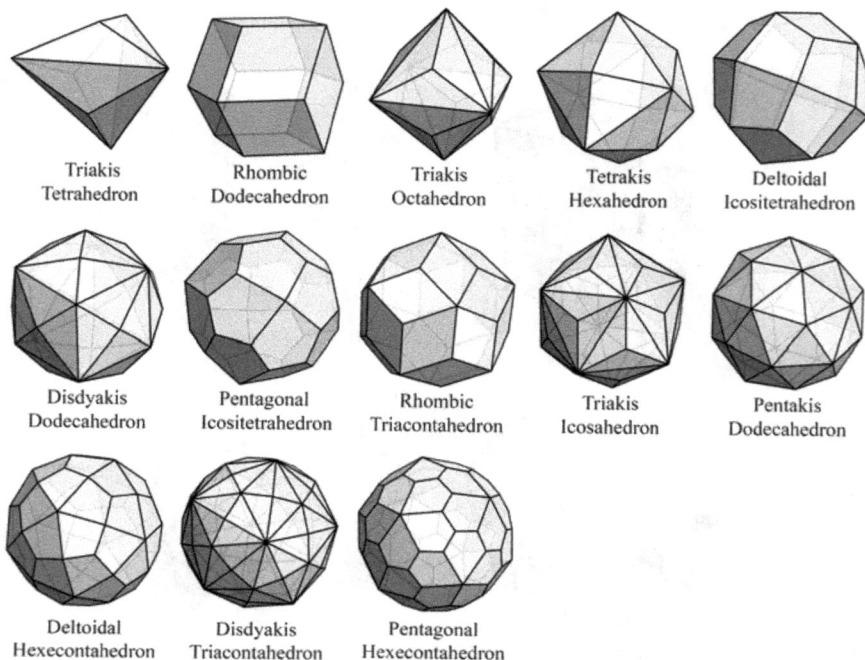

Triakis Tetrahedron	Rhombic Dodecahedron	Triakis Octahedron	Tetrakis Hexahedron	Deltoidal Icositetrahedron
Disdyakis Dodecahedron	Pentagonal Icositetrahedron	Rhombic Triacontahedron	Triakis Icosahedron	Pentakis Dodecahedron
Deltoidal Hexecontahedron	Disdyakis Triacontahedron	Pentagonal Hexecontahedron		

Fig. 6.25. The 13 Catalan solids (dual Archimedean solids).

14 vertices (the number of faces of the cuboctahedron) and 12 faces (the number of vertices of the cuboctahedron). Because of its 12 faces, it is a dodecahedron, and because each of its faces is a rhombus, it is called *rhombic dodecahedron* (see Figure 6.26). In Figure 6.27, we show the rhombic dodecahedron and its dual, the cuboctahedron, as a composite solid.

The rhombic dodecahedron consists of identical faces, all of which are rhombi. Remember, a rhombus is a quadrilateral whose four sides have the same length. The two diagonals are different and have the ratio $\sqrt{2}:1$. That means, the length of the longer diagonal is $\sqrt{2}$ times the length of the shorter diagonal. The 24 edges of the rhombic dodecahedron have the same length, but the vertices are not

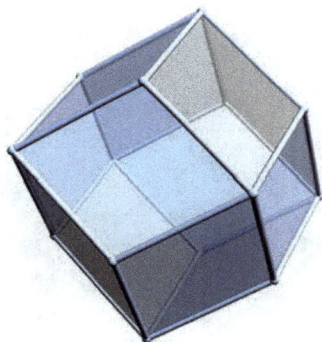

Fig. 6.26.   The rhombic dodecahedron.

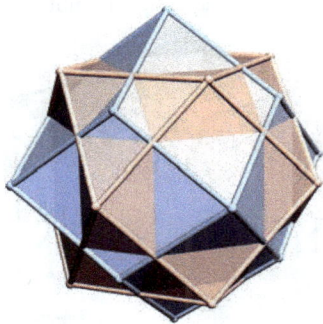

Fig. 6.27. Composite of the cuboctahedron with its dual, the rhombic dodecahedron.

congruent. There are six vertices shared by four rhombi, and eight vertices, at which only three rhombi come together.

The rhombic dodecahedron can be derived from the cube, by adding six pyramids with a slant angle of 45° to each of its six sides, as indicated in Figure 6.28.

The volume of the rhombic dodecahedron is twice the volume of this "inner cube". To prove this, we need to understand that the six pyramids on the six faces of the cube together have exactly the volume of another cube. However, this is fairly easy to see if we turn each of the pyramids upside down so that they point into the cube. From the 45° inclination of their faces they fit together perfectly, and they meet with their tip vertex in the center of the cube, thus, completely filling the cube, as shown in Figure 6.29. Therefore, the six pyramids

Fig. 6.28.   The cube inside the rhombic dodecahedron.

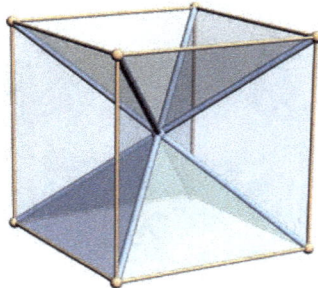

Fig. 6.29.   Six pyramids inside the cube are formed by the four spatial diagonals.

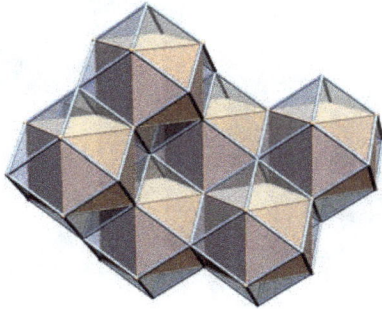

Fig. 6.30. Rhombic dodecahedra can be stacked to form a space-tiling without gaps.

together have precisely the volume of the cube, and the whole rhombic dodecahedron has exactly twice this volume.

As a consequence of the fact that the rhombic octahedron can be interpreted as a cube with protruding pyramids on its faces. Furthermore, we also notice that six of these pyramids would again fill a cubic region, and the rhombic dodecahedron has an unexpected property. It is a space-filling solid. That means, identical copies of these solids can be stacked together in such a way that they completely fill the three-dimensional space, without any gaps between them. We illustrate this procedure in Figure 6.30. The reason why this works is that the cube itself is a space-filling solid. The center-cube of the rhombic dodecahedrons would occupy every second cubic region in space, while the cubic gaps in between would be filled by the pyramids protruding from the sides of the cubes.

Filling space with a pattern of regular shapes has important applications to so-called packing problems, which is the topic of Chapter 9. In Chapter 7, we are going to explore the applications of the theory of polyhedra to the design of balls used in various sports.

Chapter 7

How to Sew a Ball –
A Polyhedral Approach

7.1 Ball Games

While some athletic activities, such as running or swimming, can be undertaken without any specialized equipment, most competitive sports require some kind of hardware. Perhaps the most common non-clothing item required in sports is a ball. For the wide majority of sports, it is desirable for such a ball to be able to move with complete symmetry, both flying in space and rolling on the ground, and for that reason, most sports require such a ball to be as spherical as possible. Naturally, American football is one exception!

For some sporting activities, like bowling, bocce, or jai alai, such spheres were originally just rocks, or carved pieces of wood or solid pieces of rubber if bouncing was required. Balls of this type, however, were not suited for all types of games. A sport like soccer or volleyball, for instance, could not be played in any reasonable way with such an object. Any game of this type requires something larger, more elastic and lighter.

In the early days of sports, there were all sorts of interesting methods used to make balls. In ancient Greece, pig's bladders were inflated and heated in the ashes of a fire. In ancient Rome, there were balls made of leather and stuffed with feathers. Early (medieval) soccer balls were sometimes made of a leather shell filled with pieces of cork. Even in modern times, it is not unheard of for a group of enthusiastic kids in impoverished circumstances to gather a lump of rags, and tie them together in a spherical shape, in order to pursue their sport.

For many types of games, inflated balls made of leather or rubber seem to be suited especially well. Such balls have just the right elasticity, weight and stability for such games as soccer, basketball, volleyball or handball, just to name a few popular examples. In order to make a leather ball as round as possible, it seems to be quite obvious that it should ideally be sewn together from symmetrical patches of material that are all of a uniform size and that come together in more or less the same way at all common edges and corners, producing a maximum amount of symmetry in the finished product. Inflating such a ball should then result in a good approximation of a sphere. A nice example of such a situation can be the "classic" soccer ball.

The ball shown in Figure 7.1 is sewn together with patches in the shape of regular pentagons and hexagons. Each vertex of such a ball is the common corner of a pentagon and two hexagons, which gives the ball a great measure of regularity because of the consistency of the angles.

For many people, this is the iconic example of what a ball is meant to be. In fact, this specific type of ball, that we would now consider

Fig. 7.1. Classic soccer ball.

to be so overtly typical, was not introduced widely until fairly recently. Although such a shape was certainly not unheard of before that, it was not until its use at the 1970 World Cup that this particular type of ball became practically synonymous with soccer. Nevertheless, this ball is an excellent starting point for our discussion of balls and their association with specific polyhedra.

In this chapter, we will consider some options for polyhedra, each of which can be used to approximate a sphere in such a way as to make it useful as the geometric basis of a ball. We will then take a look at some common types of balls, and how their respective geometric structures are derived from specific solids of this kind.

7.2 Platonic Solids and Archimedean Solids

If we are searching for polyhedra with the highest possible amount of spherical symmetry, we can start by inspecting polyhedra whose faces are all identical regular polygons. Furthermore, if we want all of their vertices to be identical, we can require the same number of identical faces to meet at each vertex. These restrictions bring us right back to the Platonic and Archimedean solids, as discussed in Chapter 6.

The three Platonic solids with triangular faces are the tetrahedron, octahedron, and icosahedron, which are shown in Figure 7.2.

As we can see, the first two do not appear at first glance to be good candidates for balls, as they do not seem to be particularly "round" in any reasonable sense of the word. The third, the icosahedron, does,

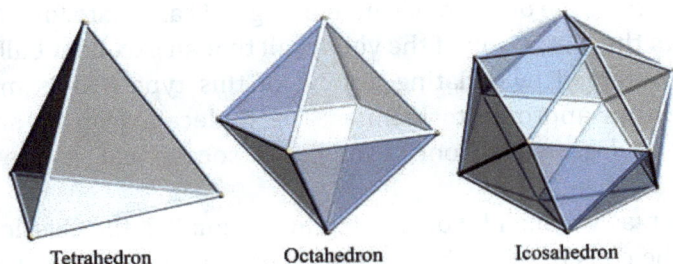

Tetrahedron   Octahedron   Icosahedron

Fig. 7.2.   Platonic solids with triangular faces.

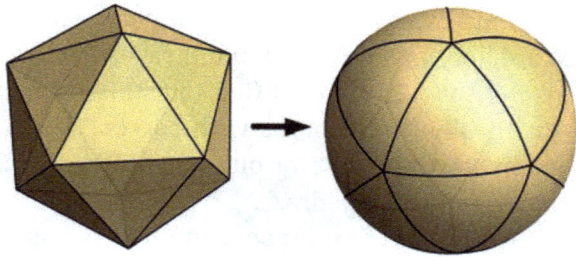

Fig. 7.3.   Inflating an icosahedron.

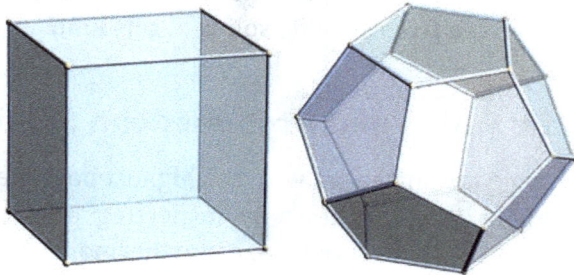

Fig. 7.4.   Cube (hexahedron) and dodecahedron.

however, seem to be a good candidate, and it is easy to imagine that such a shape can be sewn together from leather material and inflated as a useable ball for many sporting activities, see Figure 7.3. We shall be returning to this polyhedron in our discussion of the classic soccer ball.

The other two Platonic solids are the cube and the dodecahedron, which are shown in Figure 7.4.

While the cube does not seem to be a good candidate for a ball, we will see in the discussion of the volleyball that an excellent ball can, in fact, be obtained by inflating a solid of this type if it is modified slightly in an appropriate manner. The dodecahedron is, however, rounder, and seems to look a lot like a soccer ball, as we see in Figure 7.5.

In fact, as we shall be discussing in Section 7.3, the classic soccer ball can be considered to be something of a cross between the icosahedron and the dodecahedron.

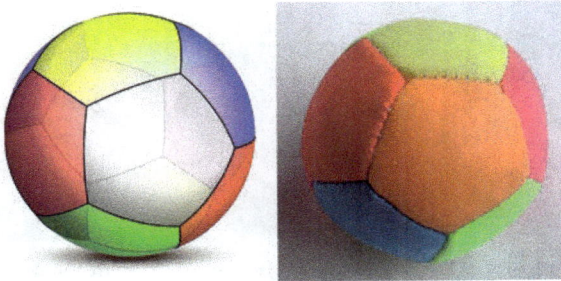

Fig. 7.5. Inflated dodecahedron. Mathematical model and a real ball.

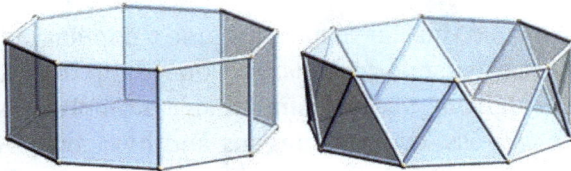

Fig. 7.6. Prism and anti-prism.

Of course, Platonic solids are not the only candidates for highly symmetric balls. We already know of other types of solids having most of the properties we required, namely, faces that are all regular polygons and vertices that are all mutually congruent. Two examples are the regular prisms and the regular anti-prisms shown in Figure 7.6.

These are certainly candidates for round balls from the standpoint of the vertices, but it seems quite obvious that the global symmetry we require of a good ball is not quite to be expected from them. They appear, at first glance, to be much too thin and flat to be useful, and increasing the number of sides of the top and bottom polygons, while keeping all the faces regular, will just make them thinner and flatter. Nevertheless, we will be encountering a useful application of polyhedra of this type later on in our discussion of beach balls.

Next, we have the *Archimedean solids*. Some of the "rounder" ones are shown in Figure 7.7.

In principle, reasonably round balls based on any of these polyhedra could be designed and manufactured, but the truncated icosahedron, which we shall be considering in more detail in the next section,

Icosidodecahedron   Truncated Octahedron   Truncated Dodecahedron

Fig. 7.7.   Icosidodecahedron, truncated octahedron and truncated dodecahedron.

is the only such solid thus far to have become a popular starting point. There are several reasons for this. For one thing, manufacturing a classic soccer ball from pieces of similar size is an obvious advantage in the sewing process, using pentagons and hexagons with sides of equal length, and of similar area. If, as an example, we were to use a polyhedron with both triangular and hexagonal faces, the hexagonal faces would have six times the area of the triangular ones, making the sewing process quite cumbersome. Furthermore, the fact that the sum of the face angles in each vertex of the soccer ball is equal to $2 \times 108° + 120° = 336°$, and therefore, quite close to the full angle of 360°, means that the vertices of the polyhedron will not be too "bumpy". Comparing this to a vertex of the truncated octahedron, for instance, we see that a square and two hexagons meet at each vertex of this object, giving an angle measure total of $2 \times 120° + 90° = 330°$, which is not quite as good as the previous version (although still better than the value we would obtain for most solids).

We could speculate some more about less popular balls, or balls that have yet to be developed, but right now it seems more productive to direct our attention to varieties of balls that are actually in wide use. There are four balls, each derived from regular or semi-regular polyhedra, which are in such common use, that their shapes are deeply associated with a specific sporting activity, even though they also find use in other contexts. These are the four types of balls we shall take a closer look at in the following section.

7.3 The Four Common Balls and Their Associated Solids

As was already mentioned, there are four balls derived from specific polyhedra whose superficial structures are so common that they have each become virtually synonymous with a specific activity. These are the soccer ball, the beach ball, the basketball and the volleyball (see Figure 7.8).

7.4 The Soccer Ball

We are now ready to take a closer look at the most common of all polyhedral forms typically associated with the structure of a ball, namely, the truncated icosahedron. This polyhedron, while also in wide use in other contexts, such as Olympic handball, has become virtually synonymous with soccer. So much so, in fact, that when it came time to create a character for soccer in Unicode, the character that was designed and introduced as U+26BD looked like a variant of the current soccer ball emoji shown in Figure 7.9.

SOCCER BALL   BEACH BALL   BASKETBALL   VOLLEYBALL

Fig. 7.8.   Some common types of balls.

Fig. 7.9.   Variants of the soccer-ball emoji (Unicode character U+26BD).

Fig. 7.10. Ball with which Pelé scored his 1,000th goal in 1969 (Maracanã Stadium Museum in Rio de Janeiro).

This came to pass despite the fact, as already mentioned, that this shape did not come into standard use as a soccer ball until the advent of the 1970 World Cup, where this polyhedron was used as the basic structure of the so-called *Telstar* ball. Before that year's tournament, balls used at the top levels of play were often made of a structure we would now more commonly associate with a volleyball, and before that even more complex structures were in common use. As a noteworthy example of a ball used before the introduction of the *Telstar*, shown in Figure 7.10, we can see the original ball now on display at the Maracanã Stadium Museum in Rio de Janeiro, with which the famous soccer star Pelé scored his 1,000th goal in 1969.

After just a few decades, the famous basic shape of the *Telstar* was already abandoned for World Cup play with the introduction of the *Teamgeist* ball in 2006. Since then, modern technology has made it possible to develop balls with ever-improving game-play related properties, derived from other basic geometric principles. An example of a later ball is the *Brazuca Rio* ball shown in Figure 7.11. This ball, which is also on display at the Maracanã stadium museum, was the ball used at the World Cup final between Germany and Argentina at the Maracanã stadium in 2014.

Despite this, the basic shape of the truncated icosahedron has simultaneously become ever more deeply entrenched in the public

Fig. 7.11.   Brazuca Rio ball (Maracanã Stadium Museum in Rio de Janeiro).

consciousness as the symbol of the game. Having completed this historical aside, we are now ready to return to more geometric matters.

We have already seen that the truncated icosahedron closely approximates many of the properties we would like a ball to have. The vertices are quite flat, allowing the ball to roll evenly and fly through the air with a minimum of wobble. Also, the faces are of similar size, allowing for relatively easy manufacture. Furthermore, we have already noted that this specific Archimedean solid can be considered as something of a cross between the two "roundest" Platonic solids, namely, the regular icosahedron and the regular dodecahedron. But what does this actually mean?

The relationship to the icosahedron is easily explained by recalling our earlier discussion on dual solids. As introduced in Chapter 6, the concept of "truncating" refers to the idea of cutting off a vertex in such a way so as to keep symmetry intact as much as possible. Starting with a regular icosahedron, this can be done in two ways that yield an Archimedean solid. If the vertices of the icosahedron are cut off in such a way that each edge of the icosahedron is cut in half, the resulting polyhedron will have a pentagonal face in the spot where each of the cuts was made (since five equilateral triangles meet in each vertex of the icosahedron), that is, there will be a pentagonal face replacing each of the vertices of the icosahedron. Furthermore, the triangular faces of the icosahedron will be replaced by smaller triangles, whose vertices are at the midpoints of the sides of the original triangular faces of the icosahedron. The resulting polyhedron will, therefore, have as many pentagonal faces as the icosahedron had vertices, namely, 12, and the same number of triangular faces as the icosahedron, namely, 20. The result is that the solid known as the icosidodecahedron (with *icosi-* referring to the 20 triangular faces and *-dodeca-* referring to the 12 pentagonal faces), shown in Figure 7.12.

On the other hand, if the vertices of the icosahedron are cut off in such a way that each edge is divided into thirds, the resulting polyhedron will still have a pentagonal face, where each of the cuts was made, but will have a hexagonal face replacing each of the triangular faces of the original icosahedron, as can be seen in Figure 7.13.

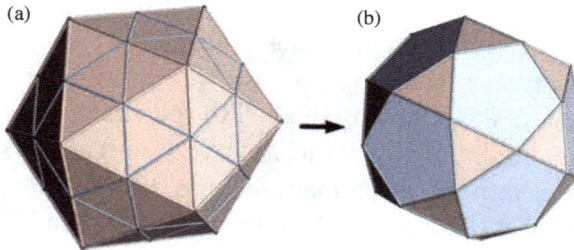

Fig. 7.12. Icosahedron prepared for cutting off vertices (a) and the resulting icosidodecahedron (b).

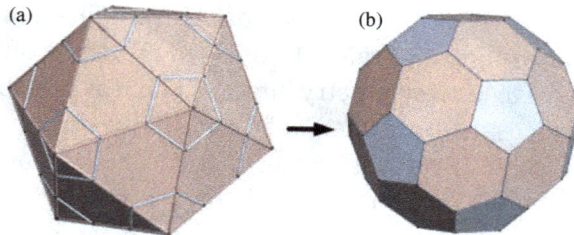

Fig. 7.13. Icosahedron prepared for cutting off vertices (a) and the resulting truncated icosahedron (b).

This is precisely the truncated icosahedron, and we now see why this is quite an appropriate name for the resulting object.

But what do we mean when we say that the truncated icosahedron can be considered something of a connection midway between the icosahedron and the dodecahedron? In order to understand this idea, it is best that we take another look at the duality of the two objects. As we know, if we join the midpoints of the faces of a dodecahedron with the midpoints of all neighboring faces, we obtain the edges of a regular icosahedron, as shown in Figure 7.14(a).

This also works the other way around. If we join the midpoints of the faces of an icosahedron to the midpoints of the adjoining faces, we obtain the edges of a regular dodecahedron, which is shown in Figure 7.14(b).

Starting with either of these situations, we can now produce the truncated icosahedron by simply slightly enlarging the inner polyhedron. In Figure 7.15, we start from the icosahedron inscribed in the

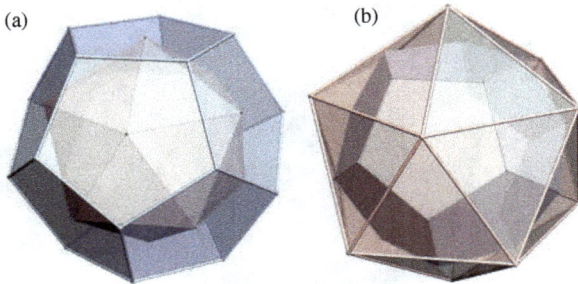

Fig. 7.14. (a) Icosahedron inside a dodecahedron. (b) Dodecahedron inside an icosahedron.

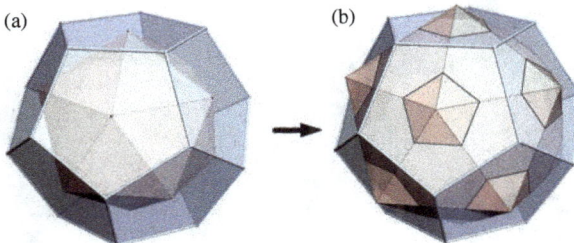

Fig. 7.15. An icosahedron grows inside a dodecahedron.

dodecahedron, and then make the icosahedron just a little bit larger, with the center of the expansion in the common midpoint of the two polyhedra.

If we properly enlarge by just the right factor, as is the case in Figure 7.15(b), the intersection of the two will yield the truncated icosahedron shown in Figure 7.16.

The process of inflating a truncated icosahedron into a soccer ball is shown in Figure 7.17.

If we enlarge the icosahedron inside the dodecahedron a little bit further, as shown in Figure 7.18, the intersection would become the icosidodecahedron of Figure 7.12.

Inflating the icosidodecahedron into a rounded form would give a somewhat unusual and unfamiliar, but nevertheless quite appealing, type of "soccer ball" shown in Figure 7.19.

Fig. 7.16.   Truncated icosahedron.

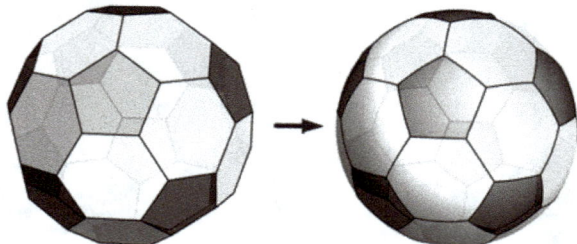

Fig. 7.17.   Inflating a truncated icosahedron.

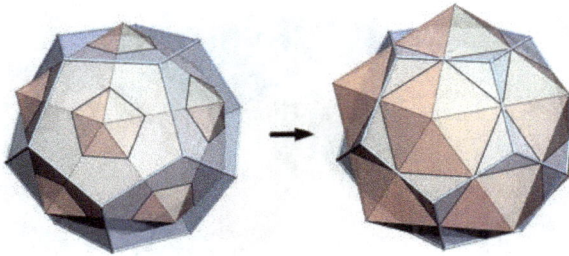

Fig. 7.18.   Growing and icosahedron inside a dodecahedron.

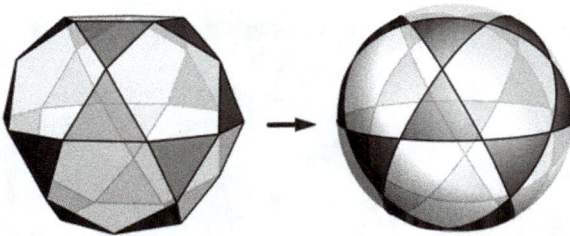

Fig. 7.19.   Inflating an icosidodecahedron.

7.5 The Beach Ball

Another very common type of ball is the beach ball. Usually made of plastic panels that have been either glued or welded together. Such balls are most commonly composed of six identical strips, whose ends are held together with circular patches. An example of such a beach ball is shown in Figure 7.20.

If we think of the colored strips as being simplified as rectangles, and let our imaginations magically transform the circles into hexagons, the basic geometric shape of the ball can be then seen to be that of a hexagonal prism, as shown in Figure 7.21.

Of course, there is no good reason to restrict ourselves to hexagonal prisms. In Figure 7.22, we see a somewhat-unusual example of an athletic ball based on a heptagonal prism. This ball is also based on a regular prism, which in this case has seven sides.

There are also beach balls with more than six sides, but these are quite unusual. The next time you find yourself on the beach, you may want to check out the various beachballs.

Fig. 7.20.   Beach ball.

Fig. 7.21.   Geometric shape related to the beach ball.

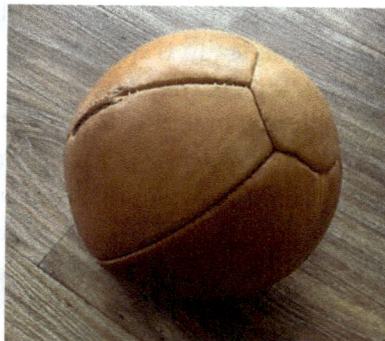

Fig. 7.22.   Athletic ball based on a heptagonal prism.

7.6 The Basketball

A quick glance at a standard basketball will not yield as obvious a polyhedral structure as is the case for the soccer ball or the beach ball (see Figure 7.23).

In fact, the surface of a standard modern basketball is not sewn from patches at all but is rather a uniform rubberized surface. The curves traditionally produced on the surfaces of the balls, however, still delineate areas on the ball recalling the shapes of the leather patches that were once sewn together to make basketballs, back in the days when such balls had visible seams.

Taking a closer look at the picture in Figure 7.23, we see that there are a total of eight regions. Each of these shapes covers the same area on the surface of the ball. The fact that the surface of the ball is divided into eight equal-area sections, is analogous to the regular octahedron. The basic geometric structure of the basketball can be considered to be derived from that of the regular octahedron in the following way. First, we see that there are three "vertices" (i.e., points in which four panels meet, two of each type) on each of the two sides of the ball. In this sense, the basketball has six vertices, each of which is a common corner of four equal-area panels, as is the case for the octahedron.

Fig. 7.23.   Basketball.

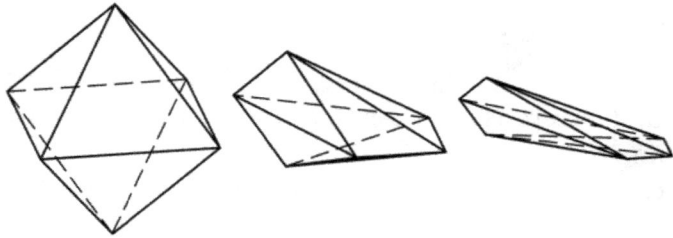

Fig. 7.24. Transformation of an octahedron.

In Figure 7.24, we see how we can imagine two steps of a transformation of a regular octahedron. Approximating a polyhedron, we can then inflate this structure to produce an approximation of the basketball. In each step, two of the vertices move along an edge, closer to a central vertex. The triangular faces are transformed from equilateral triangles to quite thin triangles in the process. Each of these thin panels can be imagined to be fleshed out to one of the panels of the basketball, and inflating the polyhedron (under the assumption that the panels are cut appropriately and made of a sufficiently elastic material) then yields a spherical form.

7.7 The Volleyball

A typical volleyball is made up of six groups of three parallel strips, as shown in Figure 7.25.

Grouping together the "parallel" strips, the surface of the volleyball is composed of six somewhat rectangular sections, which remind us immediately of the sides of a cube. The basic structure of this type of ball can, therefore, be considered to reflect the geometric properties of the cube. Each face of the cube can be thought of as being cut into three rectangular parts, with the cuts oriented in the symmetric way shown in Figure 7.26.

On the actual ball, each face of the "cube" is, indeed, cut up into three strips, but these are not quite rectangular. Ignoring some rounding for the moment, it is true that the middle strip on each face is an almost rectangular strip of leather, but the others are more-or-less trapezoidal. A better way to think of the geometric structure is to

Fig. 7.25.   Volleyball.

Fig. 7.26.   Preparing a cube.

think of these strips not as trapezoids, but rather as hexagons. This is easily motivated by noting the fact that each such strip has six vertices; four in the corners and two more along an outer edge, where the rectangular strip of the neighboring face is attached. Such a polyhedron results by chamfering the edges of a cube in an appropriately symmetrical way. This yields a polyhedron with 12 hexagonal faces and six rectangular faces. None of the faces of this polyhedron is a regular polygon, but its overall symmetry is still very high, since the faces of the same type are all congruent and themselves symmetric. Also, the placement of the faces, being derived from the cube in this

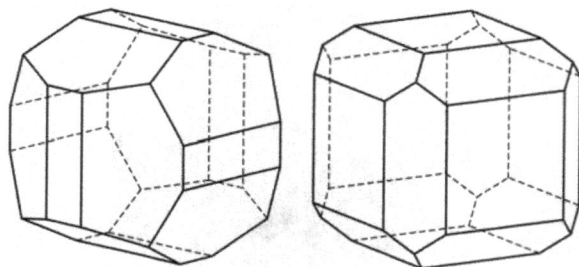

Fig. 7.27.   Chamfered cubes.

Fig. 7.28.   Golf ball.

way, yields a global symmetry that is still essentially that of the cube. Two versions of the polyhedra created as chamfered cubes are shown in Figure 7.27.

It is quite intriguing to take a closer look at the geometrical symmetries inherent in the balls we typically take for granted without focusing on their structures. In fact, we have barely scratched the surface of the geometry of ball design, despite our ruminations on regular and semi-regular polyhedra. The motivated reader might be curious enough to inspect the patterns that produce the dimples on a golf ball! (see Figure 7.28).

Chapter 8

Patterns: Voronoi, Kindergarten, Economy, and Epidemics

What is mathematics? One common answer has been: Mathematics is the science of patterns. This appropriate description stems from the English mathematician G.H. Hardy (1877–1947), one of the last century's finest thinkers. However, the world is full of patterns, though, only very few observers recognize the mathematics involved while strolling around. Throughout this chapter, we will take a journey to discover beautiful patterns and reveal the underlying mathematics.

8.1 Honeycomb Pattern

During a pleasant walk outdoors, we can often pass a lovely wild-flower meadow. Accountable for this beautiful sight is a most useful insect — the bee. According to their pollination function, bees have a great importance for many ecosystems. Estimations reveal that about one-third of the human food supply depends on insect produced pollination, where the bee plays a major role. Interestingly, these cute buzzers are full of patterns, such as the body, waggle dance, honeycomb, etc. Bees are, thus, the starting point of our voyage of discovery in this chapter. Their honeycombs provide an opening view into the patterns with which they are involved (see Figure 8.1).

Fig. 8.1.   The pattern of a honeycomb.

At first sight, we can see a perfect hexagonal tessellation of the plane. That raises the question: Why do bees produce such a tiling? Are there any deeper reasons behind this very aesthetic pattern? Bees use their honeycomb cells for two reasons, for larvae and the storage of honey. One possible explanation for this tiling is that bees are eager to keep waste as low as possible. In other words, they save on their material consumption.

If we look from above at the plane of the honeycomb, the larvae appear as roughly circular disks filling the hexagonal cells. This then raises the question: How can one arrange circles of the same size in the densest possible way on the Euclidean plane? Admittedly, bees have never asked such questions. However, the perfectly regular pattern has evolved over millions of years and evidently has advantages over other, irregular arrangements. In 1773, The Italian–French mathematician Joseph Louis Lagrange (1736–1813) provided the answer. He proved that the arrangement of circles with the highest density is the hexagonal packing, as shown in Figure 8.2. Explicitly, there are two statements in Lagrange's theorem. First, there is no packing arrangement of circles that is denser than the hexagonal packing, and second, only the hexagonal packing provides the highest density. The density of this arrangement of circles is about 90%, which means that just 10% of the plane is not covered by the circles. To calculate this percentage of coverage, we choose a representative

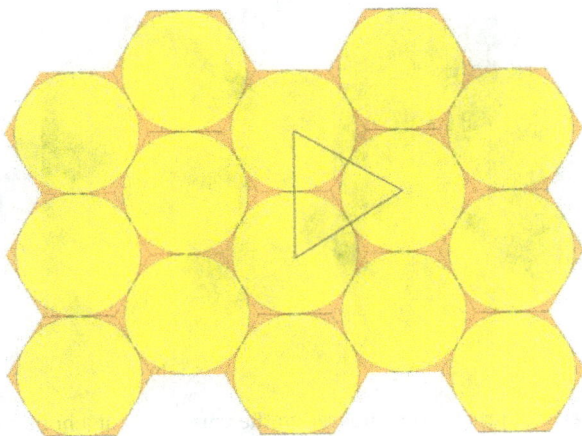

Fig. 8.2.   Hexagonal packing arrangement.

section of the tessellation, for example, an equilateral triangle, as can be seen in Figure 8.2. Let the length of a side of the triangle be $2a$, then elementary geometry tells us that has the area $a^2\sqrt{3}$. The sum of the areas of the three circle-sectors inside the triangle is one-half of the area of a circle with radius a, which is $\frac{a^2\pi}{2}$. Thus, we obtain the following proportion, which provides us with the part of the equilateral triangle covered by circles.

$$\frac{\text{area covered by circles}}{\text{area of triangle}} = \frac{\dfrac{a^2\pi}{2}}{a^2\sqrt{3}} = \frac{\pi}{2\sqrt{3}} = \frac{\pi\sqrt{3}}{6} \approx 0.90689968....$$

This computation shows, that for the hexagonal arrangement of circles shown in Figure 8.2, slightly more than 90% of the area of each triangle (and hence of the area of the whole plane) is covered by circles.

However, bees do not only save material in two dimensions, but actually in three dimensions. At first glance, one might suppose that honeycomb cells are hexagonal prisms, as the one shown in Figure 8.3(a). However, one side of the "honeycomb prism" is not flat, but it has a more complicated shape that offers the opportunity to

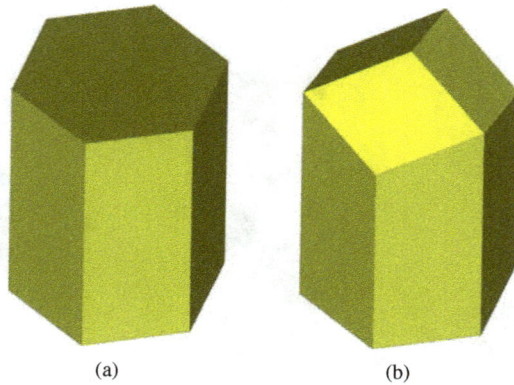

(a)   (b)

Fig. 8.3.   Hexagonal prism (a) compared to the true shape of a honeycomb cell (b).

save material without reducing the available volume. One side of the hexagonal cell ends in three rhombic faces, as shown in Figure 8.3(b). In fact, this structure can also be seen in Figure 8.1, where it shines through from the back of the honeycomb layer.

We explain how to construct this shape with the help of Figures 8.4 and 8.5. In Figure 8.4, the top face of the prism (a regular hexagon) is divided into three colored rhombuses. In order to create the true shape of a honeycomb cell, the three colored rhombuses have to be tilted around one diagonal. The green rhombus is tilted around line segment AC and for the other rhombuses it is CE and EA, respectively. This amounts to lifting the midpoint M of the hexagon, while lowering the vertices B, D, and F.

It is a basic property of any flat hexagon with side l that the diagonals of each of the rhombuses are l and $l\sqrt{3}$, as shown in Figure 8.4(a).

If one tilts the three rhombuses along their diagonals, as described above, the volume of the hexagonal cell remains constant. This can be seen as follows: Consider the rhombus $ABCM$ in Figure 8.5, and assume it gets tilted to the position of the rhombus $ATCN$. Then two tetrahedra $ABCT$ and $ACMN$ have the same volume, because the areas of the base triangles ABC and ACM are the same (due to symmetry), and also the altitudes MN and BT are equal. Hence, the original flat prism and the honeycomb cell with the tilted rhombuses on top have the same volume.

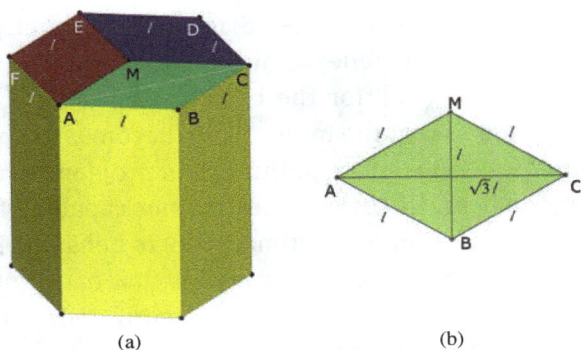

Fig. 8.4.   An incorrect honeycomb cell in the form of a hexagonal prism.

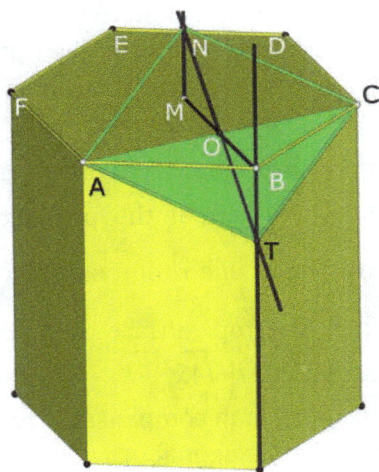

Fig. 8.5.   The tilted rhombus.

For the bees, it would be important to save material when building the honeycomb. Therefore, the shape should have as small a surface as possible without losing space inside. But is it possible to reduce the surface area compared to a regular prism just by tilting the rhombuses on the top of the honeycomb cell as described? While tilting the rhombus in Figure 8.5 around the diagonal *AC*, point *B* moves downwards on the front edge and so the area of the lateral surface decreases, as desired. On the other hand, the inclination of the rhombus makes the initially-shorter diagonal *NT* longer and, there-fore, the top surface of the honeycomb cell increases.

Hence, it is a tough call for the bee: To what extent should the rhombuses be tilted so that the total surface becomes as small as pos-sible? Mathematically, it is a nice optimization problem, which is actu-ally solved by the bees. Of course, bees are not capable of doing any calculation, here evolution has optimized their behavior by giving a competitive advantage to those bee colonies that have improved the construction of their honeycombs. For the motivated reader, we are now going to demonstrate a mathematical solution to this optimiza-tion problem.

In Figure 8.5, the rhombus *ABCM* is tilted by a certain angle to the position of the rhombus *ATCN*. We denote the length of the line seg-ment *MN* by x. It is our goal to determine the optimal value of x, for which the surface of the honeycomb cell becomes minimal.

By applying the Pythagorean theorem, the diagonal *NT* has the length $\sqrt{(2x)^2 + l^2}$, therefore, the area of the rhombus becomes $A_{ATCN} = l\frac{\sqrt{3}}{2}\sqrt{(2x)^2 + l^2}$, which is half the product of the diagonals. Hence, the increase of area of one rhombus is $l\frac{\sqrt{3}}{2}\sqrt{(2x)^2 + l^2} - l^2\frac{\sqrt{3}}{2} = l\frac{\sqrt{3}}{2}\left(\sqrt{4x^2 + l^2} - l\right)$. Then the top surface area, consisting of three rhombuses, increases by $3\frac{\sqrt{3}}{2}l\left(\sqrt{4x^2 + l^2} - l\right)$. As we shall see, this increase in area can be more than compensated for by the decrease in the areas of the faces of the hexagonal prism.

The segment *BT*, which is symmetric to *MN*, also has the length x. The segments *AB* and *BC* have the length l, therefore, the area of one of triangles *ABT* or *BCT* is $\frac{xl}{2}$. Hence, the sum of the areas of both tri-angles *ABT* and *BCT* is xl, which is the amount by which the vertical surface area decreases, if we take into account only one rhombus. This situation appears 3 times — one time for each rhombus, hence, the vertical surface area decreases by $3xl$. So, the total loss of area of the honeycomb cell — due the tilting process — equals the loss of the side surface area minus the gain in the area of the three rhombuses, that is, $3xl$ minus $3\frac{\sqrt{3}}{2}l\left(\sqrt{4x^2 + l^2} - l\right)$. Therefore, we define the total decrease of

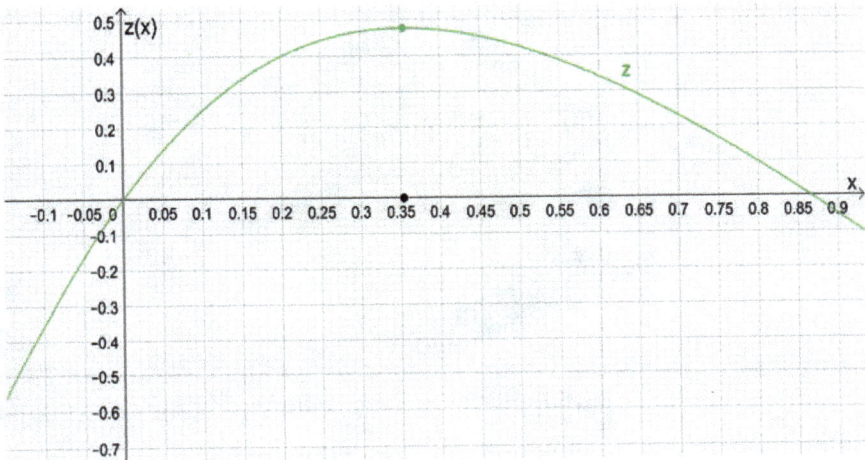

Fig. 8.6.   Graph of the function z.

area as a function of x given by $z(x) = 3l\left[x - \frac{\sqrt{3}}{2}\left(\sqrt{4x^2 + l^2} - l\right)\right]$. We are looking for that value of x, for which this area reduction is maximal. For any given length l, we can draw the graph of the function z. For $l = 1$, the graph is shown in Figure 8.6.

The maximum of this function, which is the maximal reduction of surface area, can be found at approximately 0.353, as indicated with the black dot on the x-axis in Figure 8.6. The rigorous calculation, using methods of calculus, gives $x = \frac{\sqrt{2}}{4}$. And in the case of the function z with arbitrary values of l, we obtain $x = l\frac{\sqrt{2}}{4}$.

It turns out, that in the optimal configuration, the rhombuses on top of the honeycomb cell are arranged in exactly the same way as three rhombuses of a rhombic dodecahedron, which is a dual-Archimedean solid that has been encountered earlier (see Section 6.9). The complete solid is shown in Figure 8.7.

As described in Section 6.9, the rhombic dodecahedron has the characteristic property that the long diagonal of each rhombus face is exactly $\sqrt{2}$ times the length of the shorter diagonal. This is because the rhombic dodecahedron can be derived from a cube by adding six pyramids with a slant angle of 45° to each of its six sides (see Figure 6.28).

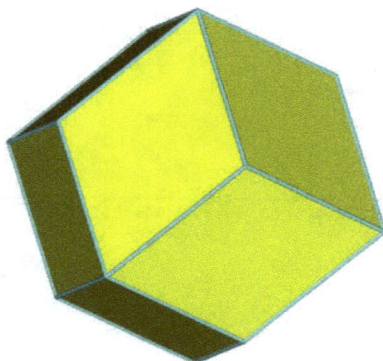

Fig. 8.7.   Rhombic dodecahedron.

(a)   (b)

Fig. 8.8.   Honeycomb-cell.

The characteristic ratio of the lengths of the diagonals can be veri-fied from our computation above. Recall that the length of the longer diagonal has not changed and is still $l\sqrt{3}$. The length of the other diagonal is $\sqrt{(2x)^2 + l^2}$. We can plug in the surface-optimizing value $x = l\frac{\sqrt{2}}{4}$, and we obtain $\sqrt{(2\frac{\sqrt{2}}{4}l)^2 + l^2} = l\sqrt{\frac{3}{2}}$. We then see that the length of the long diagonal $l\sqrt{3}$ is exactly $\sqrt{2}$ times of the length of the short diagonal $l\sqrt{\frac{3}{2}}$.

A honeycomb cell, thus, consists of an elongated hexagonal prism that ends on one side with three faces of a rhombic dodecahedron, as shown in Figure 8.8(b). The open end of the cell can be regarded as flat, as we can see in Figure 8.8(a).

We are now approaching the surprising and useful aspect, namely, the shape is such that two opposing honeycomb layers that perfectly fit into each other, as shown in Figure 8.9. Closer analysis allows one to notice that each facet is shared by another cell. This optimal

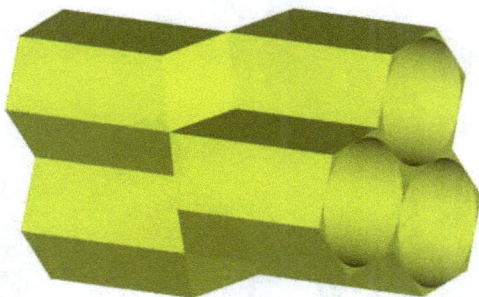

Fig. 8.9. Space-filling tessellation in Euclidean 3-space.

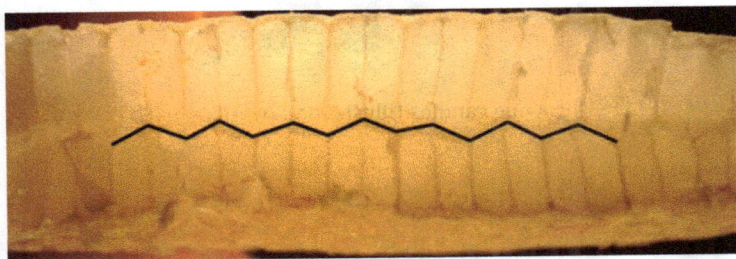

Fig. 8.10. Cross-section of a honeycomb.

construction is, of course, a consequence of the space-filling property of rhombic dodecahedra, discussed in Section 6.9.

In Figure 8.1, we notice that the cells of the far side show through because of the translucency and we can see the three rhombuses in each cell. By slicing through a (real) honeycomb, we can see how our geometric optimization has been realized in nature. The black line in Figure 8.10 indicates the non-flat ends of the honeycomb cells, the tilted rhombuses, and it also shows the perfect match of the two honeycomb layers.

The honeycomb is also a special case of another mathematical pattern, called *Voronoi diagram*, which is important in many practical applications and which we are going to explain next.

8.2 Voronoi Diagrams

We are already familiar with the concept of Voronoi diagrams since our earliest childhood — quite likely without being aware of it.

Fig. 8.11.   Throwing candies (illustration by Alessandro Lorenz).

Fig. 8.12.   Top-view of the situation.

Imagine, somebody throwing some candies into a kindergarten group, as we illustrate in Figure 8.11. Every child has to make a decision which candy to pursue. Clearly, this would be the one closest to the person. Let's take a look at the situation in an abstract top-view, which we depict in Figure 8.12.

We label the candies with the letters *A*, *B* and *C*. We consider the initial situation, when the candies were just thrown into the room and the girl (represented by a yellow circle) is going to decide which candy to pursue. Thus, she has to decide which candy is closest to her. The perpendicular bisector between the points *A* and *B* divides the plane into two regions as indicated by the dashed line in Figure 8.13.

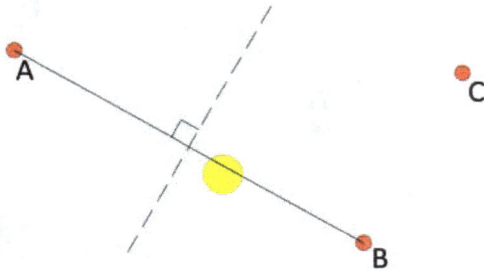

Fig. 8.13. Perpendicular bisector between A and B.

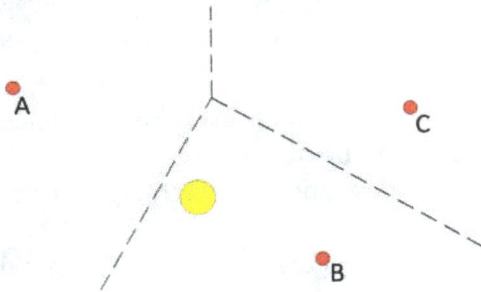

Fig. 8.14. Candy B is the favorite.

As we know, points on the bisector have the same distance to point *A* as to the point *B*. Every point on the side of the bisector where point *A* is situated is closer to point *A*, and all the points on the other side of the bisector are closer to point *B* than to the point *A*. Thus, the girl should rather run for candy *B* than for candy *A*.

Analogously, by drawing the perpendicular bisector of BC, as we have done in Figure 8.14, we once again can conclude that she should also run for candy *B* instead of *C*, since she is closer to candy *B*. However, in reality, running to the closest candy doesn't always guarantee success, as in Figure 8.11, where the boy in front is even closer to candy *B*.

If we have a given set of points in a plane and we construct all the perpendicular bisectors between all pairs of points and extend them to meet the other perpendicular bisectors, we get a division of the plane into regions as shown in Figure 8.15. If we choose any point in

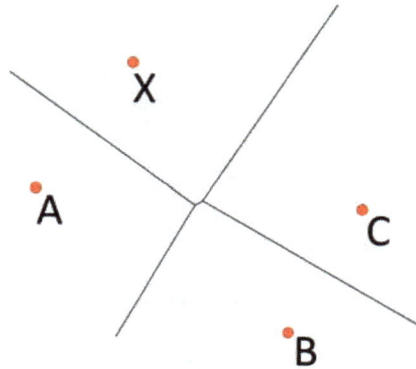

Fig. 8.15.   Four points and their Voronoi diagram.

one of the regions, say for example, in the region which contains point *X*, this selected point will be nearer to *X* than to all the other points. The diagram which describes this strategy is called a *Voronoi diagram*.

The literature backdates the first use of Voronoi diagrams to the 17th century. The famous French mathematician René Descartes (1596–1650), while studying the solar system and decomposing it into regions, essentially developed the concept of Voronoi diagrams. However, he did not provide a mathematical definition and his structure did not receive any further considerations. In 1840, these diagrams were rediscovered as an appropriate mathematical concept by the German mathematician Carl Fredrich Gauss (1777–1855), as well as in 1850 by the German mathematician Gustave Lejeune Dirichlet (1805–1959) and in 1908 by the Russian mathematician Georgi Feodosjewitsch Voronoi (1868–1908), after whom these diagrams are named. Alternatively, these diagrams are also sometimes named Dirichlet tessellations or Thiessen polygons. In 1911, the American meteorologist Alfred Henry Thiessen (1872–1956), used this kind of pattern within a geometric method for dividing land areas for weather predictions. However, many people rediscovered Voronoi diagrams and apply them in various other fields. Since his school days, Voronoi was dedicated to mathematics. As a secondary school student, he published his first paper on factoring polynomials. Although Voronoi authored many valuable contributions in the field of number theory and algebra, he is known to a larger audience for his paper on n-dimensional considerations of the Voronoi diagram. Voronoi died of a severe: Gallbladder attack in 1908 while deeply involved in his mathematics research. Yet, his reputation for having furthered the Voronoi diagrams remains to the present day.

8.3 Voronoi Diagrams in Three Dimensions

Let's take a closer look at the mathematical definition of a Voronoi diagram. We can start by considering distances in the plane. The Euclidean distance d is just the straight-line-segment distance between two points, in other words, it is the ordinary distance that we are accustomed to. More mathematically, we can consider two points in a Cartesian plane as for example, $A = (a_1, a_2)$ and $B = (b_1, b_2)$. The distance is calculated by using the Pythagorean theorem, as we can see in Figure 8.16.

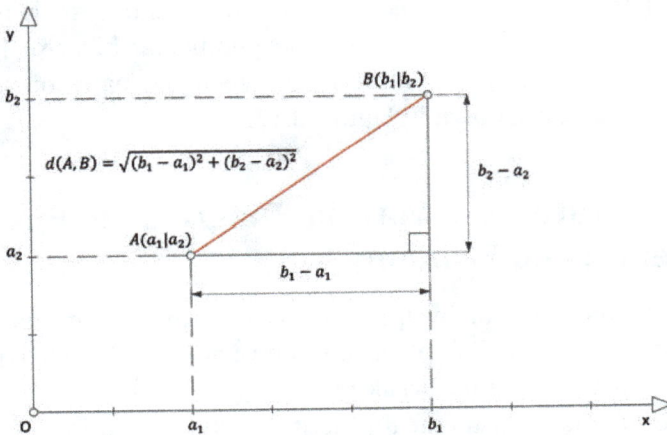

Fig. 8.16.   The Euclidean distance between A and B is the length of the line AB.

Let's consider the two points A and B in the plane. We can define the set of points P, which are closer to point A than to point B, using the mathematical notion $AP \leq BP$. This set of points P is the *Voronoi cell* of point A.

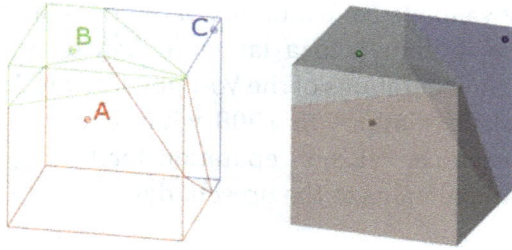

Fig. 8.17.   Two visualizations of the three-dimensional diagram.

If there are n points $X_1, X_2, X_3, \ldots, X_n$ in the plane, then the Voronoi cell of X_1, is the set of all points P that are closer to X_1 than to any other point in the list. Therefore, the following conditions must be true for the points P in the Voronoi cell of X_1: $X_1P \leq X_2P$, and $X_1P \leq X_3P$, and ... $X_1P \leq X_nP$. In this way, we can define a Voronoi cell for each of the n given points. The set of all these Voronoi cells for the given points X_1, $X_2, X_3, \ldots, X_n$ is called the *Voronoi diagram*.

Constructing a planar Voronoi diagram in a two-dimensional space can be easily generalized for higher dimensions, such as for the three-dimensional space. Instead of perpendicular bisector lines we would use the planes of symmetry between the pairs of points to divide the space, as shown in Figure 8.17.

8.4 Applications of Voronoi Diagrams to Public Health and Economy

One of the most noteworthy applications of Voronoi diagrams appears in the work of the English physician John Snow (1813–1858), which deals with the cholera outbreak in London in the year 1854. At this time, people and even medical practitioners believed in the miasma theory, which stated that cholera (and other diseases) was caused by miasma ("pollution") — in other words a kind of toxic air. The Greek physician Hippocrates of Kos (ca. 460–370 BCE) is regarded as the founder of this dubious theory. However, the ingenious John Snow was not taken in by miasma. He carefully listed the number of deaths due to cholera and marked each on the map of London. The black bars

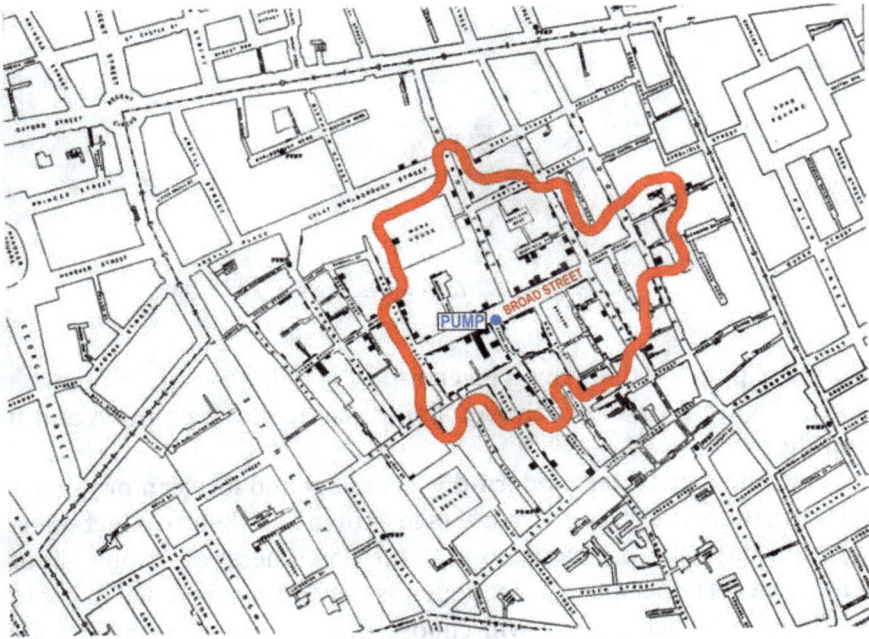

Fig. 8.18. Voronoi cell of the source of Broad Street.

in Figure 8.18 represent the number of deaths at each place. In the next step, he took the sites of the sources of drinking water into account and encircled the area on the map, which represents the boundary of equal distance between the Broad Street pump and other pumps in town. This border line drawn on the map turned out to be the Voronoi cell of the source on Broad Street (see red line in Figure 8.18). This supported his hypothesis that impurities of the source were due to human waste and not those miasma. Finally, he convinced the city government to remove the pump on Broad Street and consequently the cholera outbreak ended quickly.

One further particularly nice thing about Voronoi diagrams is that they unite the ancient Greeks and the modern economy. The Greek geometer Apollonius of Perga (ca. 250 BCE) is noted for theories about conic sections, and his definitions of parabola, hyperbola and ellipse are still used today. However, we will focus on his definition of a circle. According to Apollonius it can be defined as follows: a circle

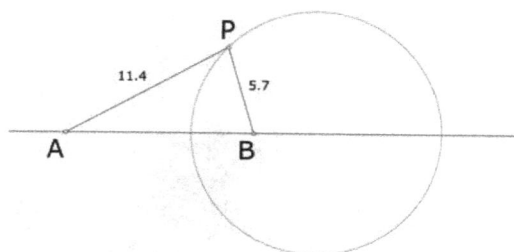

Fig. 8.19.   Apollonius circle.

is a set of points P that have a given ratio $\lambda = \frac{AP}{BP}$ for two given points A and B. Figure 8.19 shows the Apollonius circle of the points A and B and the ratio $\lambda = \frac{11.4}{5.7} = 2$.

Let's now consider modern-day economy and location problems. Assume, there are six supermarkets in a town and their product offerings are basically the same so that the costumers' decisions about going to a certain supermarket depends only on the distance to get to it. Therefore, each person will choose the nearest supermarket for his/her shopping trip. In Figure 8.20, we have divided the town into six regions that mark the catchment areas of each supermarket. If one lives on a division line, then the choice of supermarket is equally distant to each of the two. It is also possible that you may find the distance to three different supermarkets insignificant, if you live at the point where lines meet.

Obviously, the lines represent the Voronoi diagram for the set of the six given points (the supermarkets). These Voronoi patterns apply to many situations in everyday life. For instance, we could have taken the landing sites of rescue helicopters instead of supermarkets and consider their operational areas, the shape of the boarders is the same as in Figure 8.20. There are many more such applications, for example, when students are assigned to their nearest school.

Though, actually, supermarkets attract people differently. For example, larger stores with a greater selection of items draw in more customers than small grocery stores that may be just round the corner. In order to take such a feature into account mathematically, one has to weight the Euclidean distance function d of Figure 8.16. We can

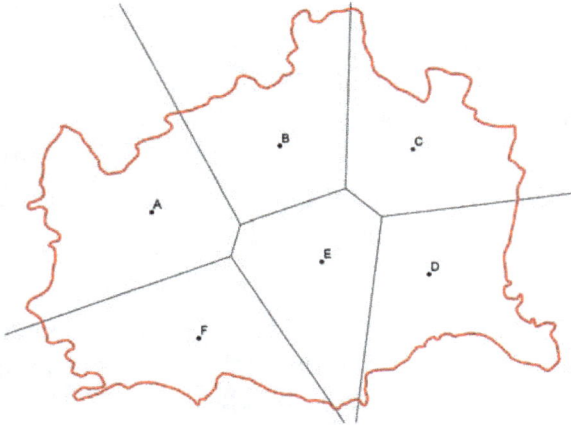

Fig. 8.20.   Catchment areas of supermarkets.

define this function as follows: $d_w(A,B) = \frac{1}{g_A}\sqrt{(b_1 - a_1)^2 + (b_2 - a_2)^2}$, and in doing so, we assume that $A = (a_1, a_2)$ and $B = (b_1, b_2)$ are points in a two-dimensional coordinate system. Consider that A and B represent locations of two supermarkets (in a two-dimensional coordinate system). The bigger the weight g_A is, the smaller is the weighted distance function d_w. Costumers always choose that supermarket, which is nearest to them in the sense of the weighted distance function. Considering the lines of indifference, again, if you live on such a line, then the choice of supermarkets is equally one or the other, in this case, one may ask about the shape of those lines. Let's take a look at the following situation in a coordinate system: The supermarket A is placed at the origin of the coordinate system and the supermarket B has the coordinates (4,0), as shown in Figure 8.21. Assume, due to some market research, the supermarket A is more popular than supermarket B, in more detail g_A values 3 and g_B is 1. Now, we want to determine the shape of the set of points $P = (x, y)$ at which one is indifferent towards these two supermarkets. Mathematicians translate these characteristics into an equation:

$$d_w = (A, P) = d_w (B, P),$$

$$\frac{1}{g_A}\sqrt{(x-a_1)^2+(y-a_2)^2}=\frac{1}{g_B}\sqrt{(x-b_1)^2+(y-b_2)^2}.$$

Let's insert numbers $(a_1 = a_2 = b_2 = 0, b_1 = 4)$:

$$\frac{1}{3}\sqrt{x^2+y^2}=\frac{1}{1}\sqrt{(x-4)^2+y^2}.$$

That leads us to the equation of a circle with radius 1.5 and center (4.5,0), which represents the border of the catchment areas of the two supermarkets A and B (see Figure 8.21):

$$(x-4.5)^2+y^2=2.25.$$

A person, who lives inside the circle, will shop at supermarket B. Unlike a person, who lives outside the circle, who will shop at the supermarket A. Lastly, a person on the border is equally placed towards both supermarkets. The attentive reader will have recognized that we have returned to an old acquaintance — the circle of Apollonius. Actually, the mentioned border can be interpreted as the Apollonius circle of the points A and B at ratio $\lambda = 3$.

This can also be regarded as the Voronoi diagram of the points A and B with respect to the weighted distance function, which takes

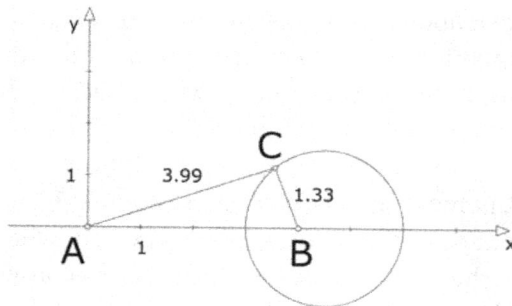

Fig. 8.21. Catchment areas of the supermarkets A and B.

the attraction of a certain supermarket into account. Accordingly, the shape of the Voronoi diagram depends on the definition of distance. Considering the weighted distance function, the Voronoi diagram becomes round, in contrast to the Euclidean function, where the Voronoi diagram consist of straight lines. Further improvements can be made by modelling the distance function correspondingly. For example, one can take the road network into account because only in very rare cases are you able to take the beeline.

8.5 Algorithms for Creating Voronoi Diagrams

Apropos bees, the construction of Voronoi diagrams is not that easy, as one has to be "busy as a bee" to create such a diagram; in particular, in case of a great number of points. The use of a computer appears to be appropriate for such an endeavor, but even so, somebody must tell the computer how it should proceed. Telling the computer about a certain procedure means implementing an algorithm. Here, we present an algorithm for the construction of a Voronoi diagram. This algorithm uses a "brute-force approach" that is not very efficient, but it is sufficient for the purpose of demonstration. Let us start by describing the algorithm in the two-dimensional plane — referring to Figure 8.22:

1. Choose a point from the list of points. Label it point A.
2. Choose another point from the list. Let us call it B.
3. Draw the perpendicular bisector of AB. Mark the area (half-plane) that contains all points that are closer to point A than to point B.
4. Remove point B.
5. Repeat the steps 2–4 until no points remain.
6. Take the intersection of all areas or half-planes, respectively, which have been created in step 3.
7. Repeat this procedure with every other point in the list that was not chosen in step 1.

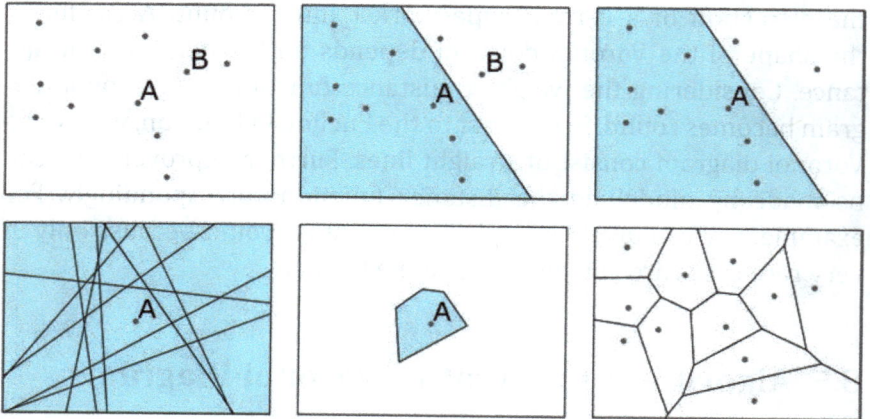

Fig. 8.22. Brute force algorithm for the construction of the Voronoi diagram of a set of points in the plane.

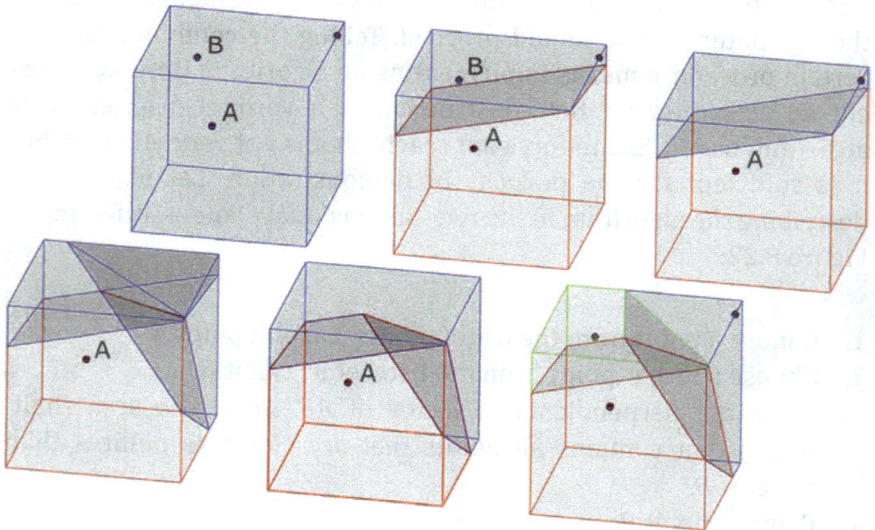

Fig. 8.23. Brute force algorithm for the construction of the Voronoi diagram of a given set of points in space.

In order to expand this algorithm to three dimensions, we have to take regions in space (half-spaces) instead of half-planes, see Figure 8.23.

However, the performance of this brute-force algorithm is rather poor. Let us assume, we have a list with n points. For every point on this list, we must consider the $n - 1$ other points to find the Voronoi cell of this particular point. This is aggravated by the fact that there are n points on the list. Finally, this algorithm requires us (or the computer, respectively) to construct $n(n - 1) = n^2 - n$ half-planes. For large numbers of n, this term equals almost n^2, which means that, if we have 1,000 points, we have to construct about 1,000,000 half-planes. Indeed, there are better, albeit more complex methods to construct a Voronoi diagram. We will now discuss one particular method, because of its relation to another important concept, the Delaunay triangulation.

8.6 Delaunay Triangulation

The Russian mathematician Boris Nikolaevich Delone (1890–1980), better known as Delaunay, which is a French translation of his name, invented a special kind of triangulation, called the Delaunay triangulation. Delaunay was influenced by Voronoi, and within this field of research, he is famous for his general mathematical model of crystals. Regarding the Delaunay triangulation, consider a given set of points in the plane, which are connected to triangles in such a way that every circumcircle of a certain triangle of the triangulation only includes the vertices of this triangle and no other vertices of other triangles. In Figure 8.24(a), we see a triangulation that is not a Delaunay triangulation because the circumcircle of triangle ABC contains point D in its interior. In this case, the Delaunay triangulation has to be organized differently, as we show in Figure 8.24(b). Here each of the circumcircles of the two triangles contains no other points of the given set.

Such a triangulation can be implemented in a three-dimensional space as well, where triangles turn into tetrahedra and circumcircles into spheres. Analogously, a sphere is determined by four points. The circumcircle condition of the triangles becomes the condition, that the circumsphere of each tetrahedron must not include a fifth point.

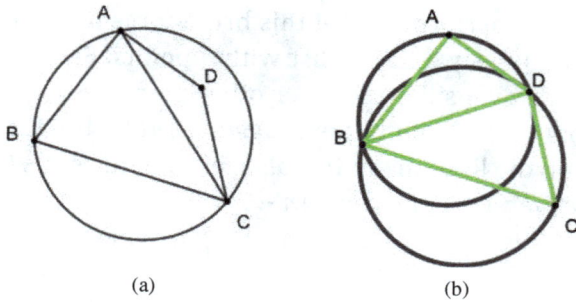

(a)   (b)

Fig. 8.24.   Delaunay triangulation of a set of four points (green lines) and a counter example (a).

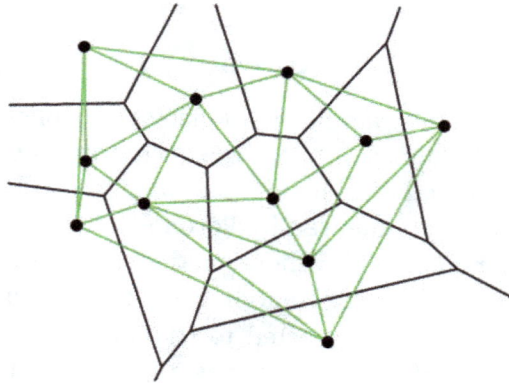

Fig. 8.25.   Voronoi diagram in black and Delaunay triangulation in green.

Up to this point, the reader may wonder why this triangulation is mentioned with respect to an algorithm for Voronoi cells. Interestingly, there is a relationship between the Voronoi diagram, and the corresponding Delaunay triangulation of a given sets of points. At first, we have to mention that the relationship holds true only, if no three points are collinear and no four points are on a circle. If these properties hold true, the vertices of the cells of a certain Voronoi diagram are the circumcenters of the triangles of the Delaunay triangulation. Conversely, we obtain Voronoi cells out of the Delaunay triangulation by constructing the perpendicular bisectors of the edges of the triangles. Figure 8.25 shows this relationship.

Certainly, this relationship holds in space as well. However, we would need additional conditions to assure that no four points are on a plane and no five points lie on a sphere.

If we know the Delaunay triangulation of a given set of points, we can easily construct the corresponding Voronoi diagram. Fortunately, there already exist algorithms for such a triangulation. One of these is called Bowyer–Watson algorithm, which we will now demonstrate for the interested reader. We consider a list of points for which we construct the corresponding Delaunay triangulation.

1. Construct a random triangle that contains all points of the list.
2. Delete all points of the list from the drawing plane.
3. Add one (new) point of the list to the drawing plane.
4. Identify all triangles whose circumcircles include the new point.
5. Delete the sides of these triangles.
6. Construct for each pair of vertices of the deleted triangles (which are at the margin of the starshaped area) the triangle that consists of the two vertices and the added point of step 3.
7. Repeat steps 3–6 until there is no more new point of the list to add to the drawing plane.
8. Delete all sides of the triangles that contain at least one vertex of the triangle of step 1 and delete the vertices of the triangle of step 1.

Then, one can construct the Voronoi Diagram. To see how this and the algorithm work, we shall consider an example. We begin with a list of points that are plotted in the drawing plane, see Figure 8.26-1. Executing the first step of the algorithm, we draw in a random triangle, which contains all plotted points, in this case $T_1 T_2 T_3$, as shown in Figure 8.26-2. Next, we remove all points of the list from the drawing plane and then, we add one point, say it is point A, as shown in Figure 8.26-3. In step 5 of the algorithm, we delete all sides of triangles, whose circumcircles encompass the added point, hence,

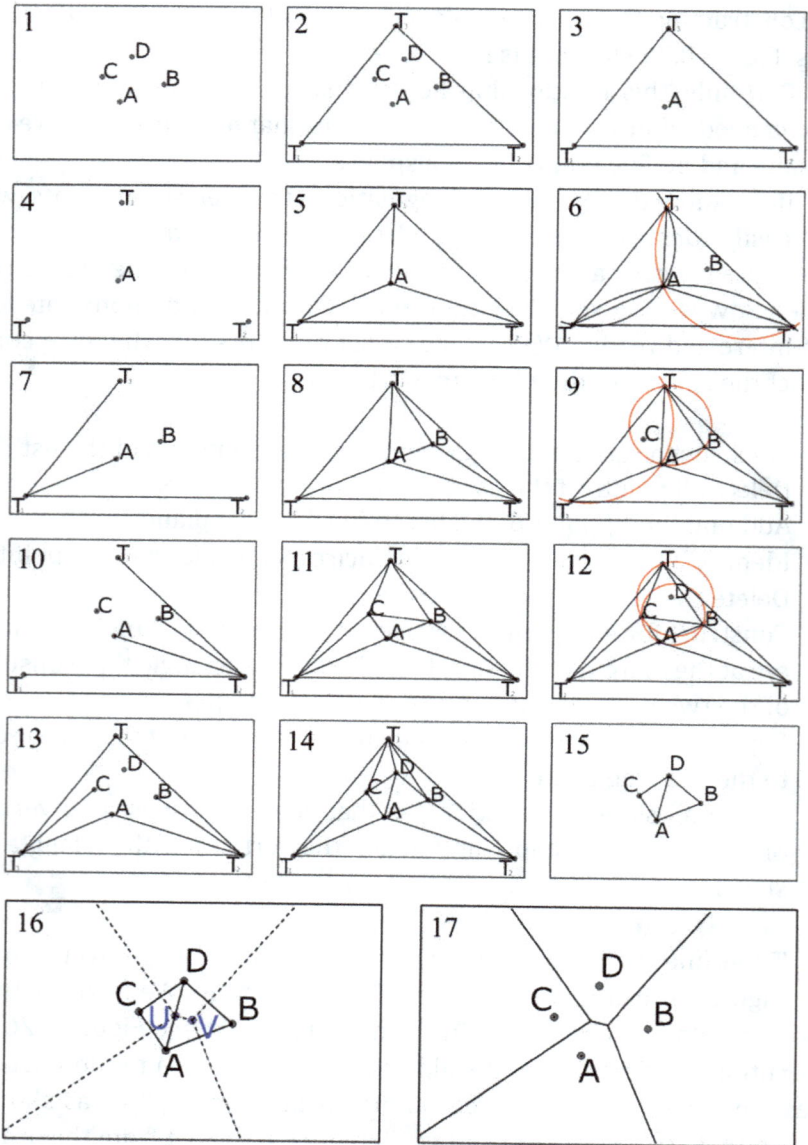

Fig. 8.26.   Bowyer–Watson algorithm for creating a Delaunay triangulation and the Voronoi diagram.

we remove the sides of the triangle $T_1T_2T_3$, as shown in Figure 8.26-4. For step 6 of the algorithm, we construct the three triangles T_1T_2A, T_2T_3A and T_3T_1A, as can be seen in Figure 8.26-5. In the next step, Figure 8.26-6, we add point B. The circumcircle of T_2T_3A encompasses point B, therefore, the sides of this particular triangle are removed (Figure 8.26-7), and we establish three new triangles ABT_2, T_2T_3B and T_3AB, as we see in Figure 8.26-8. Now, we add point C and see that the circumcircles of the triangles ABT_3 and AT_3T_1 encompass point C, shown in Figure 8.26-9. Hence, all sides of those two triangles are removed (Figure 8.26-10), and four new triangles arise: T_1CT_3, T_1AC, ABC and BT_3C (Figure 8.26-11). We repeat the steps 3–6 of the algorithm until we have added all points, then we delete all sides of the triangles that are connected with the vertices of the initial triangle $T_1T_2T_3$ and the three vertices (Figure 8.26-15), This gives the Delaunay triangulation of the four points A, B, C, and D.

We can now construct the Voronoi diagram, which is given by the perpendicular bisectors and circumcenters of the triangles obtained by Delaunay triangulation. In Figure 8.26-16, we construct the bisectors of the line segments AB, BD, DA, which intersect in the circumcenter V of the triangle ABD, and with the additional bisectors and DC and CA we obtain the circumcenter U of the remaining triangle ADC. These bisectors together with the circumcenters finally give the Voronoi diagram of the points A, B, C, and D, as shown in Figure 8.26-17.

8.7 Dynamic Construction of Voronoi Diagrams

Another, more dynamic way of constructing a Voronoi diagram is to use the given set of points as the centers of uniformly–growing circles, shown in Figure 8.27. Such dynamic procedures are not easy to implement as algorithms. Hence, we will just illustrate the procedure.

We can clarify this procedure for constructing a perpendicular bisector as we did in Figure 8.13. The procedure used in Figure 8.27 is analogous to the results shown in Figures 8.13 and 8.15. To locate all points equidistant from two given points A and B, which are on the perpendicular bisector of AB, we can draw circles of equal radii

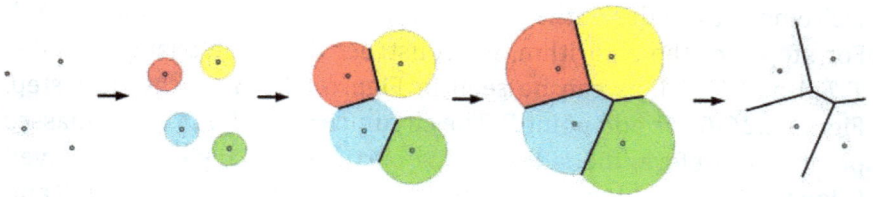

Fig. 8.27.   Growing circles.

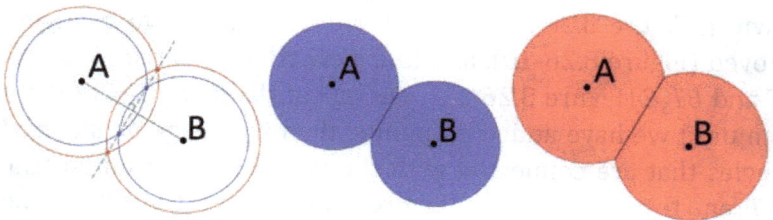

Fig. 8.28.   The link between constructing a perpendicular bisector and the meeting of uniformly growing circles.

around points A and B (Figure 8.28). Where they intersect, we get pairs of points of the perpendicular bisector. Thus, the uniformly growing circles around the points of our Voronoi diagram continuously intersect at its line segments.

Analogous to the two-dimensional case, where increasing circles meet on the perpendicular bisectors, we can construct a three-dimensional Voronoi diagram by using the given set of points as the centers of uniformly growing spheres, which meet on the planes of symmetry of the given set of points, see Figure 8.29. We can observe spheres with the same radius around three given points A, B and C. They intersect pairwise in the plane of symmetry related to their centers, which are the set of given points. This again leads to the same diagram as already shown in Figures 8.17 and 8.23.

8.8 Voronoi Diagrams in Nature and Art

Using Voronoi diagrams, many natural shapes can be described, such as soap bubbles, bone cells, animal skins, crystal structures, etc.

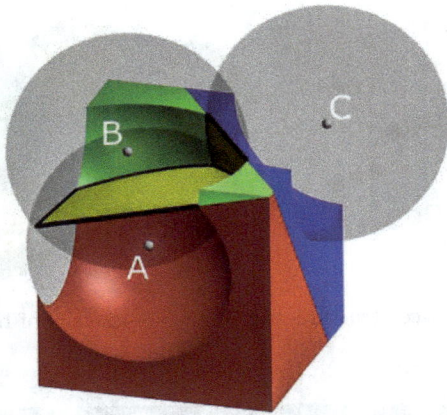

Fig. 8.29.   Growing spheres.

We are going to turn our focus on these shapes, now. The two American physicists Eugene Paul Wigner (1902–1995) and Frederick Seitz (1911–2008) developed a special cell, which is named after both inventors, Wigner–Seitz cell. Generally, crystals are arranged in a regular three-dimensional array, called a lattice. The crystal properties or features are defined by the lattice. In order to examine the characteristics of a crystal, one uses the Wigner–Seitz cell, which can be regarded as mathematical examination tool for crystals. Specifically, the Wigner–Seitz cell is a special example of a primitive cell of crystal structure, which only contains one lattice point and it is defined as the locus of points in space that are closer to that lattice point than to any other lattice point. In fact, that is the definition of a Voronoi cell. Figure 8.30 shows a set of points arranged in a three-dimensional lattice and the corresponding Voronoi diagram can be seen on the right side. In fact, this Voronoi diagram represents a space tessellation of Wigner–Seitz cells, which are, in this case, truncated octahedra — Archimedean solids with the surface of eight hexagons and six squares Section 6.8.

We can discover Voronoi diagrams in many places in nature. We have seen that Voronoi diagrams can emerge when things grow in a special way. The fur covering of a giraffe, the shell of a turtle, or the

Fig. 8.30.   Crystal lattice structure and Wigner–Seitz cells of body-centered cubic lattices.

Fig. 8.31.   Voronoi diagrams in nature: Animals (https://pixabay.com/de/photos/giraffe-wildtier-flecken-2222908/).

human skin are just a few examples of the Voronoi pattern with respect to cell growth, as we can see in Figures 8.31 and 8.32.

Moreover, the dehydration process of soils obeys the pattern of Voronoi. An explanation for this phenomenon may be that every seed tries to gain as much water as possible, and if the seeds have the same power of attraction, the soil becomes a Voronoi diagram with linear bisectors (see Figure 8.33).

Inspired by nature, Voronoi diagrams are used in arts to obtain an organic-looking design (see Figure 8.34).

When seeing a certain pattern, one cannot immediately decide whether the presented examples are Voronoi diagrams or not,

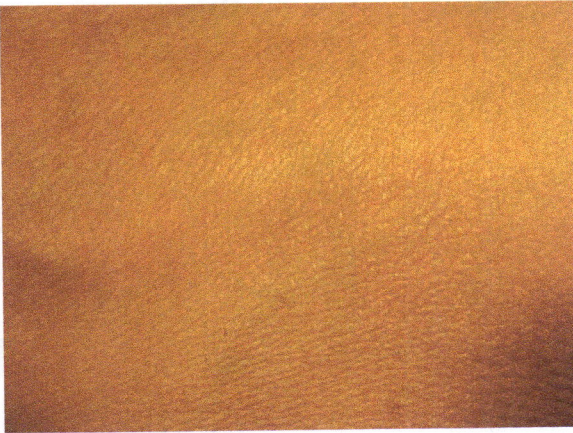

Fig. 8.32. Voronoi diagrams on the human skin.

Fig. 8.33. Voronoi diagrams in nature: Desert (https://pixabay.com/de/photos/w%C3%BCste-trockenheit-composing-279862/).

especially, if the corresponding set of points is not visible, as you might have experienced while viewing the Figures 8.31–8.34. As a first step, we may check the Voronoi cells for convexity. A polygon is convex (Figure 8.35(a)), if it contains no two points, which when

Fig. 8.34.   Jewelry art.

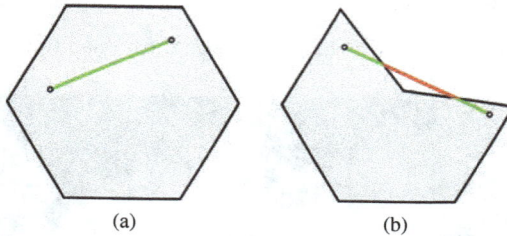

| (a) | (b) |

Fig. 8.35.   Illustration of a convex and a non-convex set.

connected with a straight-line would find a part of the line outside the polygon. A pattern containing non-convex cells (as in Figure 8.35(b)) is not a Voronoi diagram.

However, this condition is necessary but not sufficient. Hence, there are patterns that fulfill the convexity condition but they are not Voronoi diagrams. In this case, the Delaunay triangulation supports us in deciding on the Voronoi issue. Recall that the vertices of the cells of a certain Voronoi diagram are the circumcenters of the triangles of the Delaunay triangulation. Within this triangulation, the points are connected to triangles in such a way that the circumcircle of any triangle includes only the vertices of the same triangle. At worst, in some cases, if there are three or more collinear points, or if there are four or more points on a circle, the Delaunay triangulation does not exist. However, the pattern could be a Voronoi diagram, nevertheless.

8.9 Architecture and the Weaire–Phelan Structure

Even in architecture, we can find the patterns of Voronoi as we can see in Figure 8.36, which shows the National Aquatics Center that was built in the years 2004–2007 for the 2008 Summer Olympics in Beijing, China.

Although the building has the nickname "Water Cube", it is a rectangular box, or more specifically, a cuboid where the ground plane is a square of 584 square feet with a height of 102 feet. The outer wall is based on the Weaire–Phelan structure, which we show in Figure 8.37, and which is a complex three-dimensional structure representing an idealized form of equal sized bubbles. This is to date the best-known solution for what is known as the Kelvin problem. In the year 1887, the British mathematical physicist William Thomson Lord Kelvin (1824–1907) posed the following problem: Which arrangement of cells of equal volume is the optimal with the restriction that the total surface area is minimal? Kelvin's solution was a structure made of

Fig. 8.36. Beijing National Aquatics Center (Andrey Belenko, CC BY 2.0, free commercial use).

Fig. 8.37.   Weaire–Phelan structure.

copies of truncated octahedra, shown in Figure 8.30. For more than 100 years it was believed to be the best solution for this problem. However, in 1993, the Irish physicist Denis Weaire (1942–) and his student Robert Phelan discovered a better solution by using computer simulations. These two solutions do not differ greatly. Actually, the difference is merely 0.3%, which means the surface of Weaire–Phelan structure is 0.3% less than the surface of the truncated octahedra.

In more detail, the Weaire–Phelan structure consists of two kinds of cells, which have the same volume, they are the pyritohedron and the tetrakaidekahedron. The first one is a dodecahedron with 12 pentagonal faces, which are not constrained to be regular, as shown in Figure 8.38. The edges can be divided into two sets of equal length, one consisting of 24 edges and one consisting of 6 edges. Although the regular dodecahedron does not occur in crystals, the pyritohedron does. Maybe, it was an inspiration for the discovery of the regular dodecahedron.

The second solid is a truncated hexagonal trapezohedron, which is shown in Figure 8.39. It has two hexagonal and 12 pentagonal faces and it has only two reflection planes.

The Weaire–Phelan structure can be built by sticking together the so-called translation-group of six trapezohedra and two

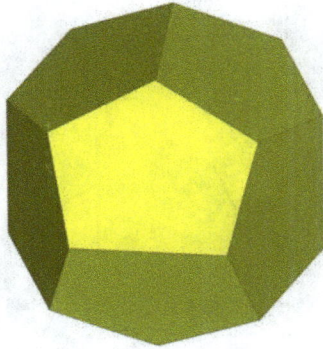

Fig. 8.38.   Pyritohedron.

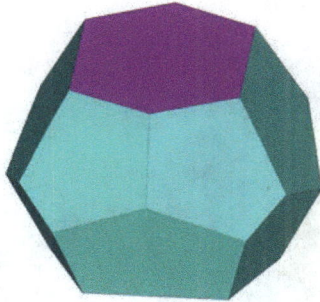

Fig. 8.39.   Truncated hexagonal trapezohedron.

pyritohedra, as shown in Figure 8.40. We can copy and translate this whole translation-group of eight polyhedra (the one in the lower left corner is covered), to build the whole structure.

The impressive surface of the "Water Cube" was achieved by obliquely slicing through the Weaire–Phelan structure, as shown in Figure 8.41.

The cross-sectional area of the slice or the surface, respectively, are further examples of Voronoi diagrams. Actually, the genesis of the appearance stems from a collaboration of Chinese and Australian architecture firms that partly refer to cultural symbols. The Chinese partners suggested the square layout, which is kind of symbolic to the Chinese culture and represents the earth, and the Australian partners

Fig. 8.40.   Translation group.

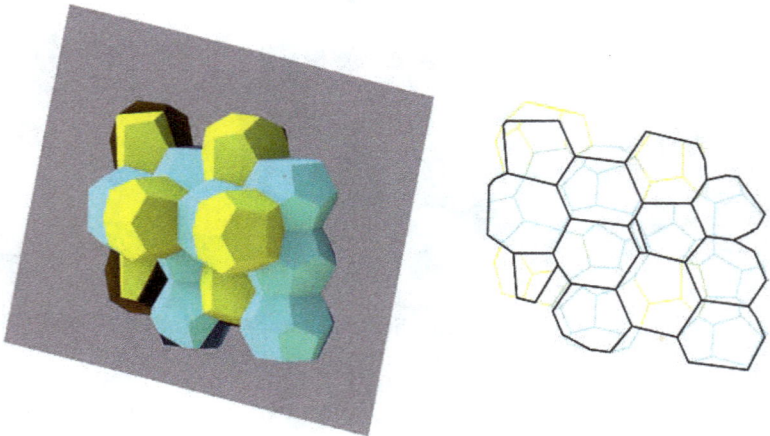

Fig. 8.41.   A slice through the Weaire–Phelan structure.

came up with the idea of bubbles symbolizing water. The result was the "Water Cube" with its impressive surface, which is pictured in Figure 8.36.

As we have seen in this chapter, the world is full of patterns. By revealing the underlying mathematics, we attempted to show that the world is full of mathematics. One must only appreciate the visible.

In more detail, we studied Voronoi diagrams in different contexts, such as bees, kindergarten children, epidemics, economy, human skin and architecture. Although the concept of Voronoi diagrams seems to be rather easy to comprehend at the first glance, it can be a tough challenge to create such a diagram. For example, if one uses different notions of distance, as we have seen in the context of catchment areas of supermarkets. Even the implementation of an efficient algorithm can become rather tricky.

Many natural, economical and even social phenomena can be described with Voronoi diagrams. They occur more often than one may think. Let's keep our eyes open to discover the pattern of Voronoi diagrams in our environment.

Chapter 9

Packing Problems

Modern container ships are giants of the sea, approximately 1,300 ft (400m) long and 200ft (60m) wide and equipped with diesel engines producing 100,000 kW of power. Container shipping is the basis for world-wide trade and a lot of mathematics is involved in the transport logistics as well as in the basic problem of finding the best way of packing objects into a container in order to maximize its full capacity. After a brief historical account of the origin of the standardized inter-modal shipping container, this chapter presents various packing problems in three-dimensional geometry, leading to interesting mathematical questions such as Kepler's conjecture for the optimal packing of equal spheres and the related "kissing problem" of finding the maximum number of equal spheres that can be arranged in such a way that they touch ("kiss") a common center sphere of the same size. We will also discuss three-dimensional packing puzzles as well as certain aspects of real-world packing problems as they appear in the loading of shipping containers.

9.1 History of Containerization

Sometimes, a simple invention can change the world. In the early 1950s, the American transport entrepreneur Malcom McLean (1913–2001) had an idea that would revolutionize international trade and facilitate the creation of a global economy. At that time, almost all commonplace items were produced close to the consumers, since shipping over large distances was very expensive. To transport goods from one place to another, they were packed into bags, boxes, drums, or barrels that were handled manually as break bulk cargo. This means they had to be loaded onto a vessel individually, piece by piece. Loading and unloading a ship at a port was a time-consuming process. Moreover, cargo was loosely placed in the holds of the ship, using the available space not very efficiently. Watching dock loaders unloading freight in odd-sized wooden boxes from trucks and transferring them to ships, McLean recognized that a standardized process of cargo transfer would have the potential to dramatically increase efficiency and thereby, reduce shipping costs. He started experimenting with more efficient ways of transferring cargo and came up with a very simple, but brilliant concept. With the help of engineer Keith Tantlinger (1919–2011), he adopted a transport system, developed by the US Army, for commercial use and designed robust corrugated steel boxes together with some trucks and a ship that would seamlessly hold these boxes. The boxes were 35 ft (10.7 m) long, 8 ft (2.44 m) wide, 8 ft 6 (2.59 m) high. The length of 35 ft was determined as the maximum length of trailers for trucks then allowed on Pennsylvanian highways. A twist-lock mechanism atop each of the four corners of a box allowed them to be easily secured and lifted using cranes. This was the birth of the intermodal shipping container (see Figure 9.1).

McLean bought two World War II tanker ships and converted them to carry containers on and under the deck. On April 26, 1956, the *SS Ideal-X* was loaded and sailed from the Port Newark–Elizabeth Marine Terminal, New Jersey, to the Port of Houston, Texas. It was the first time in history that a ship had its cargo packed in containers. Between 1968 and 1970, ISO standards for containers were published by the International Maritime Organization. ISO containers are one of the rare examples of normed objects all countries in the world agree upon. By greatly reducing the cost and increasing the speed of international trade, containerization was a decisive factor for worldwide economic growth after the World War II. Today approximately 90% of all non-bulk cargo worldwide is transported by containers

Fig. 9.1.   Intermodal shipping container, by IPLManagement/CC BY-SA.

stacked on ships. Large container ships can carry more than 20,000 containers, with up to eight containers stacked upon each other (see Figure 9.2).

9.2 What is a Packing Problem?

Mathematics plays an important role in container shipping in several ways, such as in optimizing the loading and unloading process, finding optimal routes to distribute the cargo, and of course packing objects optimally together into containers with a minimum of wasted space. The latter belongs to a class of mathematical optimization problems known as *packing problems.*

In a packing problem, the task is to pack a single container as densely as possible or to pack a certain collection of objects minimizing the number of containers. These objects may have different sizes or shapes, or they may be all identical. In addition, although this is not the typical case in the real world, the container need not be a cuboid, it could have other shapes as well. Even if one only considers regular geometric shapes like platonic solids, spheres, cylinders, or cuboids,

Fig. 9.2.   Container ship, image by JoachimKohlerBremen/CC BY-SA.

there is a large variety of different situations and possible arrangements and it should not be too surprising that packing problems, in
general, turn out be rather complex. In the following, we will only
present some special cases and their solutions, including famous
problems that have received the attention of many of the greatest
mathematicians in history. Applications of dense packings of various
three-dimensional shapes are by far not limited to transportation and
logistics. In fact, they represent very useful models for the structure
of matter in liquid, glassy and crystalline states.

9.3 Packings in Infinite Space

If the size of the container is much larger than the size of the objects,
the problem essentially boils down to the problem of packing
objects as densely as possible in infinite Euclidean space. Cubes and
cuboids can easily be arranged so that they fill the whole space. In
geometry, a tiling of space by polyhedral cells such that there are no
gaps or overlaps, is called a honeycomb. Among the platonic solids

Fig. 9.3. Cubic honeycomb.

Fig. 9.4. Tetrahedron packing, viewed from different angles.

(tetrahedron, cube, octahedron, dodecahedron, icosahedron), only the cube can fill the whole space on its own. The most natural arrangement is the cubic honeycomb as shown in Figure 9.3.

The proportion of space occupied by an arrangement of shapes is called its density. Imagine a three-dimensional lattice of shapes, such as the tetrahedra as shown in Figure 9.4. Finding the densest packing for a given set of shapes means to maximize the (average) density. With regular tetrahedra, the currently densest known packing is a double lattice of triangular bipyramids. It fills 85.63% of space, which we can see in Figure 9.4.

If we use both tetrahedra and octahedra, we can fill all of space (density 1) in an arrangement known as the tetrahedral–octahedral honeycomb pictured in Figure 9.5.

Fig. 9.5. Tetrahedral–octahedral honeycomb.

Twenty-eight convex uniform honeycombs are known to exist. A convex uniform honeycomb is made up of non-overlapping convex uniform polyhedral cells. Figure 9.6 shows two further examples. The rectified cubic honeycomb (Figure 9.6(a)) consists of octahedra and cuboctahedra. A cuboctahedron is a polyhedron with eight triangular faces and six square faces. It can be obtained by cutting the vertices of a cube. The gyrated triangular prismatic honeycomb (Figure 9.6(b)) is made up of identical triangular prisms (a polyhedron with two equilateral triangular bases, joined by three quadrilateral surfaces). It is called gyrated, because the prisms in each layer are rotated by a right angle to those in the next layer. In Figure 9.6, the lower layer can be obtained from the upper layer by a reflection with respect to the plane defining the boundary between the two layers and then rotating the lower layer by 90° around an axis perpendicular to this plane. The colors indicate which units correspond to each other.

Obviously, curved objects such as cylinders or spheres cannot form a convex uniform honeycomb, since there will always remain gaps between them. However, the problem of finding the densest possible packing is still very interesting and has caught the attention of many great mathematicians. For identical cylinders, such as the cylindrical barrels as shown in Figure 9.7(a), the problem of finding the densest packing reduces to finding the densest packing of circles

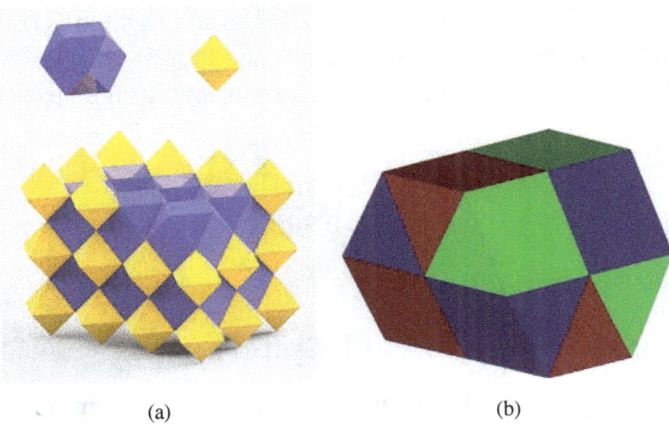

Fig. 9.6. (a) Rectified cubic honeycomb. (b) Gyrated triangular prismatic honeycomb.

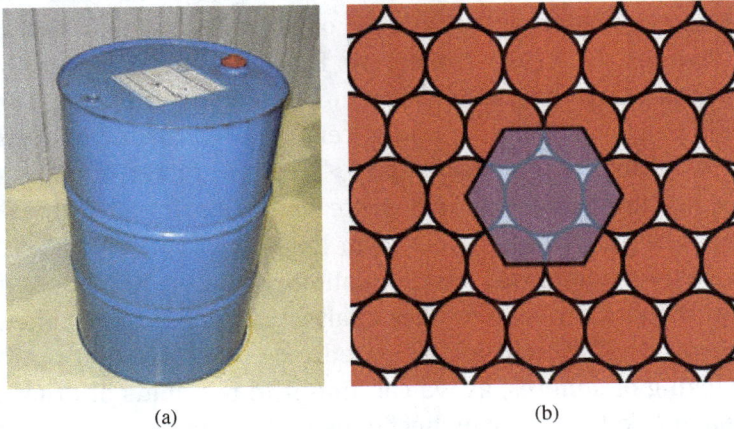

Fig. 9.7. (a) Cylindrical barrel, image by Meggar at English Wikipedia/CC BY-SA. (b) Hexagonal circle lattice, image by Inductiveload/Public domain.

in the plane. It is intuitively clear that this must be the arrangement shown in Figure 9.7(b), where the centers of the circles lie on the vertices of hexagons which fill the plane. Indeed, in 1773, it was proved by Joseph Louis Lagrange (1736–1813) that the hexagonal lattice has the highest density of all circle packings. Its density can be calculated as the ratio of the part of the area of the hexagon which is covered by circles to the total area of the hexagon (see Figure 9.7). We omit this calculation here, but the ambitious reader is encouraged to verify that the result is

$$\frac{\pi\sqrt{3}}{6} \approx 0.9069.$$

9.4 The Best Way to Pack Spheres

The problem of finding the densest packing for identical spheres has an interesting history that begins with Sir Walter Raleigh (1552–1618), an English explorer and one of the most prominent figures of the Elizabethan era. Around 1587, Raleigh posed a question to his assistant and friend, the mathematician Thomas Harriot (1560–1621), that became the origin of a famous mathematical problem known as the Kepler conjecture. Raleigh wanted Harriot to find out how to stack the cannonballs on the decks of his ships in the most efficient way. Harriot, who is remembered for introducing the symbols "<" for "is less than" and ">" for "is greater than", studied various stacking patterns and wrote about his investigations in a Letter to the German mathematician Johannes Kepler (1571–1630), who is best known for his laws of planetary motion. Kepler got interested in the problem and further analyzed it. In his 1611 paper, "On the six-cornered snowflake", he stated the conjecture that the "typical" stacking of spheres, as we can find it in pyramids of oranges or apples at markets, has the highest density that can be achieved, which we can see pictured in Figure 9.8.

At first glance, it might appear as if there is only one way of stacking spheres, layer upon layer, in such a way that the density is maximized. However, there are in fact infinitely many possible arrangements, differing in the relative positioning of successive layers. The two simplest regular patterns are shown in Figure 9.9. These arrangements are known as the hexagonal close-packed (hcp) lattice and the cubic-close-packed (ccp) or face-centered cubic (fcc) lattice,

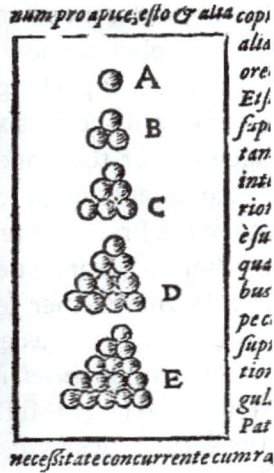

(a)   (b)

Fig. 9.8.   (a) Image by Adoscam/CC BY-SA. (b) One of the diagrams from Kepler's paper "On the six-cornered snowflake".

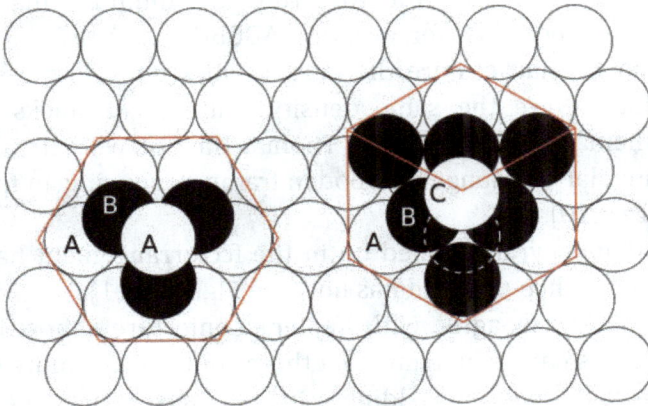

Fig. 9.9.   Comparison of hcp and fcc lattices of spheres, image by en:User:Twisp/ Public domain.

based on their symmetry. In the following example, one should think of an infinite number of layers, each extending to infinity. Put in another way, we are only interested in the relative positioning of the layers.

What is the difference between the two arrangements shown in Figure 9.9? It is obvious how to build the first layer, since it corresponds to the densest packing of circles in the plane. Each sphere in the first layer is touched by six other spheres in this layer. We denote the arrangement of the spheres in the first layer by A. We then build the second layer by putting the spheres into the gaps between the spheres of the first layer and call this arrangement B. For the next layer, two different arrangements with respect to the existing layers are possible. We may either repeat positioning A, meaning that each sphere of the third layer lies exactly above a sphere of the first layer, or we may place the spheres in such a way that the positioning of the spheres in the third layer is different from the positionings of both the first and the second layer. We call this new arrangement C. In the latter case, the fourth layer is inevitably a repetition of the first or the second layer, that is, it must be arrangement A or B. The hcp lattice is obtained by the sequence AB AB AB AB ..., that is, every other layer is the same. In the fcc lattice, every fourth layer is the same, that is, we have the sequence ABC ABC ABC ABC ... . Infinitely many other sequences are possible, for example ABCBCA ..., ABAC ..., etc., but the fcc and hcp lattices are the most regular ones. However, all of these lattices have the same density. Cannonball stacks as they were mathematically analyzed by Thomas Harriot, were usually piled in a rectangular or triangular wooden frame, producing an fcc lattice (see Figure 9.10).

Note that a pyramid piled up in the fcc arrangement has sharp edges, while an hcp pyramid has not (see Figure 9.11).

Intuition lets us agree with Kepler's conjecture almost immediately, since it is hard to imagine that there could be any more efficient way of packing spheres, yielding a higher density than the lattices shown in Figure 9.11. And indeed, in 1831, the famous German mathematician Carl Friedrich Gauss (1777–1855) proved that among all regular arrangements, these packings have the highest density. An example for a lattice packing with a lower density is the simple cubic lattice, where the centers of the spheres sit at the vertices of a cubic honeycomb. Table 9.1 shows the densities of cubic lattices and a schematic drawing of their structure.

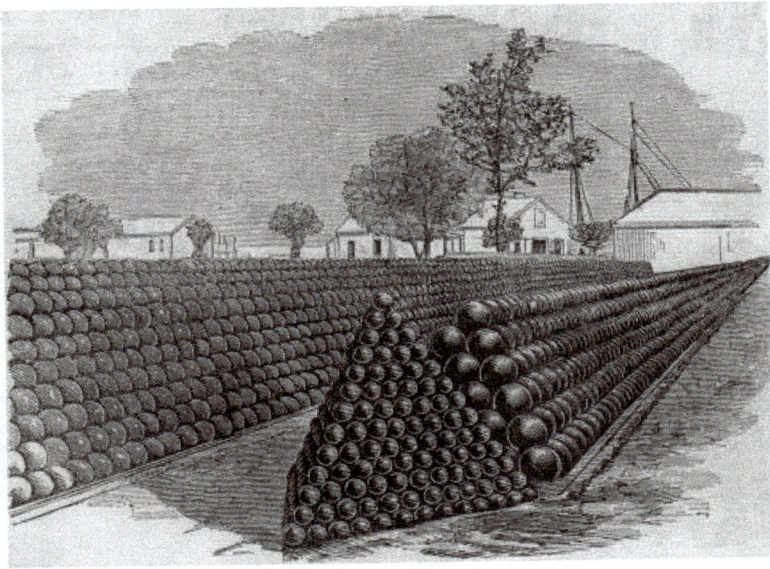

Fig. 9.10.   Cannonball stacks in an fcc lattice, image by STACY/Public domain.

(a)   (b)

Fig. 9.11.   (a) hcp lattice, (b) fcc lattice.

However, Gauss' result did not prove the Kepler conjecture, since there might be irregular arrangements with an even higher density. An irregular packing occurs, when spheres are randomly thrown into a container and then compressed. Although experiments show that irregular packings will have a density of only about 63%, this does not rule out the possibility that there exists a certain irregular arrangement with a higher density than that of an fcc lattice.

Table 9.1.   Cubic lattices and their densities.

Name	Density	Structure
Face-centered cubic (fcc)	$\dfrac{\pi}{3\sqrt{2}} \approx 0.74$	
Body-centered cubic (bcc)	$\dfrac{\pi\sqrt{3}}{8} \approx 0.68$	
Simple cubic (sc)	$\dfrac{\pi}{6} \approx 0.52$	

9.5 The Kissing Problem

Without a mathematical proof that there is also no irregular arrangement that is denser than the fcc or hcp lattice, Keplers's conjecture
remained open. Again, looking at the fcc and hcp arrangements in
Figure 9.11, it seems impossible that there could be any irregular
arrangement with an even higher density. However, there is an interesting observation that may make one doubt this conclusion. In the
fcc and hcp lattice, each sphere is touched by 12 other spheres.
Imagining layers as shown in Figure 9.9, where each sphere is surrounded by six other spheres in the same layer, by three spheres in
the next layer and by three spheres in the previous layer, giving $6 + 3$

+ 3 = 12 surrounding spheres. Is this the largest number of spheres that can be arranged in such a way that each of them is in contact with a common sphere in the center? This question belongs to a mathematical problem that is very interesting by itself; it is known as the "kissing problem", where the kissing number is defined as the greatest number of non-overlapping unit spheres that can be arranged in such a way that they each touch ("kiss") a common unit sphere. The kissing problem is the problem of determining the kissing number in n-dimensional Euclidean space. In two dimensions, the Kissing number is obviously 6, since at most six congruent circles can be in touch with a common circle (see Figure 9.12).

In three dimensions, however, the problem is already not at all trivial. In fact, it was the subject of a disagreement between the English physicist and mathematician Isaac Newton (1642–1727), well-known for his axioms of classical mechanics and his law of gravitation, and the Scottish mathematician David Gregory (1659–1708). Newton thought that the maximum number of spheres in contact with a central sphere would be 12, but Gregory considered it possible that a thirteenth sphere would fit. A close packing of spheres as in the fcc lattice is completely rigid, meaning that the spheres cannot be moved, their places are fixed. However, if we do not consider a lattice of spheres, but simply try to bring as many spheres as possible in

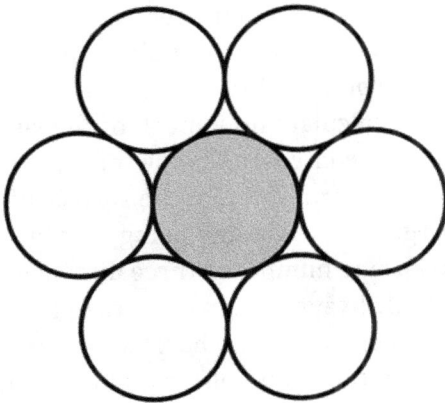

Fig. 9.12.   The kissing number in two-dimensional space, image by N. Mori/Public domain.

Fig. 9.13.   The kissing problem in three-dimensional space.

contact with a common sphere, then the configuration is not rigid. There is still some space between the surrounding spheres, allowing us to move them around without losing contact to the center sphere (see Figure 9.13).

But is there enough free space to squeeze in a thirteenth sphere? There is a nice argument showing that this is at least conceivable. Imagine that the central sphere is transparent and has a lamp in its center. Now enclose the whole configuration in a larger sphere whose center coincides with the center of the central sphere (see Figure 9.14). The lamp casts shadows of the surrounding spheres onto the inside surface of the larger sphere. Clearly, these circular shadows do not overlap, and one can calculate how much of the surface of the outer sphere is covered by the circular shadow of one of the surrounding spheres. Surprisingly, it turns out that the surface of the outer sphere is 14.9 times as large as the shadow being cast by one sphere. Does this mean that the kissing number in three dimensions is actually 14? No, because there will always remain interstices between the spheres, so not all of the available space can be used. But considering the fact that the surface area of the outer sphere is almost 15 times as large as the projected area of one of the equal spheres, it does not seem impossible that 13 spheres could simultaneously be brought into contact with the center sphere.

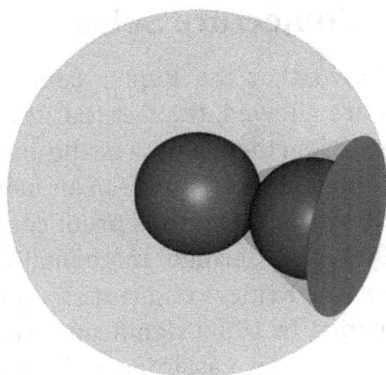

Fig. 9.14.   The surrounding spheres cast shadows on the inner surface of a larger sphere.

Yet, Newton was correct, the kissing number in three dimensions is 12. While it is relatively easy to verify this by an experiment, the first complete mathematical proof for this statement did not appear until 1953. The kissing problem can be generalized to hyperspheres in spaces of arbitrary dimension n. Although spaces of dimension greater than 3 seem to exceed our imagination, since we cannot even mentally visualize four-dimensional objects, let alone five- or six-dimensional ones, their mathematical description is not much different from ordinary three-dimensional space. Whereas, we need three numbers or coordinates to specify a point in three-dimensional space, four numbers are required to determine a point in four-dimensional space, and n numbers in n-dimensional space. The notion of distance between two points can be easily generalized to n-dimensional space, and a sphere is then simply defined as the set of points at a constant distance from a given point, just as in 2 or 3 dimensions. In two dimensions, we call it a circle, in three dimensions a sphere, and in dimension 4 or higher, it is called a hypersphere (see Chapter 15). In four-dimensional space, the kissing number (the greatest number of non-overlapping unit hyperspheres that can be arranged in such a way that they each touch a common unit hypersphere) is 24, but the kissing numbers for $n > 4$ are still not known, with the only exceptions being the cases where $n = 8(240)$, and $n = 24$ (196,560).

9.6 The Kepler Conjecture Solved

While Gauss has shown that the Kepler conjecture is true when only regular lattices are allowed, the fact that the kissing number in three dimensions is 12, further supports the intuitive answer that also no irregular arrangement of spheres can have a higher density than the fcc lattice. However, a rigorous proof could not be achieved. Furthermore, in 1900, the German mathematician David Hilbert (1862–1943) included the Kepler conjecture in his list of 23 unsolved problems of mathematics. In 1953, significant progress was made by the Hungarian mathematician László Fejes Tóth (1915–2005) who was able to show that the problem of determining the maximum possible density of all arrangements (regular or irregular) can be reduced to a finite, but very large number of calculations. Based on Tóth's work, the American mathematician Thomas Hales (1958–) announced a complete proof in 1998, which consisted of 250 pages of notes and 3 GB of computer programs and data. However, due to its complexity, the correctness of the proof could not be verified with certainty, so Hales began to create a complete formal proof that can be verified by automated proof-checking software. After almost 20 years of work, this proof was accepted by a mathematics journal and published in 2017. The proof is now widely accepted in the mathematics community, eliminating Kepler's conjecture from Hilbert's list of unresolved problems.

9.7 How Many Billiard Balls Fit into a Shipping Container?

We started our journey into the realm of packing problems with the invention of the intermodal shipping container that dramatically accelerated worldwide trade and paved the way for a global economy. Although containers will in general not be filled with equal spheres, we may entertain ourselves by applying the densest packing of spheres to estimate the maximum number of billiard balls fitting into a standard shipping container. The most common containers are 20 ft (6.1 m) long, 8 ft (2.44 m) wide, 8 ft 6 in (2.59 m) high. They differ only by their length from the containers that were transported by

Malcom McLean's SS Ideal-X in 1956. The total volume of a container is therefore $2.44 \times 2.59 \times 6.1 \approx 38.55$ m^3 or 38,550 dm^3. An international pool billiard ball has a diameter of 57.15 mm and, thus, a volume of $\frac{4\pi}{3}(\frac{57.15}{2})^3 \approx 97,734$mm^3 or 0.097734 dm^3. If we throw the billiard balls randomly into the container, they will fill approximately 63% of the container volume, amounting to $0.63 \cdot \frac{38550}{0.097734} \approx 248,500$ billiard balls. To get as many balls as possible into the container, we have to arrange them in a close-packed lattice, filling approximately 74% of the container volume. We can, therefore, increase the number of balls in the container by a factor $\frac{0.74}{0.63} \approx 1.175$, that is, by 17.5% or 42,500 balls compared to a random arrangement. Hence, the maximum number of billiard balls that can be packed into a standard shipping container is approximately 291,000.

9.8 Dense Packings Can Help Save the Climate

Although we also encounter dense packings of spheres in everyday life, for instance, when oranges are piled at market displays, typical real-world packing problems are not about arranging a large number of equal objects in a close-packed lattice. In a more realistic situation, items of different shapes and sizes must be packed into identical containers in a way that minimizes the number of containers used. In mathematics, this is known as the bin-packing problem and this problem is in general, at least from a practical perspective, much more complicated than the packing problems with identical objects we have considered thus far. Even if the number of items of different shapes and sizes is not very large, there are incredibly many different configurations possible. In principle, one could use a computer to simply calculate the occupied volume for each of these configurations. But the number of possible configurations grows so fast with the number of items that even the fastest computers would need a very long time to check all the possible configurations. Therefore, very sophisticated algorithms have been developed to find optimal solutions in a reasonable time, at least if the number of items is not too big. Moreover, there exist much faster heuristic algorithms, which, however, do not always produce an optimal solution. Considering that the worldwide container throughput in 2019 was approximately 800 million TEU (Twenty-foot Equivalent Units) and that typical costs for shipping a container lie between \$1,500 and \$2,500, a lot of money can be saved by finding more efficient packing algorithms. For example, increasing the average packing density in shipping containers with the help of a better algorithm by just 1% would correspond to 8 million less containers or a total cost reduction of about \$16 billion. Considering that modern container ships have to burn approximately 3 L of heavy oil per container and per 100 km and are responsible for a significant part of worldwide emission of CO_2 and pollutants, increasing the packing density would also contribute to climate protection.

9.9 Three-Dimensional Packing Puzzles

One of the smallest non-trivial three-dimensional packing problems is the Slothouber–Graatsma puzzle, which we show in Figure 9.15. Six $1 \times 2 \times 2$ blocks and three $1 \times 1 \times 1$ blocks have to be arranged such that they fit into a $3 \times 3 \times 3$ box. The solution to this puzzle is unique (absenting mirror reflections and rotations).

Since the larger blocks can only occupy an even number of the nine cells in each layer, each of the layers must contain one of the unit blocks. There are three mutually-orthogonal ways to decompose the cube into three layers (one horizontal and two vertical decompositions) and the only way to achieve that for each decomposition, each layer contains a unit block, is to place the three unit blocks along a body diagonal of the cube. As soon as one has realized this, the puzzle is solved. A similar solid dissection puzzle is the Soma cube shown in Figure 9.16, which was developed by the Danish polymath Piet Hein (1905–1996).

Here, seven pieces, which are comprised of unit cubes, must be assembled into a $3 \times 3 \times 3$ cube. There are 480 distinct solutions to the Soma-cube puzzle, which cannot be transformed into each other by rotations, or 240 if reflections are also excluded, but in all solutions the "T" piece has to sit at the same place. To see this, notice that this piece will always fill either two corners of the large cube or zero corners, it cannot be oriented such that it fills only one corner.

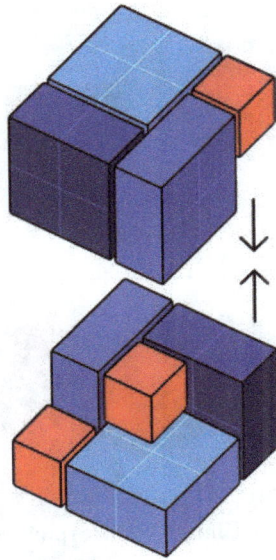

Fig. 9.15.   Slothouber–Graatsma puzzle, image by Cmglee/CC BY-SA 4.0.

Fig. 9.16.   Pieces of the Soma cube.

The "L" piece, on the other hand, can be placed such that it fills two corners, one corner, or zero corners. As we can see in Figure 9.16, each of the other five pieces can only fill either one corner or zero corners. Hence, without the "T" piece, at most seven corners can be filled. A cube has eight corners, so to fill each of them, the "T" piece must fill at

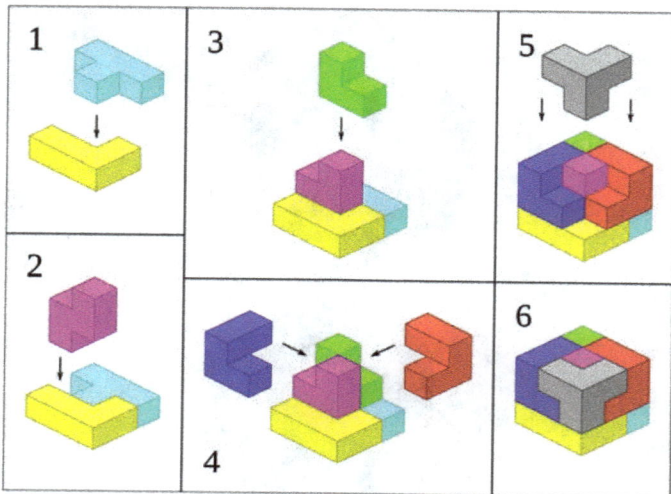

Fig. 9.17. Soma cube assembled, image by Дмитрий Фомин (Dmitry Fomin)/CC0.

least one corner. Since it can only fill two corners or no corners, it must, therefore, be placed in such a way that it fills two corners. One particular solution of the Soma-cube puzzle is shown in Figure 9.17.

9.10 Bin-Packing Algorithms

In realistic packing problems, such as they are encountered by container shipping companies, the individual pieces will not fit so nicely together as in these three-dimensional puzzles. In general, there will always remain gaps and there may be additional constraints, for instance, weight capacity limits or delivery routes. Typically, a collection of items in the shape of cuboids, but with different dimensions, must be packed into bins or containers such that the number of bins used is minimized. If the items cannot be put on top of each other, as for example, palletized cargo that cannot be stacked, the problem reduces to two-dimensional bin-packing. One of the simplest algorithms is the first fit algorithm, which places one item after the other in the first bin in which it will fit. The algorithm is more effective, if the items are first sorted by decreasing-size order, which is also known as the first-fit decreasing algorithm. More sophisticated bin-packing algorithms categorize items by their size into classes or take into account that some items should preferably be packed together.

As global commodity flows are further increasing, and thereby, also the number of containers shipped across the ocean, the development and improvement of algorithms to optimize packaging and transportation is a very active and important field in applied mathematics.

Chapter 10

Fascinating Shapes

Mathematical topology deals with those properties of shapes and surfaces that are not changed under continuous deformations. One of these properties is *one-sidedness*, and it is shared by a class of surfaces in three dimensions that paradoxically appear to have only one side. The best-known example is the *Möbius strip*, which can easily be made from a simple strip of paper.

10.1 Magic with Paper Strips

Flat paper strips can be easily turned into fascinating three-dimensional shapes. One such possibility is a delightful children's activity that involves folding a strip of paper to create a paper spring (sometimes referred to as "witch's ladder"). To make this, one would need two thin strips of paper of equal size. The first step is to glue these strips together to form a right angle as shown in Figure 10.1. Then fold the lower paper strip over the top one. Next, fold the other strip, which is now the lower one over the first strip, as further shown in the second picture of Figure 10.1. Continue to fold the lower strip over the top strip until the entire length of the paper strips has been folded into a small square shape. Glue the two ends together and you will obtain a stable structure that can be expanded into a beautiful paper spring.

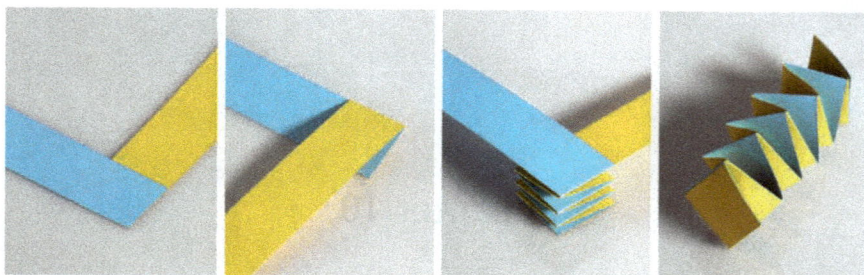

Fig. 10.1.   Folding a paper spring.

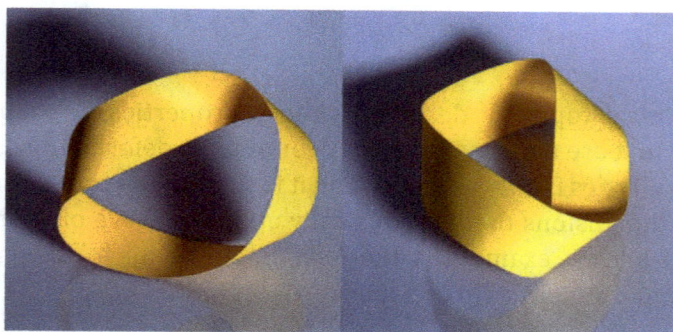

Fig. 10.2.   Two views of a Möbius strip.

Interesting and challenging experiments can even be carried out with a single strip of paper. If you glue a strip of paper together at the ends, you simply get a cylindrical ring. But if you twist one end 180° before gluing the two ends together, you get a ring called a *Möbius strip* with some unusual properties, which we shall investigate. Figure 10.2 shows two views of a Möbius strip.

The Möbius strip is an example of a one-sided surface in three dimensions. It was named after the German mathematician August Ferdinand Möbius (1790–1862). At first sight the Möbius strip appears to have two sides at every point, but from any point one can reach the other side without having to cross over the edge: For example, if you take a Möbius strip and run a pencil line along the center of the strip continuously and without lifting the pencil, you will — after completing one cycle around the strip — reach a point on

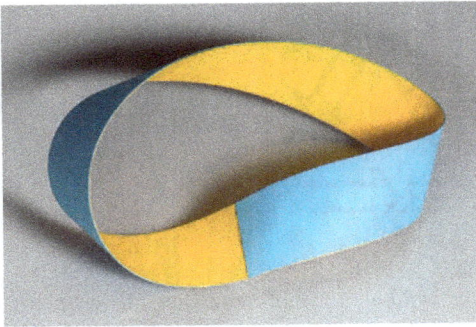

Fig. 10.3.   Möbius strip with two different colors.

Fig. 10.4.   Möbius strips with different chirality.

the "other side" of the strip. Continuing along the center of the strip, the pencil mark will eventually reach the point at which you started. Therefore, it is not possible to paint the two opposite sides with different colors without these two colors colliding somewhere on the surface, as shown in Figure 10.3.

We can actually produce two different types of Möbius strips, as shown in Figure 10.4. Each of these two strips is twisted in opposite directions and differ in what is known as chirality, which means that they are mirror images of each other. We turn one end clockwise when gluing the first strip, and we turn the end counterclockwise when gluing the second strip.

The design of the universal recycling symbol in fact uses the layout of a Möbius strip, but it is often a Möbius strip with three half-twists, as shown in Figure 10.5(a). In any case, it is a one-sided figure.

(a)   (b)

Fig. 10.5.   Recycling symbol. Examples of Unicode character U+2672 (a) and U+267B (b).

Fig. 10.6.   Sculpture inspired by the Möbius strip created by Sebastian (Enrique Carbajal).

The Möbius strip has inspired many artists. A sculpture of a one-sided ribbon inspired by the Möbius strip is shown in Figure 10.6.

10.2 Experiments with Möbius Strips

A Möbius strip is not only one-sided, it also has only one edge. If you slide your finger along the edge, you circle the Möbius strip twice before returning to the starting point. Some very interesting and

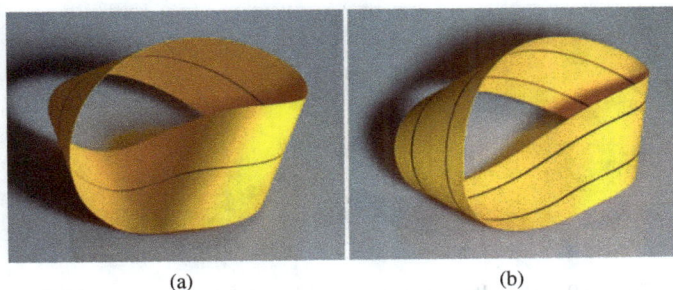

(a)   (b)

Fig. 10.7.   What happens if one cuts the Möbius strip along the line?

entertaining experiments can be carried out with Möbius bands. For example, if we cut a Möbius strip in Figure 10.7(a) along the center line, we do *not* get two strips, but — much to the surprise of those who are doing this experiment for the first time — only a single, double-twisted closed strip with twice the length of the original strip, as shown in Figure 10.8(a).

What happens when you try to cut the edge off? The lengthwise line on the Möbius strip in Figure 10.7(b) is located at a quarter of the width from the edge, and even if it looks like there are four lines, it is indeed only one line, but it is four times as long as the Möbius strip. What happens if you cut along this line, thereby removing the edge of the strip? Cutting along the line in Figure 10.7(b) gives *two* closed loops. The middle part remains a Möbius strip and the cut edge becomes a long thin double twisted strip that is twice the length of the original strip. It is interesting that the two parts are intertwined like the links in a chain, as shown in Figure 10.8(b).

Instead of twisting one end of the paper strip by 180°, this time twist one and by 360° and then glue the ends together, which will result in a double-twisted closed tape as shown in Figure 10.9. It is *not* a Möbius strip and it has two separate surfaces, which means that you can paint the two sides with different colors. What happens if you cut such a strip lengthwise, along the midline shown in Figure 10.9?

Cutting this double-twisted strip gives two double-twisted strips, which are interlocked as the links of a chain as shown in Figure 10.10.

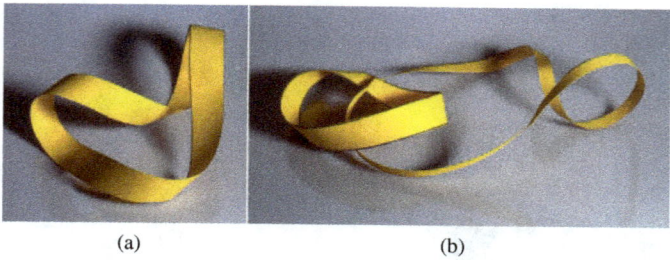

Fig. 10.8.   Result of cutting a Möbius strip as in Figure 10.7.

Fig. 10.9.   Strip with a double-twist.

Fig. 10.10.   Result of cutting a closed band with a double-twist.

For each of the following experiments we will require two strips of paper. Figure 10.11 shows two normal circular loops that were glued together at right angles. Our concern is to determine what happens when both rings are cut at the center along the dashed line shown in Figure 10.11.

When we cut one of these loops along the center line and right through the area where the two loops have been glued together, we get a "handcuffs" result, as shown in Figure 10.12(a). Our next step is to cut the long midline of the connecting strip shown in Figure 10.12(b). Amazingly, after that the structure can be opened out to reveal a square, as shown in Figure 10.12(c).

Fig. 10.11. Two ordinary circular arcs, glued together at right angles.

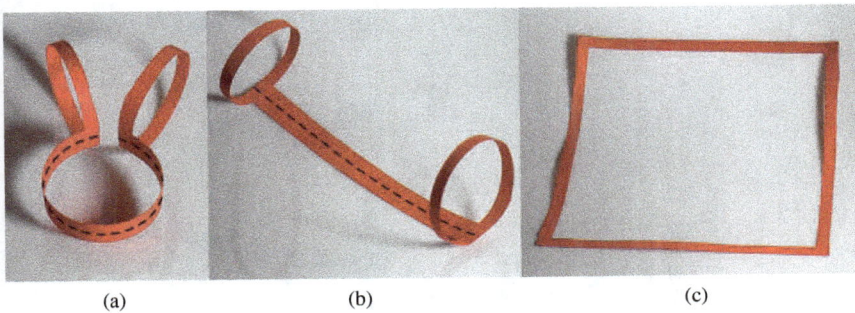

| (a) | (b) | (c) |

Fig. 10.12. Squaring the circles.

The analogous question can be asked for Möbius strips. If we glue together the two Möbius tapes with different chirality as we have shown in Figure 10.4, we get the structure shown in Figure 10.13. Now, we can again cut along the midline of these two Möbius strips, which might be a bit challenging due to the shape. Perhaps the easiest way is to make the last cut at the portion where they are glued together.

The result of this experiment is presented in Figure 10.14. After cutting the two Möbius strips along the midline and carefully opening them out, we will get the two interlocking hearts shown in Figure 10.14. Mathematical results can have romantic resulting images.

Fig. 10.13.   Two Möbius strips with opposite chirality, glued to each other at right angles.

Fig. 10.14.   Two hearts for Möbius.

10.3 The Topology of Paper Strips

A cylindrical loop is created by gluing together the short sides of a rectangular strip of paper (see Figure 10.15). After gluing the strip in Figure 10.15 to form the closed loop, the two blue points and the two yellow points are no longer distinct from each other. Each corresponding pair of points becomes one and the same point on the cylinder.

A Möbius strip is created in a very similar way, only opposite points are identified "crosswise" with each other. In Figure 10.16, the two blue dots and the two yellow dots each become one dot, when the ends of the strip are glued together. In order to achieve this, the strip has to be twisted by 180° before its ends are brought together. The right and left edges are, thus, identified in the opposite direction, as indicated by the arrows in Figure 10.16.

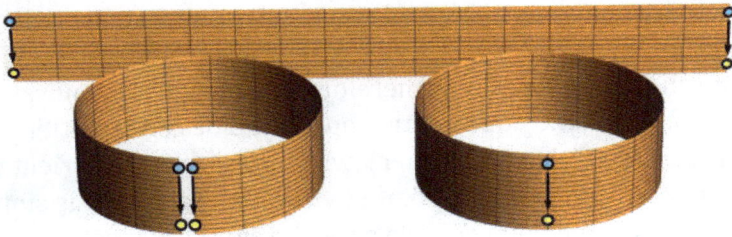

Fig. 10.15.   Identifying points on a rectangle.

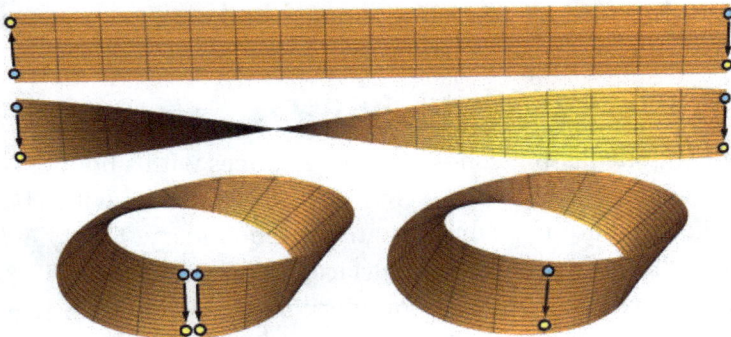

Fig. 10.16.   Gluing a Möbius strip.

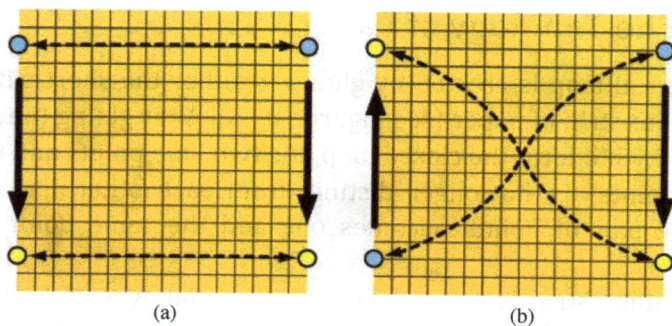

Fig. 10.17. Topology of a cylindrical ring (a) and a Möbius strip (b).

In mathematical "topology" one examines properties of surfaces that do not change when the surface is deformed (by squeezing, stretching, bending, in other words, by deformation without tearing or gluing). Therefore, from a topological point of view, it doesn't matter whether you use an elongated rectangular paper strip or a square strip at the start.

A normal, cylindrical loop, therefore, is topologically equivalent to a square in which the points of two opposite sides are identified with one another (i.e., "glued" together), while preserving the orientation. Likewise, from a topological point of view, a Möbius strip is simply a square in which the points of two opposite sides are identified crosswise, and hence, orientation of one side has to be changed. We show these two methods of identifying two opposing sides of a square in Figure 10.17.

10.4 Surfaces Without Borders

In topology, one distinguishes between surfaces with a boundary and surfaces without a boundary. A circular disk is a surface with a boundary, while a sphere is a surface without a boundary. The cylindrical loop and the Möbius strip are surfaces with a boundary. They are related to surfaces without a boundary in an analogous way as the disk is related to a sphere. This can be seen easily in case of the cylindrical loop, which can be rolled up into a torus. A torus has the shape of a donut, as shown in Figure 10.18, and it is a surface without a boundary.

The procedure to obtain a torus can also be described in the following way. We start with the square shown in Figure 10.19. The connection of the two vertical sides forms a cylindrical loop. The connection between the horizontal sides at the top and the bottom rolls it up into a torus.

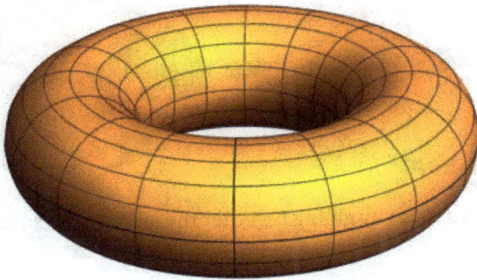

Fig. 10.18.   A torus.

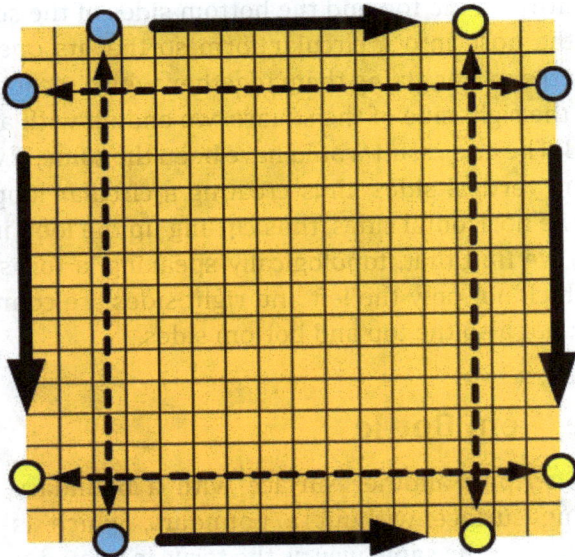

Fig. 10.19.   "Topological" view of a torus.

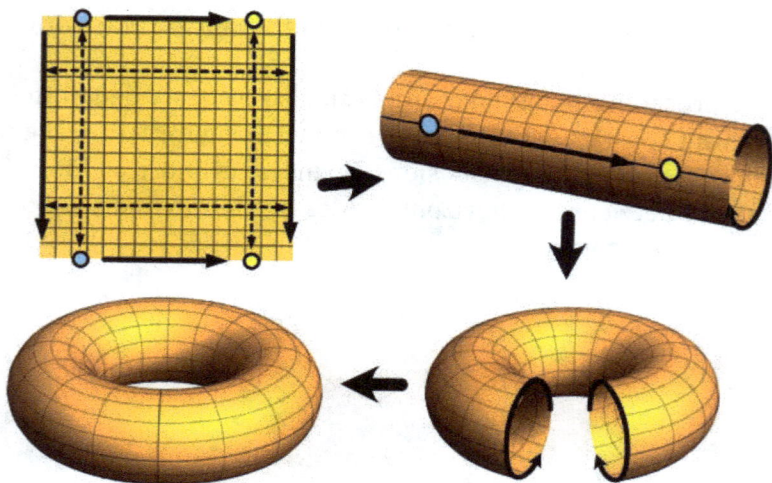

Fig. 10.20.   Forming a torus by gluing together the sides of the square in Figure 10.19.

The whole procedure is shown in Figure 10.20. We first roll up the square to form a tube or a hose (a cylinder). Mathematically, this is done by identifying the top and the bottom sides of the square. Next, one bends the hose into a circular form so that its open ends are brought together. After gluing them together, which means to connect the left and the right side of the square, we end up with a torus as in Figure 10.18. The end result would have been the same, if we had first identified the vertical sides, thus creating a circular loop, and then identifying the horizontal sides, thus curling up the loop into a torus. In any case, we find that, topologically speaking, a torus is a just a square in which not only the left and right sides are connected with one another, but also the top and bottom sides.

10.5 The Klein Bottle

The Möbius strip is another surface with a boundary. Is there a corresponding surface without a boundary, which is related to the Möbius strip in the same way as the torus is related to a cylinder?

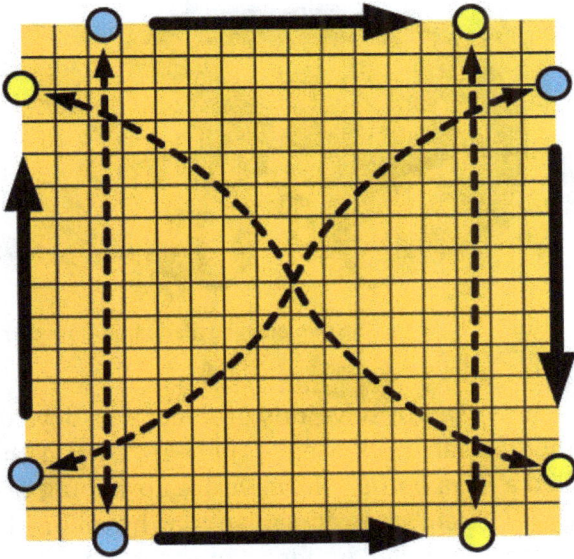

Fig. 10.21. "Topological" view of a Klein bottle.

It is very difficult to imagine how to roll up a Möbius strip into a hose or donut shape. The twist of the Möbius strip would always obstruct the process. But from the abstract point of view of topology, this can actually be done very easily. Take a square, as in Figure 10.21, where the vertical sides are identified with a change of orientation, that is, glued together with a twist, as was the case with the Möbius strip (Figure 10.17(b)). Then identify the two horizontal sides in the same way as was done with the torus (Figure 10.19), that is, without any change in orientation. This would give a surface without a boundary, which is a Möbius strip with its long edges glued together — in fact, it is rather the case that the single edge of the Möbius strip is glued to itself.

How can we imagine the result of following the prescription given in Figure 10.21? We have to glue the top and bottom side together in the same way as for the torus, while we have to glue the left side to the right side with a twist as for the Möbius strip. This leads to the problem illustrated in Figure 10.22.

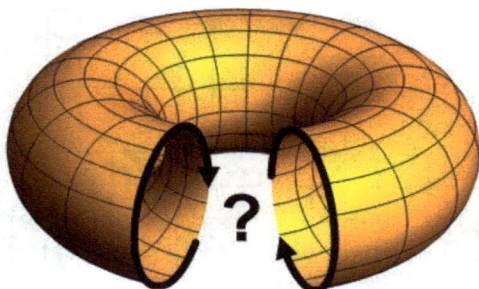

Fig. 10.22.   How can the two ends be made to fit together?

The procedure used to solve this problem was developed by the German mathematician Felix Klein (1849–1925) and the resulting surface is called a *Klein bottle*, which we are now going to describe.

Following Felix Klein's procedure, we start by rolling the square of Figure 10.21 into a cylinder by identifying the top and the bottom sides. One would then have to bend the hose into a ring and glue its open ends together. This step corresponds to the last step of the analogous procedure for the torus in Figure 10.20. But this is where the problem arises, as illustrated by Figure 10.22. But how can we glue the ends of the hose together by reversing at the same time the orientation of one end? How can the different orientations of the open ends of the hose be made to fit together?

This is obviously impossible unless we bring one end of the hose *from the inside* to the other end. This can only be done if we allow the surface to penetrate itself. Figure 10.23 shows the result, where one end of the hose is expanded to enable the other end of the hose to penetrate through the side into the interior of the hose leading to the other end. In this configuration the two ends meet with matching orientations, which are then glued together. This results in the famous shape of the Klein bottle, which is a one-sided surface without a boundary. Here one-sidedness means that if you start painting the two locally opposite sides with different colors, the colors will collide somewhere on the surface, similar to the case with the Möbius strip.

The Klein bottle can also be understood as two Möbius strips glued together along their borders. This can be seen by cutting a Klein bottle into two halves. Figure 10.24 shows lower half of the

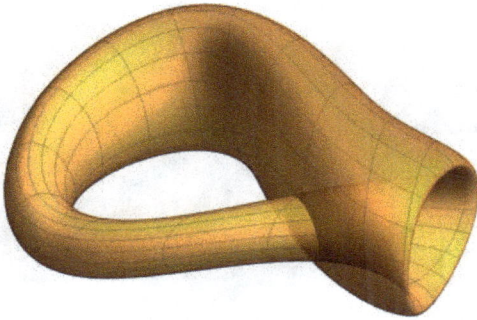

Fig. 10.23.   Klein bottle.

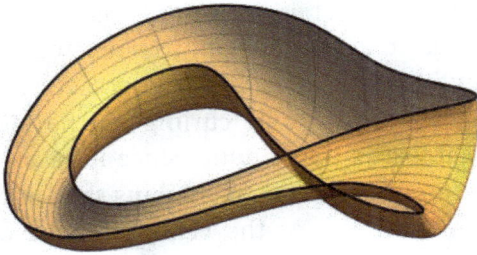

Fig. 10.24.   Lower half of the Klein bottle in Figure 10.23 — a Möbius strip.

Klein bottle. If we ignore the self-intersection (as we should), we see that this is actually a distorted Möbius strip!

It can be shown that any one-sided surface without a boundary would be self-intersecting in three-dimensional space. No matter how we twist, stretch or squeeze it, one-sidedness necessarily comes with a self-intersection. In Chapter 15, we will consider an extension of geometry to four dimensions by imagining an additional spatial direction that is perpendicular to all spatial directions familiar to us. It turns out that in four dimensions, a Klein bottle could be created as a smooth one-sided surface without any self-intersections.

There is a variant to creating a surface with the properties of a Klein bottle. This would be a three-dimensional representation of the Klein bottle with a different placement of the self-intersection. To create this, we start with a Möbius strip; that is, the first step would be to identify the vertical sides of the square in Figure 10.21 with the

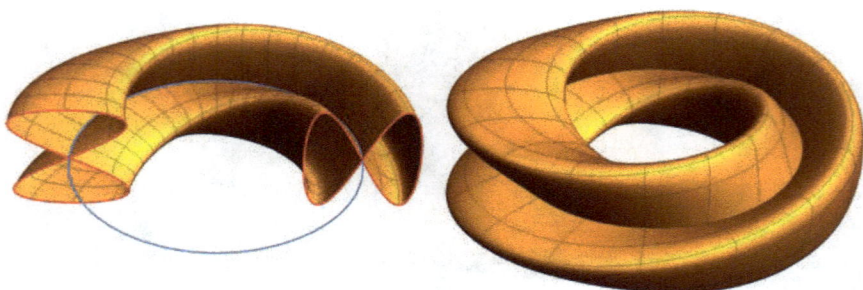

Fig. 10.25.   A one-sided surface, equivalent to Klein's bottle.

orientation reversed. The next step would be to join the other two sides of the original square. It is difficult to imagine how this could be done with a Möbius strip that has only a single edge. And indeed, it cannot be done without creating a self-intersecting surface. But we can imagine that we would start curling up the Möbius strip and bending its edge towards the midline. Since it has actually only one edge, we can work our way along the Möbius strip and bend the edge to the midline everywhere. At the centerline, the edge meets the opposite part of the edge, which has also been bent towards the center line. The cross-section would be something resembling the figure "8". This is shown in Figure 10.25.

In fact, this surface can be created by transporting the figure "8" (lemniscate) around a circle, thereby, rotating it about 180°. It is a torus with a twisted "8"-shaped cross-section. On the topological level, this figure is represented by the same square as the Klein bottle (Figure 10.21). Hence, the shape in Figure 10.25 is topologically equivalent to the Klein bottle in Figure 10.23. It is interesting that the first visualizations of this type of "Klein bottle", similar to that shown in Figure 10.25, did not appear before the last quarter of the 20th century.

10.6 Further Examples of One-Sidedness

There are other one-sided surfaces. Figure 10.26 shows the creation of a so-called cross-cap surface. One could start with a

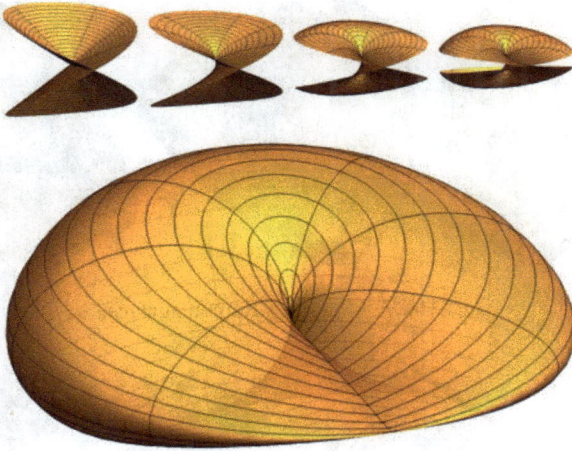

Fig. 10.26.   Creating a cross-cap surface from a self-intersecting disk.

self-intersecting disk (Figure 10.26, top-left side) and bend the edges of this disk together, as shown in Figure 10.26, finally arriving at the surface called cross-cap.

The tree-dimensional image of the cross-cap has a straight line where the surface intersects with itself. At the end of that line is a singular point, where the surface is not smooth.

Steiner's Roman surface is another example of a one-sided surface, which is more symmetric than the cross-cap. It was discovered in 1844 by the Swiss mathematician Jakob Steiner (1796–1863), when he was in Rome. Figure 10.27 shows three views of Steiner's Roman surface.

This surface is closely related to a self-intersecting polyhedron, the tetrahemihexahedron. (The name comes from Greek, 4 = "tetra" and 3 = 6/2 = "hemihexa".) This polyhedron has the six vertices of an octahedron. Its faces have four triangles and three intersecting squares. Figure 10.28 shows how to create this shape step by step.

The three squares are perpendicular to each other. We note that the lines of intersection are ignored, that is, they are not regarded as additional edges and the center is not an additional vertex. We are already familiar with this point of view, as it relates to a similar

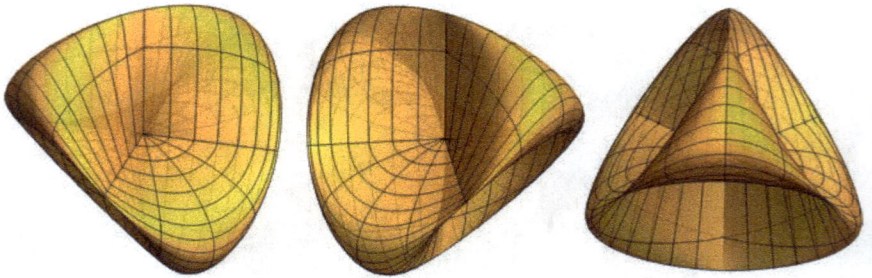

Fig. 10.27.   Steiner's Roman surface.

Fig. 10.28.   Constructing a tetrahemihexahedron.

Fig. 10.29.   Three views of a tetrahemihexahedron.

situation with the great dodecahedron, shown in Figure 6.17. The three squares do not yet make a polyhedron, because we need two faces meeting along each edge. For this, we need the four triangles in an alternating pattern, completing the tetrahemihexahedron. Figure 10.29 shows the tetrahemihexahedron from different sides. Comparing these with the views of the Roman surface in Figure 10.27, we see that the Roman surface is just a smooth version of the tetrahemihexahedron.

10.7 The Real Projective Plane

It can be shown that the tetrahemihexahedron and the Roman surface (as well as the cross-cap) are three-dimensional representations of the so-called real projective plane. Topologically, the real projective plane can be defined as a square, where both the horizontal and the vertical sides are identified "crosswise", that is, in the opposite direction, as indicated in Figure 10.30.

The real projective plane is, thus, a square, where both the horizontal sides and the vertical sides are glued together in the fashion of a Möbius strip. Another, equivalent topological representation would be a disk, where along the circumference, all antipodal points are identified.

It is not at all easy to understand, what the tetrahemihexahedron of Figure 10.29 (and thus, the Roman surface) has to do with the real projective plane of Figure 10.30. For the interested reader we provide an argument with the sequence of images in Figure 10.31. The first step shows an attempt to unfold the polyhedron and lay its faces flat on the plane. This figure (top left in Figure 10.31) is called the net of the polyhedron. In the case of a convex polyhedron (e.g., a cube),

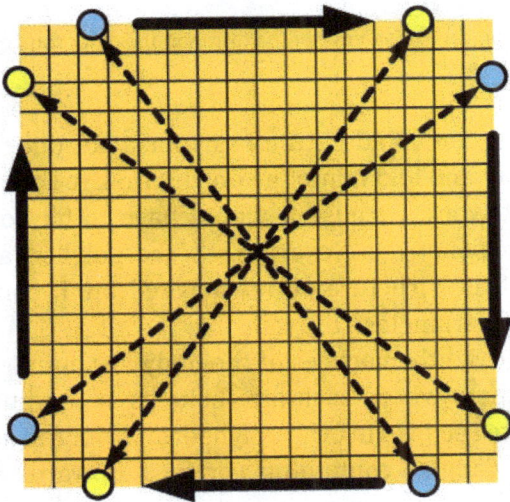

Fig. 10.30.   The projective plane.

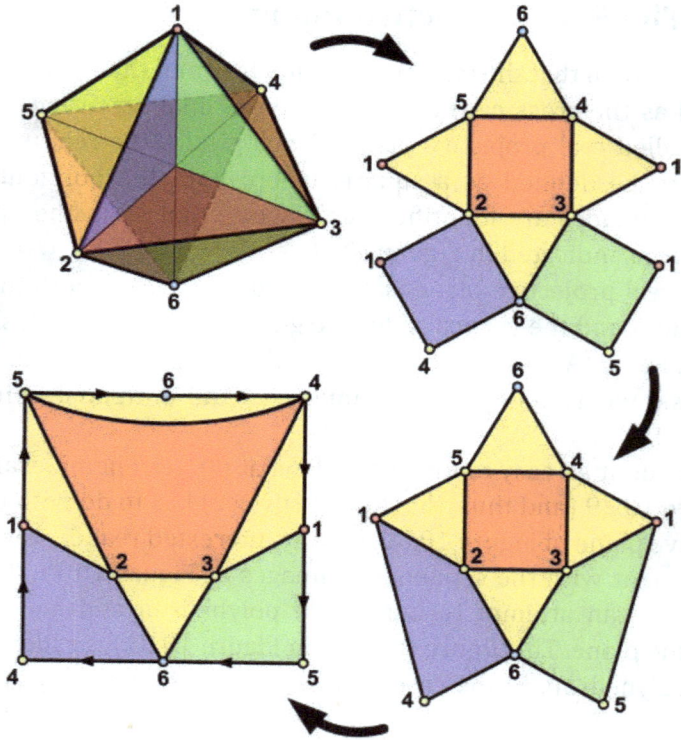

Fig. 10.31. Topological equivalence of the tetrahemihexahedron and the real projective plane.

a corresponding net could be the basis for a cut-out sheet, because one can cut it out and fold it into the polyhedron. In this case, however, this would not work, because it would have to be possible for the squares to interpenetrate. Nevertheless, the polyhedral net gives us an overview of the structure of our polyhedron. It consists of three squares and four triangles.

In Figure 10.31, the vertices of the body are numbered from 1 to 6 so that we know where each face belongs at the finished polyhedron. In order to reconstruct the polyhedron from the net, we have to fold the two triangles containing vertex 1 upwards from the red square. The other two triangles (containing vertex 6) have to be folded down.

Then the squares hanging on the lower triangle 2–3–6 have to be folded upwards into the interior of the resulting figure. We have to imagine that they can penetrate each other without damaging themselves. In this case, for example, the side 1–4 of the blue square comes to lie on the side 1–4 of the right yellow triangle. There are a few more pairs of edges in the flat network that overlap with each other.

We remind the reader that for topological comparisons we can deform, bend, stretch and compress surfaces as long as we do not tear them apart. We now use this "freedom" to simplify the drawing of the polyhedral net. In the first step, we bring together the two adjacent edges on the right yellow triangle and the green square. They both connect the vertices 1 and 3 and, hence, they are identical anyway. This can be done without any spatial folding, we just have to distort the green square. We do the same with the two edges connecting the vertices 1 and 2 on the left side of the figure and arrive at the net at the bottom right, which is somewhat distorted but still has the same topology as the original figure.

In the last step, we continue to distort this figure until it becomes a square. To do this, we only have to pull the upper corner points 4 and 5 outwards and push the lower corner points 4 and 5 into position. The resulting square at the bottom left is still topologically equivalent to the tetrahemihexahedron, provided that we ensure the correct identification of the corresponding edges. For example, the edge 15 of the yellow triangle at the top left is identical to the edge 15 of the green quadrilateral at the bottom right. Closer inspection shows that each side of the square is identified crosswise with the opposite side. This is precisely the situation already described in Figure 10.30. The net of the tetrahemihexahedron is the real projective plane. This shows us that the tetrahemihexahedron (and, therefore, also Steiner's Roman surface) can be seen as a three-dimensional representation of the projective plane, to which it is topologically equivalent.

Chapter 11

Manipulating Two Dimensions in a Three-Dimensional World — Origami in Space

In our intrinsically three-dimensional world, there is nothing that better approximates a two-dimensional sub-space than an ordinary sheet of paper. For all practical intents and purposes, paper has width and height, but not depth; the thickness of a sheet of paper is a measure we generally ignore.

It is quite noteworthy, however, that origami models — the results of folding such a nominally two-dimensional sheet of paper — are, for the most part, three-dimensional. Most origami models, be they geometric or representational, are not themselves flat, see Figure 11.1.

There are many interesting aspects of the three-dimensionality of origami models, too many, in fact, to be fully considered in the context of this book. In this chapter, we will consider three sub-topics relating to origami models and what makes them three-dimensional (or not!). Then, in the final part, we will look at some real-world technological applications of this idea — three-dimensional things that are produced practically from very flat materials through the application of origami methods.

Fig. 11.1.   Regular tetrahedron model and traditional frog model, each folded from a square of paper without cutting or pasting.

11.1 Flat-Folding and Non-Flat-Folding Origamis

If we are planning to take a close look at the majority of origami models that do not fold flat, it will certainly be of interest to first gain a bit of deeper understanding of those special ones that do. What specific properties of combinations of folds allow an origami model to fold flat? Obviously, just folding over a piece of paper once will always yield a flat result, as we see in Figure 11.2.

In this section, we will use some of the standard symbols of origami, namely, dashed lines for *valley folds* (i.e., folds "up"), dot-dashed lines for *mountain folds* (i.e., folds "down"), thick lines for edges, and thin lines for creases already present in the paper. Dotted lines will be used to represent "X-ray" views of edges under the top layer of paper, when it seems useful to present them. The resulting crease is assumed to be a straight line for now; we will be considering some of the consequences of non-straight creases later on.

Any folding process can be repeated over and over, until the paper becomes too thick to fold. As a brief aside, we can note that the thickness of real paper, the very thickness we are so used to ignoring in everyday situations, is what makes it impossible to fold paper indefinitely often in practice. For regular paper, a rule of thumb states that seven successive folds pretty much limit what can be achieved.

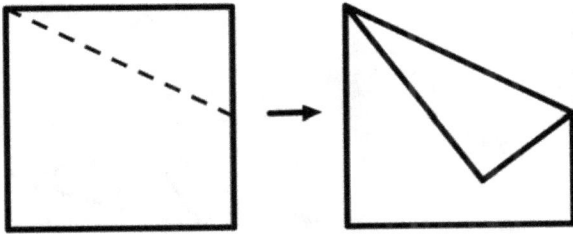

Fig. 11.2.   One single fold.

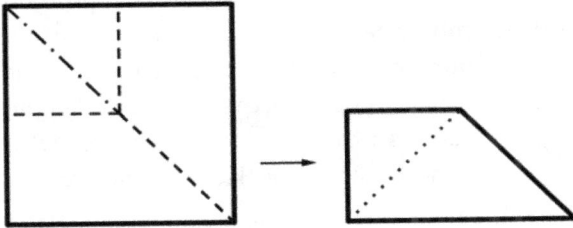

Fig. 11.3.   A simple flat-folding crease pattern.

You may want to experiment with a piece of paper to convince yourself of this feature. You will find that getting seven folds is not at all easy to do! Due to the exponential nature of successive doubling, the thickness of an actual sheet of paper will certainly become prohibitively large for any further folding after a limited number of folds, no matter how thin the paper being used is. After all, each fold doubles the number of layers. This means that seven folds of the same paper yields $2^7 = 128$ layers of paper, and it is quite easy to imagine how hard it is to fold a 256-page book! (Don't forget that there are two printed pages on each sheet of paper in a book.) By the way, the world record for successive folds of a really long single sheet of paper as of this writing stands at 12 folds.

Returning now to the matter of flat-folding, we find that a more interesting question than simply creating a single fold is the matter of flat-folding under the assumption that a number of creases meet in a single point in the interior of a sheet of paper. Such a situation is illustrated in Figure 11.3.

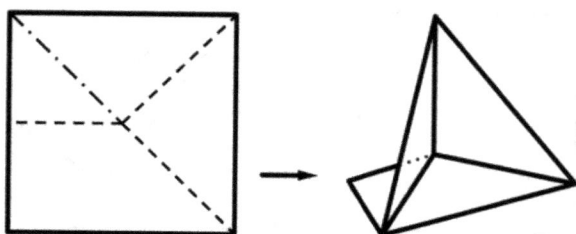

Fig. 11.4.   A very similar crease pattern that does not fold flat.

We can think of this folding process as leaving the bottom trapezoid in place, while the upper corners of the square are lifted up and folded down, with both the upper left-hand corner and the upper right-hand corner coming to lie over the bottom left-hand corner.

Just changing one small thing in such a crease pattern immediately results in a situation that is not flat-foldable. Such a situation is illustrated in Figure 11.4.

We can think of this folding process as leaving the bottom trapezoid in place again (note that the right-hand part of Figure 11.4 is now a three-dimensional representation), with the left-hand corner again lifted up and folded down, coming to lie over the bottom left-hand corner. In this case, however, unlike the folding pattern from Figure 11.3, this does not bring the upper right-hand corner to the same spot. In fact, it does not allow us to fold this corner into the plane of the trapezoid at all, and the crease in this corner stands up horizontally with respect to the plane of the trapezoid.

So, what is the essential difference between the two crease patterns? In both cases, we have three valley folds and one mountain fold, meeting at the midpoint of the folding square. The clockwise order of these folds is also the same. The only difference lies in the angle of a single fold relative to the others, namely, the fold that points straight up in Figure 11.3, but to the upper right-hand corner in Figure 11.4.

If we want to find a general rule that describes the conditions under which a grouping of folds in a crease pattern will fold flat, it seems clear that we will have to take a closer look at the angles between the folds.

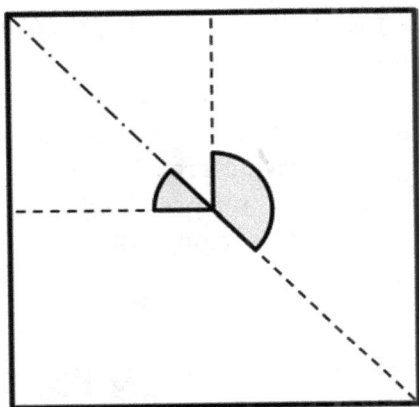

Fig. 11.5.   Alternate angles between creases converging in an interior point.

In fact, it turns out that the rules governing the flat foldability of a crease pattern around a single inner point of a sheet of paper are not too complicated. If such a crease pattern is given, and it is known to fold flat, it can be shown that the number of mountain folds and the number of valley folds must differ by exactly two (which means that the total number of folds must necessarily be even), and that the sum of alternate angles must be equal to 180°. Taking another look at the crease pattern from Figure 11.3 as seen in Figure 11.5, we see that the sum of the marked angles is equal to the sum of the unmarked angles, each of which is, therefore, equal to half of the full angle, or 180°.

In a sense, these properties are also sufficient for a crease pattern to fold flat. If we are given an even number of creases converging in a unique interior point of a folding medium such as a piece of paper, and the sums of alternating angles between these creases are equal, it is always possible to determine a classification of these folds as mountain and valley folds in such a way that the resulting model will fold flat. We know in advance that the number of mountain and valley folds will differ by two, but the specific arrangement of these depends on the angles between them.

Of course, things become much more complicated when the crease pattern has creases meeting in more than one interior point, thereby, creating a more intricate pattern in the paper. One reason for this is the fact that the creases containing the smallest angle must necessarily be oriented in opposite directions, which is actually quite clear if you think about it a bit. A consequence of this is the existence of crease patterns in which each interior point of the patterns would fold flat, even though the total patterns do not.

If you are interested in a more detailed description of the mathematics behind all of this, along with the appropriate proofs, you might want to take a look at one of the many sources available.[1]

11.2 Modular Origami and Polyhedra: Nets and Folding Units

While the more classic style of origami traditionally involves the creation of models from a single piece of paper, there is a whole sub-division of origami models, in which folding methods are used to create building blocks, which can then be assembled to create three-dimensional models. This style of origami is known as *modular*, or *unit origami*. Two examples of models created in this style are shown in Figure 11.6.

Unlike single sheet models, most origami models of this type are not representational creations like animals or masks, although such things are certainly possible. It is far more common for origami artists, working in this style, to create abstract structures with a high degree of symmetry. These can be decorative objects, such as boxes or stars, or more mathematical objects, such as polyhedra. Specifically, Platonic and Archimedean solids are interesting in this context because of their high degree of symmetry. Identical or symmetric building blocks can be used to create such things in any number of interesting ways.

A very simple example of a unit origami model is the cube shown in Figure 11.7.

---

[1] Hull, T. (1994). On the mathematics of flat origamis. *Congressus Numerantium*, **100**, 215–224.

Fig. 11.6.   Some modular origami models.

Fig. 11.7.   Cube composed of squares folded in half.

Each of the "building blocks" used to create this cube is simply a square piece of paper folded in half once, parallel to its sides, at a right angle, as illustrated in Figure 11.8.

By placing each of 12 squares in such a manner that the crease comes to lie in an edge of the cube and alternating the order of the

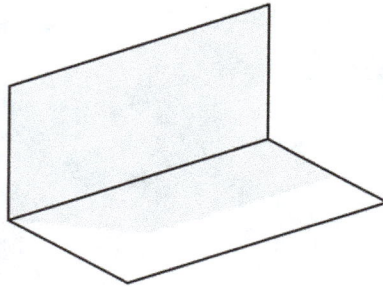

Fig. 11.8. Basic unit used to build the simple cube.

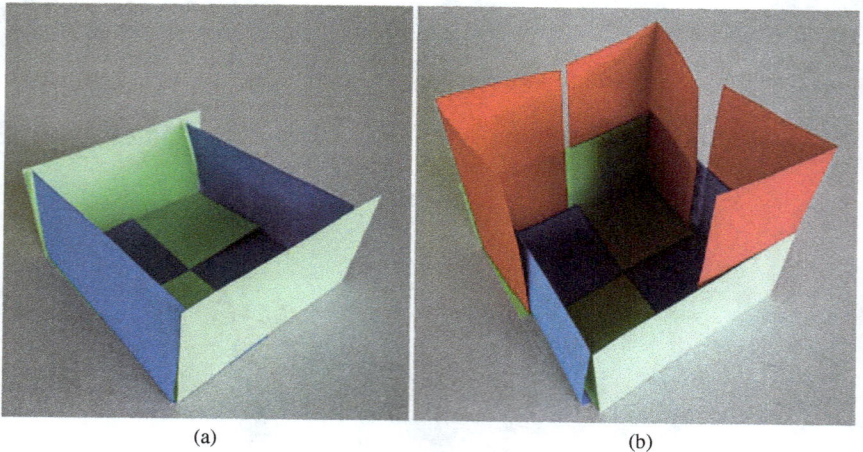

(a) (b)

Fig. 11.9. The first steps in assembling the simple cube.

sheets appropriately, as shown in Figure 11.9, the cube results. In Figure 11.9(a), we see four such units put together to form the bottom of the cube. In Figure 11.9(b), three of the four vertical edges have been added in a second step of assembly. This can then be completed to form the cube shown in Figure 11.7.

Of course, this model is not very stable. While the friction between the sheets of paper gives the model a minimal measure of stability, it will tend to fall apart in the slightest breeze, as the sheets are essentially just lying on top of each other. In order to avoid this degree of frailty, origami units are generally designed with extruding flaps and pockets to contain them. Such units can be combined by inserting the flap of one unit into the pocket of another, or simply into the space between some other units. This greatly increases the stability of the finished model, both because of the improved friction between the units and because the units will often tend to support each other symmetrically. A simple example of an application of this idea is the cube shown in Figure 11.10.

Each unit used to assemble this cube is folded as shown in Figure 11.11.

Six such units are required to assemble the cube, as each unit covers one face of the cube. The small half-squares standing at right angles to the square in each unit are the flaps, which are inserted into the interior of the cube. The assembly process is shown in Figure 11.12. Note that the pockets in this model are not actually part of the units but result by simply orienting the units appropriately.

Fig. 11.10.   Cube composed of face units.

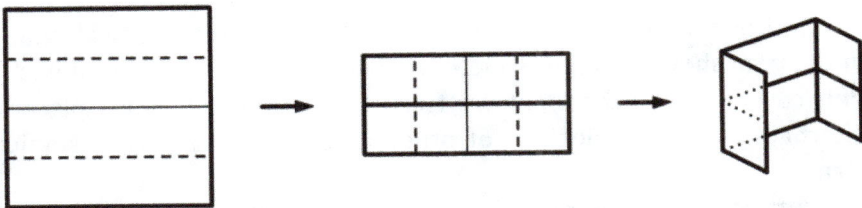

Fig. 11.11.   Folding the units for the face-unit cube.

Fig. 11.12.   The first steps in assembling the face-unit cube.

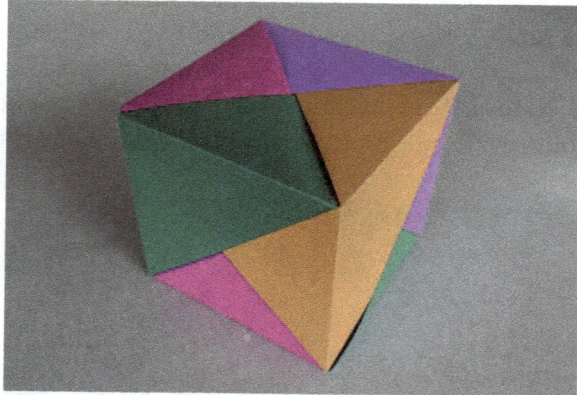

Fig. 11.13.   Cube composed of eight vertex units (Kunihiko Kasahara).

This is a very simple model, yet already quite stable, as a motivated reader can readily check out by doing the actual construction. The cube shown in Figure 11.13 is even more stable, and much more sophisticated from the point of view of the structural process required to develop the units.

In order to create this cube (Figure 11.13), eight identical units of the type shown in Figure 11.14 are placed in such a way that each unit covers one of the eight vertices of the cube, with flaps placed alternately into the pockets of other units, creating a very stable structure.

Putting the pieces together properly is a bit of a puzzle and is left to the interested reader. A source that can provide further

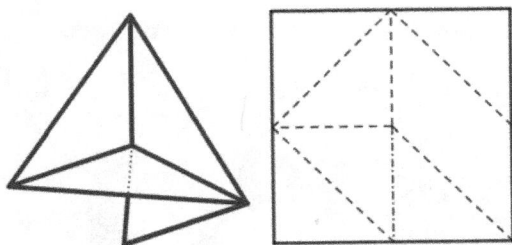

Fig. 11.14.   One vertex unit of a Kasahara cube and its folding pattern.

assistance about this particular model is the recommended book by Kasahara.[2]

These three cube models are examples of the three basic ideas available for creating origami units that can be assembled to form polyhedral structures. The first model (shown in Figure 11.7) used 12 units to create the cube, with one unit covering each of the 12 edges of the cube. The second model (shown in Figure 11.10) used six units to create the cube, with one unit covering each of its six faces. Finally, the third model (shown in Figure 11.13) used eight units to create the cube, with one unit covering each of its eight vertices.

It is possible to apply variations of these ideas in creating various origami units. A unit can cover two or more faces instead of just one. However, unit origami models can generally be categorized according to the roles of each unit in the assembled model, covering either a face, an edge or a vertex. In Figure 11.15, we see yet another cube model, which is shown in Figure 11.15(a), and is composed of six units of the type shown in Figure 11.15(b).

We see that each unit has a square with four pockets, and two flaps that fit into such pockets in adjoining faces of the cube. This has the added feature of creating an interesting triangle pattern on the faces of the cube. The units in modular origami are often specifically designed with such aesthetic effects in mind, along with the geometric stability of the resulting polyhedra. Readers interested in folding

---

[2] Kasahara, K. (2004). *The Art and Wonder of Origami.* Gloucester, Massachusetts: Quarry Books.

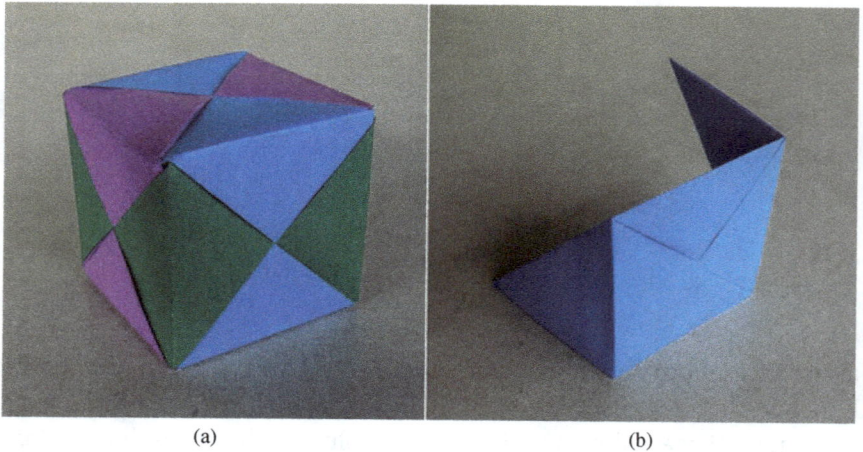

Fig. 11.15.   Face-model of a cube and a unit of the type it is composed of.

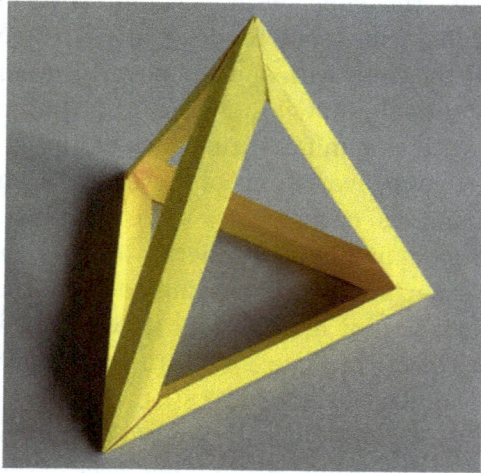

Fig. 11.16.   Edge-model of a regular tetrahedron.

this traditional model themselves are encouraged to search out the folding patterns in various books that specialize on the subject.[3]

In Figure 11.16, we see an edge-model of a regular tetrahedron.

---

[3] Biddle, S. and Biddle, M. (1993). *The New Origami*. London: Ebury Press.

In this model, which is composed of six units corresponding to the six edges of the tetrahedron, where the faces are not closed. Such open frame models are not only possible as edge-models but can also be designed as vertex-models. This specific model was originally designed by Francis Ow.[4]

Finally, in Figure 11.17, we see a classic "skeleton"; a vertex-model of an octahedron, composed of six units corresponding to the six vertices of the octahedron.

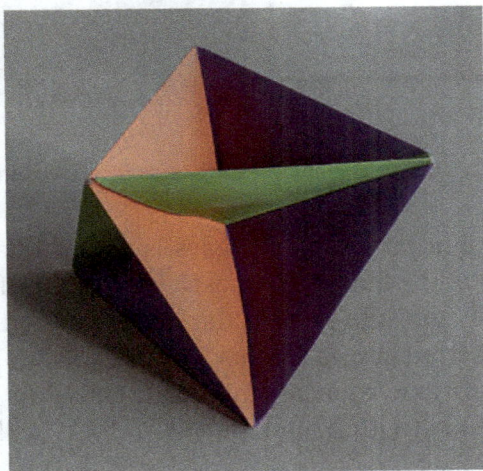

Fig. 11.17.   Classic octahedron skeleton.

Such a model is called a skeleton, because the plane faces of the model are not the faces of the octahedron, but rather the triangles joining the edges with its midpoint, creating a view of the polyhedron's interior support.

This model is especially interesting, as it is very easy to fold, and reasonably easy to assemble, but astoundingly stable once assembled. Each unit is simply a square in which the diagonals and the mid-parallels of the sides are folded in opposite directions, as shown in Figure 11.18(a), and the result pushed together, as shown in Figure 11.18(b).

---

[4] Ow, F. (1986). Modular origami (60° unit). *British Origami*, **121**, 30–33.

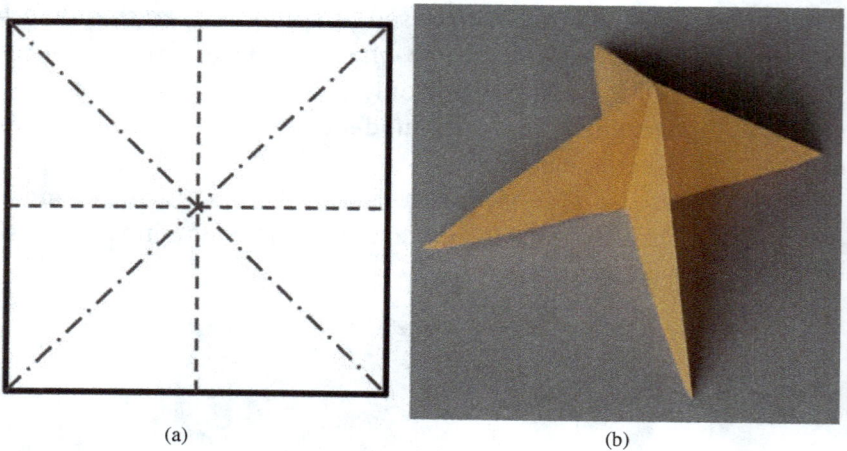

(a)   (b)

Fig. 11.18.   The unit for the octahedron skeleton.

Each of the four extruding triangles is either a flap or a pocket, and the stability of the finished model stems from the fact that the six units all tend to converge toward the mid-point of the octahedron once they are inserted into each other cyclically. This is another model interested readers may enjoy assembling themselves. Note that consideration of the alternating colors in Figure 11.17 will help you with your assembly.

There is a great deal of literature on unit origami available, but any reader interested in pursuing this subject further is well-advised to start by getting a copy of the classic book on the subject.[5]

11.3 Folding Curves and the World of Developable Surfaces

We will now take a closer look at an idea already alluded to in the earlier section on flat folding, namely, the question of what it means to fold something other than a straight-line crease. An example of an origami square folded along a curved crease is shown in Figure 11.19.

---

[5] Fusè, T. (1990). *Unit Origami.* New York and Tokyo: Japan Publications.

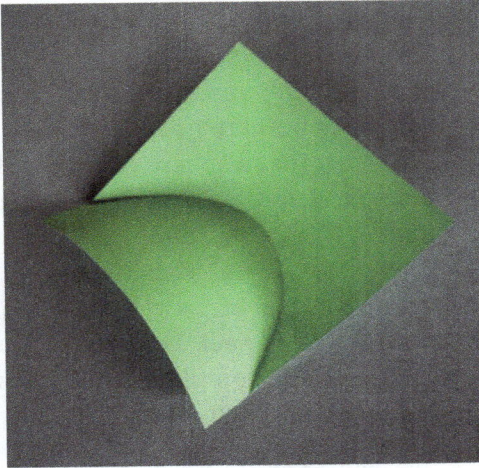

Fig. 11.19.   Example of a curved crease.

It is immediately obvious that the crease separates the paper into two sections, neither of which is flat.

In fact, introducing any kind of continuous fold in a flat piece of paper, while keeping one section of the paper flat, will always imply that the resulting crease must be a straight line. This seems to be obvious, but it is not quite so simple to prove in a precise way. Still, we can be certain that introducing any kind of a curved fold will always cause both sections of the paper to warp. Folding curved creases is, therefore, a confident way to create three-dimensional objects from two-dimensional sheets of paper.

In order to deal with this idea more deeply, we will need to introduce the concept of a *developable surface*. Essentially, a developable surface is anything that can be created by somehow bending or twisting a plane without stretching, creasing or tearing it. From a purely descriptive standpoint, this is exactly what each of the sections of the paper will tend to, no matter how the nonlinear crease is distorted, as long as it remains a smooth curve, without any corners. Such surfaces are called developable, because they can be "developed", that is, evened out into a plane without any stretching or tearing or creasing. Examples of developable surfaces are cylinders and cones, but not every developable surface is a cylinder or a cone.

In addition, not every simple curved surface is developable. Donuts (tori) and balls (spheres) are good examples of surfaces that are not developable. The fact that a sphere is not developable is actually the reason that maps of the globe can never accurately reflect every geometric aspect of the Earth, so this is quite a deep-reaching concept. Just imagine trying to flatten out an orange peel on the surface of a table. There is no way you can ever get it completely flat, even if you ignore the thickness of the peel. This means, for instance, that it can never be possible to draw a completely accurate map of a whole continent that shows all distances and all directions with complete accuracy. See Chapter 4 for more details on cartography and various methods to create maps.

It will prove useful to introduce a few classic concepts of differential geometry to our discussion. In 1828, in his *Theorema Egregium* (the "remarkable theorem"), the famous German mathematician Carl Friedrich Gauss (1777–1855) proved that there is a measured quantity that remains invariable in any point of a surface when the surface is "developed". The precise definition of what it means to "develop" a general surface is a quite technical matter in differential geometry, but for our purposes, it is sufficient to think of it as already described, an action of the type we would expect to be able to apply to a sheet of paper, changing its form without introducing creases. This would include any kind of rolling or bending that does not tear or stretch the surface, and is, therefore, in a way, an application of the qualities that differentiate a relatively rigid medium like paper from a slightly more malleable one, like a sheet of rubber. This invariable quantity is known as the *Gaussian curvature* of the surface and is usually denoted by a capital letter K (short for *Krümmung*, the German word for curvature).

In order to determine the Gaussian curvature of a surface in a specific point, we must first assume that the surface meets certain conditions of differentiability, or put more colloquially, that the surface is sufficiently smooth at the point in question. If this is the case, the value of K can be calculated quite simply as the product of the principal curvatures K_1 and K_2 of the surface in that point. Now, of course, this begs the question of what is to be understood by the term "principal curvature".

In order to get an idea of the meaning of this concept, we can imagine the unique line perpendicular to a (bent, but smooth) surface in a specific point P. The planes containing this line cut the surface in various curves, each of which has a specific curvature in the point P. The principal curvatures in P are simply the largest and smallest of these, noting that curvatures in opposite directions are given opposite signs. The Gaussian curvature will, therefore, be a negative number if the surface has some kind of saddle shape that contains curves bending in opposite directions, see Figure 11.20(a). For surfaces such as spheres, with the property that all curves bend in the same direction, all curvatures will be positive, and therefore, the Gaussian curvature at each point of such a surface will also be positive, as in Figure 11.20(b).

If the surface under consideration is a plane, it is obvious that the curves we obtain in this way are all straight lines, and their curvature is, therefore, always zero. The value of the Gaussian curvature in each point of a plane is also always zero.

This implies that the surfaces that we can bend a sheet of paper to are precisely the surfaces with Gaussian curvature zero in every point. As we have already mentioned, cylinders and cones are

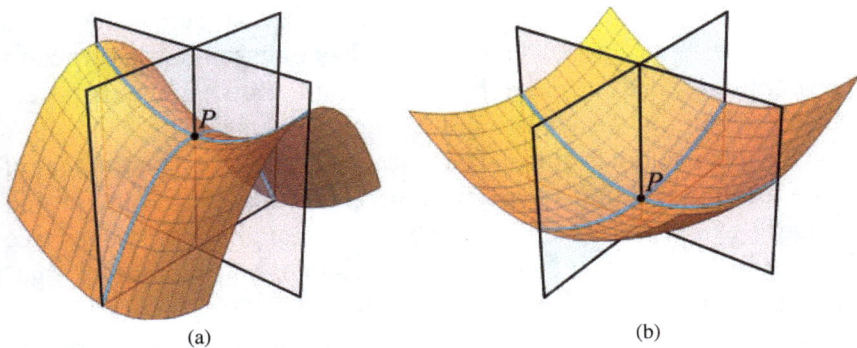

Fig. 11.20. A surface with negative Gaussian curvature (a) and positive Gaussian curvature (b) in a point P.

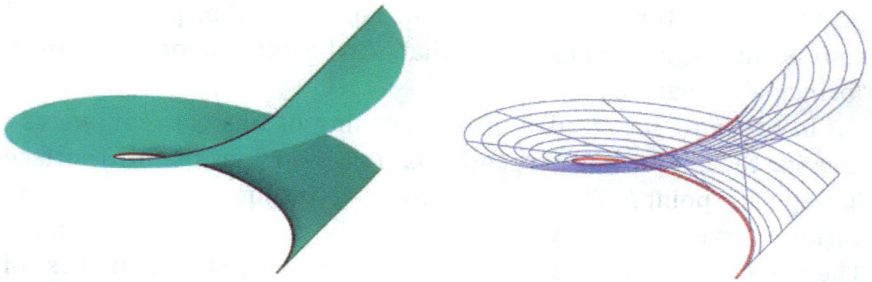

Fig. 11.21.   A tangent-developable surface.

examples of such surfaces, and it is not difficult to imagine bending a sheet of paper to form a cylinder or a cone. The only other develop-able surfaces are the so-called *tangent developable* surfaces. Each such surface is made up of the tangents of a specific spatial curve. As it turns out, twisting a sheet of paper in any way that does not result in a cylinder or a cone will always yield a surface of this type. Figure 11.21 shows an example of a tangent-developable surface.

This means that all developable surfaces are *ruled* surfaces, built up of straight lines. Cylinders are, of course, surfaces composed of parallel lines, while cones are surfaces composed of lines through a common point. Since tangent developables are surfaces composed of lines tangent to a common curve in space, all developable surfaces are composed of lines.

However, the converse is not true. Not every ruled surface is developable, as not every ruled surface has Gaussian curvature zero in every point. An example is the hyperboloid, shown in Figure 11.22, which is commonly used for building cooling towers.

The hyperboloid is quite a fascinating surface. It can be thought of as the result of rotating a hyperbola on its minor axis, as is illustrated in Figure 11.23(a). It can also, however, be thought of as the surface that results by joining the points on circles in parallel planes by regu-larly placed straight lines as shown in Figure 11.23(b).

In fact, since the hyperboloid is symmetric with respect to any plane containing its central axis, there is also a second, symmetrical set of lines on the surface. This means that there are actually two

Fig. 11.22.   Cooling tower in the form of a hyperboloid.

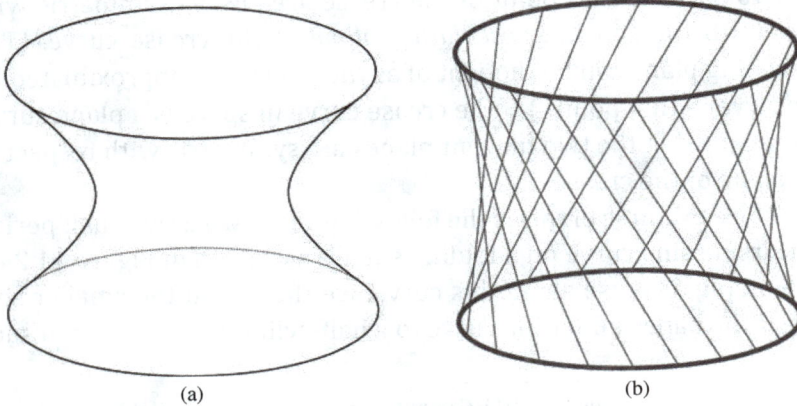

(a)   (b)

Fig. 11.23.   Two different interpretations of a hyperboloid.

straight lines passing through every point on its surface. The line perpendicular to the surface is, therefore, perpendicular to the plane containing both of these lines. However, even though there are two curves on the surface through this point with curvature zero, that is, the lines, there are also curves on the surface and on planes containing the perpendicular line oriented in both directions relative to this line. For this reason, the Gaussian curvature is negative everywhere, and certainly not equal to zero. The hyperboloid is, thus, not developable.

Having finished our brief tour of the world of developable surfaces and ruled surfaces, we now have the tools at our disposal to take a closer look at an example of what can happen when we fold a specific curved crease in a sheet of paper and form the resulting crease to fit a given shape.

We are free to choose any smooth curve on a sheet of paper, fold it in such a way that a crease is created along this curve, and then bend the crease to create any smooth curve in space. This will then cause the two parts of the paper, delineated by the chosen curve, to bend in a predetermined fashion.

If we are given the curve in the folding sheet and the spatial curve we wish to fit the crease to, it is possible to calculate the resulting curved surfaces the two parts of the paper will bend to. A useful tool in dealing with such calculations is the fact that the tangent planes of the two surfaces in a point of the crease are always symmetric with respect to the so-called *osculating plane* of the crease curve. (The osculating plane can be thought of as the plane best approximated by the curve in that point.[6]) If the crease curve in space is a plane curve, this means that the two tangent planes are symmetric with respect to the plane of the crease.

A nice example of this is the following. First, we draw a half period of a simple sine curve on a folding square as shown in Figure 11.24.

Folding a crease along this curve, we then bend the smaller section of the paper under the curve to a half-cylinder of rotation in such

---

[6] Pottman, H. and Wallner, J. (2001). *Computational Line Geometry.* Heidelberg, Berlin: Springer Verlag.

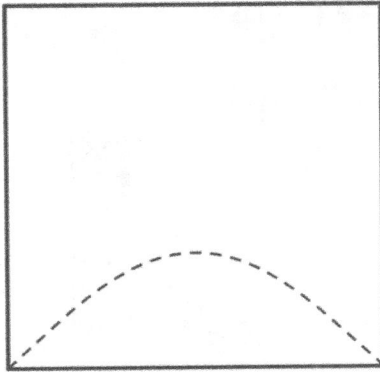

Fig. 11.24.   Sine curve as a crease.

a way that the bottom edge of the paper creates a semi-circle. In can be shown by calculation that the crease will then be a plane intersection of the resulting half-cylinder, and therefore, a section of an ellipse. The upper part of the paper will have tangent planes symmetric to the tangent planes of the smaller cylindrical section with respect to the plane of the crease and this yields a symmetric cylinder for the other section of the paper with respect to the plane of the crease. A picture of the result is shown in Figure 11.25.[7]

A number of origami artists are making use of the wonderful opportunities afforded by the folding of curved creases. Two particularly impressive models with curved folds, created by Ekaterina Lukasheva, are shown in Figure 11.26.

As a final comment on this subject, it is worth noting that folding a curve intentionally is not only relatively difficult to do, but also something that does not readily occur in a natural way. We might expect curved folds to occur more or less automatically in crumpled paper, but this turns out not to be the case. We are all familiar enough with crumpled wads of paper. After all, who of us has not played a round of trashcan basketball with the result of some failed

---

[7] Geretschläger, R. (2009). Folding curves. In R. J. Lang (Ed.), *Origami 4, Fourth International Meeting of Origami Science, Mathematics and Education*. Natick Massachusetts: A.K.Peters Publishing.

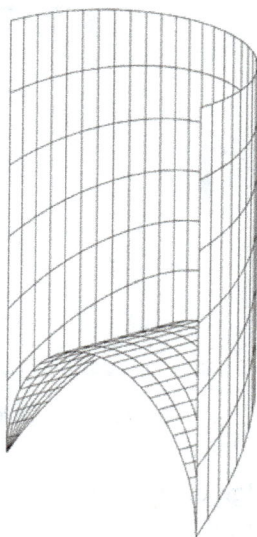

Fig. 11.25.   Result of folding along the sine curve and placing the crease in an appropriate plane.

Fig. 11.26.   Origami art by Ekaterina Lukasheva (Pavlovic) https://www.instagram.com/ekaterina.lukasheva/.

masterpiece we had just given up on? We could certainly be excused for thinking that the (roundish) crumpled paper would have mostly curved creases. In fact, there are very few curves in the crease pattern that result from such a spontaneous balling of paper.

11.4 Some Origami Applications in Modern Technology

One of the fascinating aspects of theoretical origami is its applicability in a surprisingly large number of technological contexts. The whole basic idea of origami, namely, using a flat, thin, lightweight material to create three-dimensional sturdy objects by folding means that the ideas of origami will be useful in all situations in which we are called upon to create something lightweight or small that can be easily expanded to fill a larger space in a sturdy and durable way.

One obvious application of this concept is in the area of satellite technology. For a satellite to work well, there are several components that should ideally be quite large, like solar panel arrays or parabolic antennae. These preferably large objects must be transported into orbit, and should, therefore, be as lightweight as possible. Origami to the rescue!

Obviously, paper is not a useful medium for a solar panel or a radio antenna, but there are other flat materials that will do the job. By folding them up in sophisticated ways derived from origami research, it is possible to compactify them, and thus, make their transport into orbit as efficient as possible. They can then be unfolded to their full size once they have reached their intended deployment area. An example of such an array is shown in Figure 11.27, both in its unfolded state and in a semi-folded state.

Fig. 11.27.   A folding array for satellite solar panels. Photo by Mark Philbrick/BYU Photo.

Another completely different application of the compactification of a useful technology that aids its transfer to the site of its intended use is the medical world of stents. These tubular aids help to open up anatomic vessels to allow the free flow of liquid in the body. Examples of these are coronary or vascular stents that are inserted into partially blocked arteries to ease blood flow and ureteral stents, which are placed in the ureter to insure the flow of urine from the kidney to the bladder.

Such devices should ideally be approximately cylindrical and exert some pressure from within on the walls of their host vessels. For the purposes of insertion, however, they should fold up to a smaller size such that the walls of the vessel are not harmed in the process. Once again, methods derived from origami research have proven a great help in devising such health aids, examples of which are shown in Figure 11.28.

Of course, there are as many such applications as can be imagined, and engineers are constantly searching for more such useful ideas. A more prosaic example of such an application is the collapsible drink container shown in Figure 11.29, which could someday prove useful in waste reduction.

Fig. 11.28.   Stent model designed to snap open inside an artery. By Zhong You, University of Oxford.

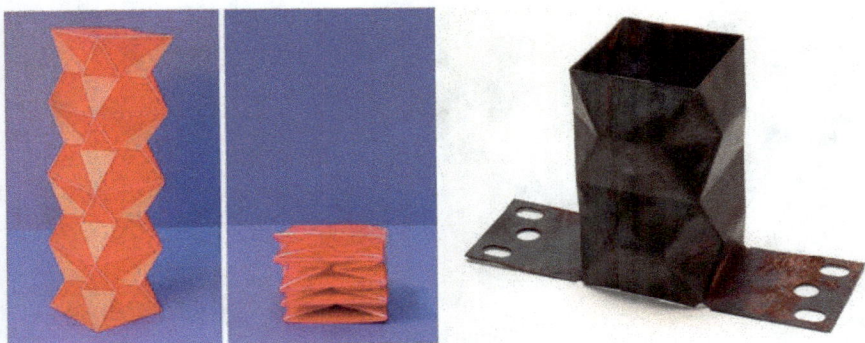

Fig. 11.29.   Collapsible drink container and a shock absorber by Zhong You, University of Oxford.

Origami methods can also be applied to such diverse areas as architecture, map folding and airbag construction. Origami science is even proving useful to research on the way DNA folds in the process of cellular reproduction. So, the next time someone tells you that origami science is cutting edge, you know they aren't just referring to paper cuts! Also, if you are interested in learning more about the geometry of origami constructions, you might find the book *Geometric Origami* useful.[8]

---

[8] Geretschläger, R. (2008). *Geometric Origami*. UK: Arbelos Publishing.

Chapter 12

Knot Theory

Knot theory is the study of knots from a mathematical perspective. Obviously, tying a knot is only possible in three dimensions, a knot cannot be formed in the plane. Thus, it is not surprising that the various knots that can be created and their geometric properties are intimately related to properties of three-dimensional space itself. In this chapter, we will first consider practical knots, some of which you may be familiar with. We will then explain the notion of a mathematical knot and briefly review the history of knot theory. As we go along, all mathematical knots with up to six crossings will be presented. The question of how to determine whether two knots are equivalent will lead us to the so-called Reidemeister moves, which is a method to simplify knot diagrams, and to knot invariants such as tricolorability.

12.1 Knots in Everyday Life

Knots are ubiquitous in everyday life. Most probably, you have already tied a knot today, perhaps the knot in your shoelaces, the knot that arranges your necktie or the knot that holds your hair together. A simple and effective way to secure the opening of a sack or bag is to put a rope around its neck and tie a knot. Knots are an ancient technology, they have already been used by humans in prehistoric times, possibly even before the use of fire. Just as the invention of axe and wheel, knot tying is usually viewed as an ability unique to humans. Although some animals have developed impressive technical skills to construct their nests, in particular, birds and the great apes, they have not been observed tying knots in the wild. However, captive orang-utans have learned to tie knots, so some scientists speculate that knot tying is not an exclusively human technique. While knots are very practical in many circumstances, we also encounter unwanted knots, for example, seemingly spontaneously forming knots in computer cords or long strands of hair. Soft ropes or wires tend to entangle themselves when they are curled up and as long as electrical cords are needed for the power supply for computers, printers, lamps and other devices, we may find knotted cords behind or under our desks. Untangling a messy ball of knotted cords can be a challenge.

12.2 The Gordian Knot

The difficulty of untying a complicated knot is best expressed in the metaphor of the "Gordian knot" for an intractable and convoluted problem. "Cutting the Gordian knot" then means to solve such an extremely difficult and involved problem. The Gordian knot is a leg-end associated with Alexander the Great and took place in Gordian, which was the capital city of ancient Phrygia (today Turkey). Following the judgment of an oracle, the Phrygians would declare as their king the next man driving into their city with an ox-cart. So, the farmer Gordias became king of Phrygia. To show his gratitude, his son Midas tied the ox-cart to a column with an intricate knot of cornel bark. The knot could not be untied and was still in place hundreds of years later. Again, according to the prophecy of an oracle, the man who would be able to untie the Gordian knot, would become ruler of Asia. And indeed, Alexander, who sliced the knot with his sword, conquered Asia (see Figure 12.1).

12.3 Practical Knots

The simplest and most fundamental knot is the overhand knot, also known as a half knot (see Figure 12.2). It is the basic element of many

Fig. 12.1.   Alexander cuts the Gordian knot. Painting by Jean Simon Berthélemy (1743–1811).

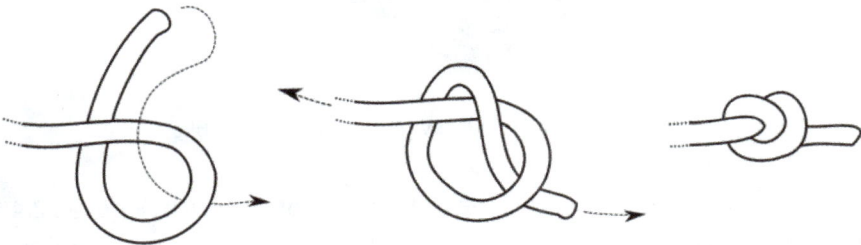

Fig. 12.2.   Overhand knot. Image by QuasarFr/CC BY-SA.

other, more complex knots. If a strip of paper is tightly folded into a flattened overhand knot, it assumes the shape of a regular pentagon (see Figure 12.3).

Only a few different knots are essential in everyday life, but a huge variety of knots is used in climbing and mountaineering. One of the

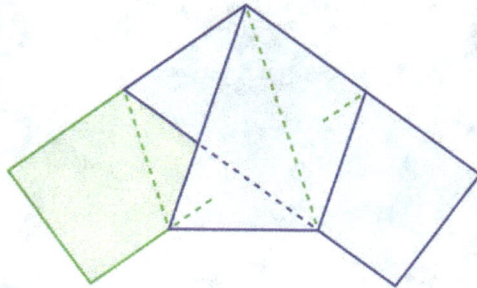

Fig. 12.3.   Strip of paper folded into an overhand knot. Image by AnonMoos/Public domain.

Fig. 12.4.   Butterfly knot, image by Mher Hovsepian, public domain.

essential knots for climbers is the Butterfly Knot, shown in Figure 12.4. It forms a secure loop in the middle of a rope to which a carabiner can be attached. There are also special knots used by anglers for tying a fishing line to a hook. Different types of knots are used depending on the thickness and elasticity of the line.

However, the sport associated the most with knots is probably sailing. The speed of ships is even measured in knots, a term dating from the 17th century. To determine the speed of a ship, sailors dropped off the stern of the ship a log attached to a rope knotted at regular intervals and counted the number of knots that unspooled from a reel in a specific time. A knot (kn) is equal to one nautical mile per hour, which is exactly 1.852 km/h (approximately 1.15078 mph). Rope knots can be roughly divided into the following categories: hitches, bends, and splices. A hitch fastens a rope to another object, for example, the barrel hitch knot shown in Figure 12.5.

A bend unites two rope ends, for example the Carrick bend, also known as the Sailor's breastplate, is shown in Figure 12.6.

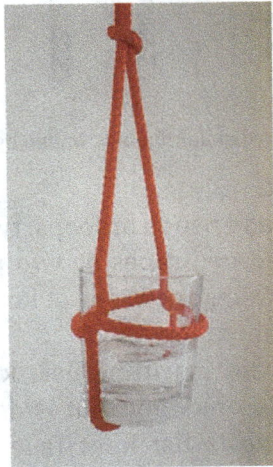

Fig. 12.5.   Barrel hitch.

Fig. 12.6.   Carrick bend. Image by David J. Fred/CC BY-SA.

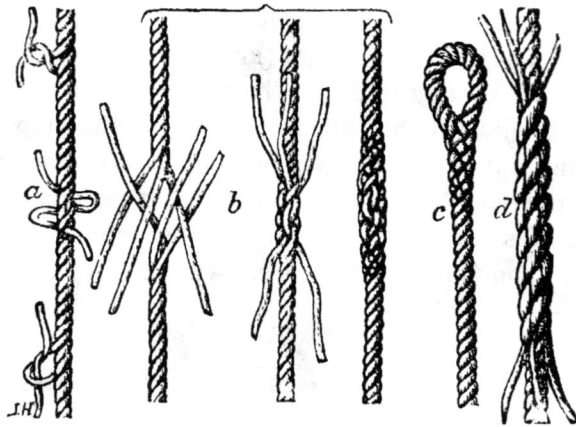

Fig. 12.7.   Splices, Nordisk familjebok. Image by NH/public domain.

Splices are multi-strand bends or loops, typically forming a semi-permanent joint between two ropes or two parts of the same rope (see Figure 12.7). Other categorizations of knots are also possible, as well as further sub-divisions.

Often originating from maritime use, knots are also used as decorations. An example of a knot that is decorative, but also has a function, is the monkey's fist knot (see Figure 12.8). Tied to the end of a rope, it serves as a weight and makes it easier to throw the rope.

Tying knots for decorative purposes was very popular in China during the Qing Dynasty (1644–1911) and developed into an accepted art form. But even among practical knots, there is a fascinating variety to be discovered. A beautiful collection of knots appeared on the front cover of Scientific American, March 18, 1871 (see Figure 12.9).

12.4 Knotting Matters

One of the most comprehensive encyclopedias of knots is *The Ashley Book of Knots*, written and illustrated by the American artist Clifford W. Ashley (1881–1947). First published in 1944 after 11 years of

Fig. 12.8. Monkey fist knot. Image by Markus Bärlocher, Public domain, via Wikimedia Commons.

work, it describes more than 3,800 different knots and contains approximately 7,000 illustrations (see Figure 12.10).

In October 1978, an article on the front page of the British newspaper *The Times* presented an allegedly newly invented knot credited to Edward Hunter (see Figure 12.11(a)) The question whether or not it really was a new knot, gave rise to discussions between knot-tying experts around the world, and thereby, initiated a process that led to the foundation of the International Guild of Knot Tyers (IGKT)[1] in 1982. It is a worldwide association for people interested in knots and publishes a quarterly newsletter, *Knotting Matters* (see Figure 12.11(b)).

12.5 Mathematical Knots

What do knots have to do with three-dimensional geometry? Obviously, tying a knot is only possible in three-dimensional space. The notion of a knot does not make sense in two dimensions. The huge variety of practical knots, described, for instance, in *Ashley's Book of Knots*, suggests a systematic classification of knots. Classifying

---

[1] www.igkt.net

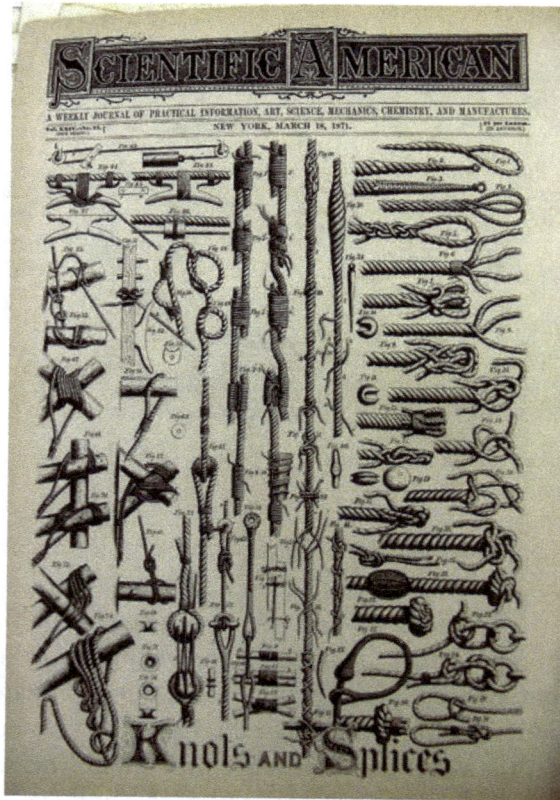

Fig. 12.9.   Cover of Scientific American, March 18, 1871.

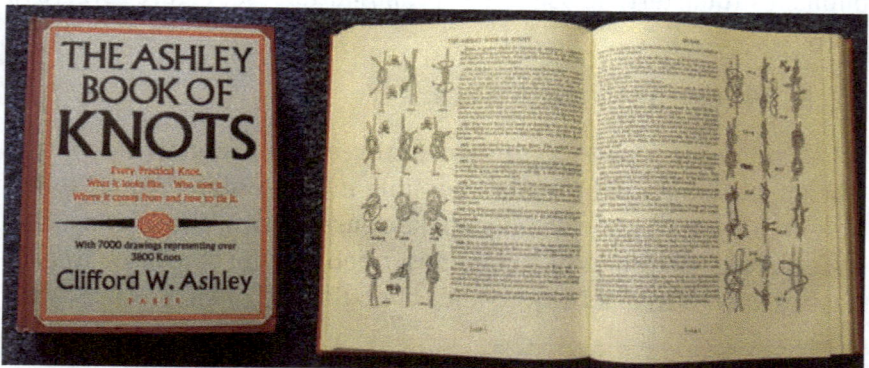

Fig. 12.10.   *The Ashley Book of Knots.* CC BY-SA 2.0 DE.

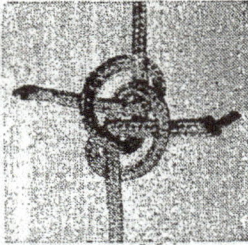

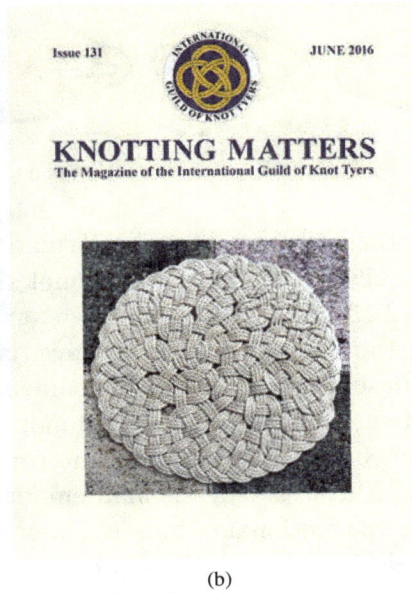

Fig. 12.11. (a) Article from *The Times*, October 1978. (b) *Magazine of the International Guild of Knot Tyers (IGKT)*, Issue 131, June 2016.

objects with respect to certain properties is a typical aspect of mathematical research. And indeed, there is a branch of mathematics devoted to the study of knots. Although the mathematical notion of a knot was inspired by the various knots we encounter in everyday life, there is an important difference between a mathematical knot and a conventional knot. If we take a piece of rope and tie a knot into it, we can always, at least in principle, untie the knot. The rope can be brought from an unknotted state into a knotted state and back again by certain manipulations such as forming a loop and pulling trough one end of the rope. Thus, it can be deformed into a knot just by bending it in a particular way, without tearing or gluing it. Topology is a branch of mathematics concerned with the properties of a geometric object that are preserved under continuous deformations. Examples for continuous deformations are bending, stretching, twisting or even crumpling, but not tearing or gluing. Two objects are considered topologically the same (equivalent), if they can be transformed into

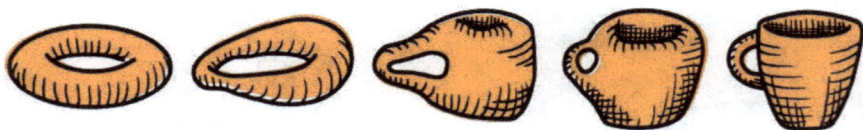

Fig. 12.12.   Transformation of a torus into a coffee mug.

one another by continuous deformations. There is no difference, topologically, between a torus (donut shape) and a coffee mug (see Figure 12.12).

This topological equivalence gave rise to a joke among mathematicians, describing a topologist as someone who cannot tell the difference between a coffee mug and a donut. Topologically, there is also no difference between the state of the rope before and after we tied a knot into it. That is why the mathematical definition of a knot differs from the conventional notion of a knot. A mathematical knot has no loose ends, it is closed. Hence it cannot be untied. However, any conventional knot formed with a single rope can be turned into a mathematical knot by joining the ends of the rope. Knot theory is a sub-branch of topology concerned with the study of mathematical knots. One of the basic problems of knot theory is determining the equivalence of two knots. Two knots are called equivalent, if they are topologically the same, that is, if one can be deformed into the other without cutting it. Being presented with two knots, a knot theorist will try to answer the question "Are these two knots the same or are they different?"

12.6 The Trefoil Knot

The simplest knot is a ring, also known as the "trivial knot" or the "unknot". The simplest non-trivial knot is the trefoil knot, named after the three-leaf clover (or trefoil) plant. Here, "non-trivial" means that it cannot be deformed into a ring, hence, it is not equivalent to the unknot. The trefoil can be obtained by joining together the two loose ends of a common overhand knot (see Figure 12.13).

It has three crossings, and there are no knots with fewer crossings. The crossing number of a knot is the least number of crossings

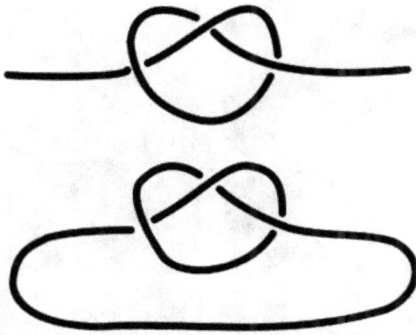

Fig. 12.13.   Creating a trefoil knot.

Fig. 12.14.   Two representations of the unknot.

that are visible on a picture of the knot or on a two-dimensional representation of the knot as in Figure 12.13, called a knot diagram. It is
important to use the smallest number of crossings that occur to
define the crossing number, since one can always produce extra crossings by twisting a given knot. For example, Figure 12.14 shows two
diagrams of the unknot, which has crossing number zero. But by
twisting the unknot, we can always generate additional crossings.

There are two configurations of the trefoil knot, which cannot be
deformed into each other: The left-handed trefoil has as its mirror
image, the right-handed trefoil (see Figure 12.15). In geometry, a
figure that is not identical to its mirror image, is called chiral. The
trefoil knot is an example of a chiral knot.

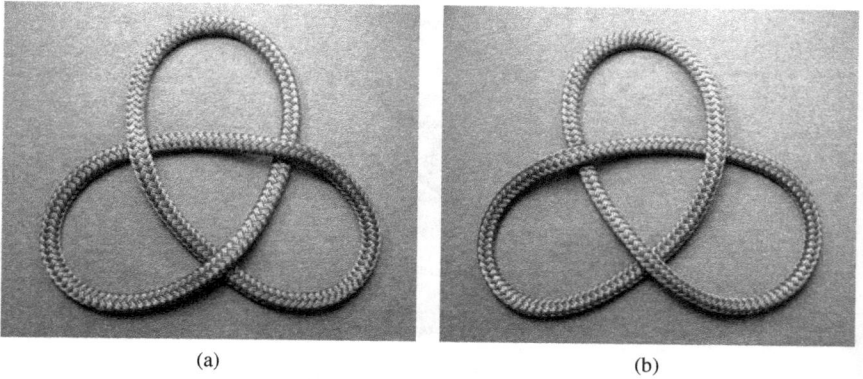

(a)   (b)

Fig. 12.15.   Left-handed and right-handed trefoil knot, image by David J. Fred/CC BY-SA.

Fig. 12.16.   A trefoil knot in disguise. Image by AnonMoos, Public domain, via Wikimedia Commons.

The knot represented in Figure 12.16 is a trefoil knot in disguise. It appears to have more than three crossings, but these additional crossings arise from simple twists as shown in Figure 12.14.

12.7 The Origins of Knot Theory

The first mathematical paper mentioning knots was written by the French mathematician Alexandre-Théophile Vandermonde

(1735–1796), who recognized that different knots are to be distinguished not by geometrical notions such as length, angle, curvature etc., but by "the manner in which the threads are interlaced", that is, by properties which would now be called topological. While Vandermonde did not pursue this much further, it was the German mathematician Carl Friedrich Gauss (1777–1855), who took the initiative to develop a mathematical theory of knots. Gauss was fascinated by knots as early as his teenage years and one of the oldest notes found among his belongings, dated 1794, is a collection of knot drawings. However, it was not until the year 1833 that Gauss achieved mathematical results on knots, thereby initiating knot theory as a field of mathematics. Apart from the left-handed and right-handed trefoil knots, there is no other knot with three crossings. Figure 12.17 shows different depictions of a knot with four crossings. It is the next simplest knot after the trefoil knot and it is called the figure-eight knot. The figure-eight knot can be continuously deformed into its mirror image, hence, it is achiral. In fact, it is also the only knot with crossing number 4.

There are two knots with five crossings, which are not mirror images of each other. These are the cinquefoil knot (see Figure 12.18(a)) and the three-twist knot (see Figure 12.18(b)). Both of them are chiral, so they are not equivalent to their mirror images. Thus, there are in total four different knots with crossing number 5.

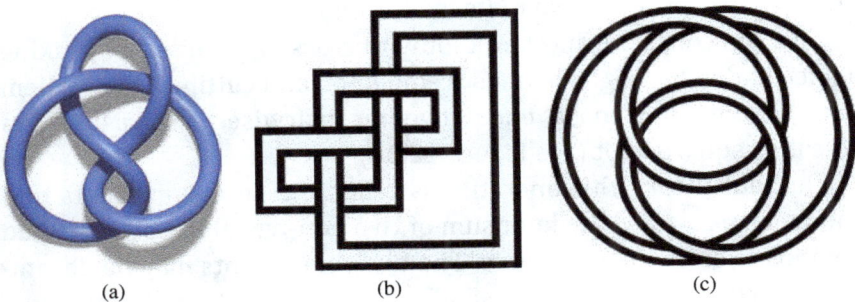

(a) (b) (c)

Fig. 12.17. Figure-eight knot. Images by Jim.belk (a) and AnonMoos (b, c)/public domain.

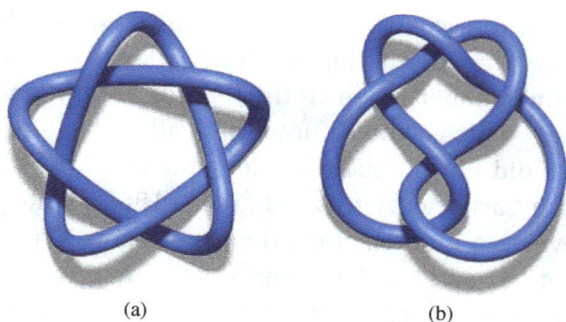

Fig. 12.18.   (a) Cinquefoil knot. (b) Three-twist knot. Images by Jim.belk/Public domain.

Fig. 12.19.   Knots with crossing number 6. Images by Jim.belk/Public domain.

How many knots are there with crossing number 6? Three of them are shown in Figure 12.19, the first two are chiral, while the last one (on the right) is equal to its mirror image.

We can construct another knot with crossing number 6 by taking a left-handed and right-handed trefoil knot, and cutting each of them and joining their loose ends together pairwise. The result is a so-called square knot (see Figure 12.20).

In knot theory, this operation is called the knot sum. Knots that can be expressed as the knot sum of two non-trivial knots, are called composite or compound knots. The two components may be decomposed further into knots with smaller crossing numbers. A non-trivial knot that cannot be written as the knot sum of two non-trivial knots, is called a prime knot. Thus, similar to the prime decomposition of

Fig. 12.20.   Square knot. Image by Jim.belk/Public domain.

natural numbers, there is also a prime decomposition for knots. The number of possible knots increases very rapidly with the crossing number. For example, if we include composite knots, there are almost 10,000 different knots with 13 crossings, and over a million with 16 crossings. The earliest tables of mathematical knots were compiled by the Scottish mathematical physicist Peter Guthrie Tait (1831–1901). In 1885, he published a table of knots with up to ten crossings (a part of the table is shown in Figure 12.21). This was quite an impressive work and Tait's efforts in this undertaking cannot be valued enough. Remarkably, Tait's tables remained mostly unchanged for almost 100 years, but in 1973, it was discovered that two of the knots with 10 crossings in Tait's tables were actually the same knot.

12.8 Are Atoms Composed of Knots?

Tait was led to his study of knots by a beautiful and tempting idea of the Scottish mathematician and physicist William Thomson, Lord Kelvin (1824–1907), who thought that atoms might consist of knotted vortex tubes of the ether, with different elements corresponding to different knots. In Thomson's theory, the unknot represented hydrogen and the trefoil knot represented carbon (at that time, most chemical elements were not yet discovered). The crossing number would then correspond to the number of the element in the periodic table, which was developed simultaneously by the Russian chemist

Fig. 12.21.   A table from P.G. Tait's treatise *On Knots*, 1885.

Dmitri Mendeleev (1834–1907). However, Thomson's theory failed, but it motivated Tait and other mathematicians to study knots and further develop knot theory, which is still on the forefront of mathematical studies today. With regard to Thomson's theory of atoms, Tait wrote about his attempts to classify knots: "The development of this subject promises absolutely endless work — but work of a very interesting and useful kind."

Indeed, the number of enumerated mathematical knots exceeds the number of conventional knots by far. While the approximately 3,800 entries in *Ashley's Book of Knots* are already quite impressive, there are 1,701,936 prime knots with up to 16 crossings (not even counting mirror images). Prime knots can be joined to form composite knots, so the number of composite knots is even much higher. Figure 12.22 depicts all prime knots up to crossing number 7, with only one entry for a knot and its mirror image, as it is common in knot theory.

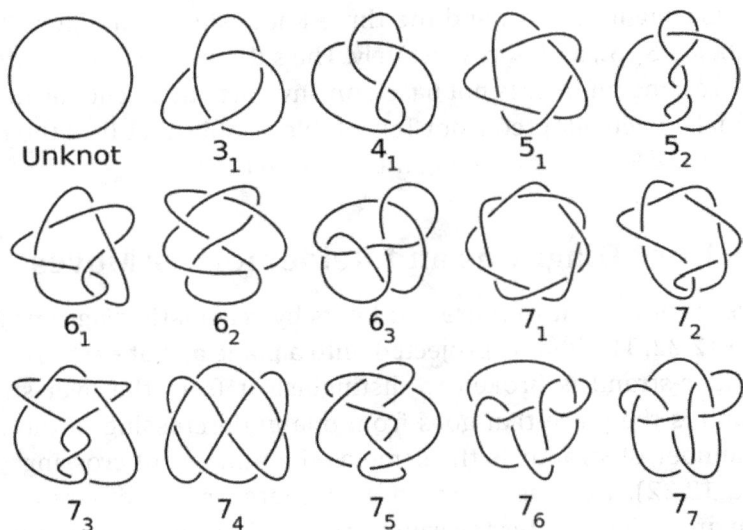

Fig. 12.22.   All prime knots with crossing number up to 7. Image by Jkasd/Public domain.

Table 12.1. Number of prime knots for crossing numbers up to 16.

Crossing number	Number of prime knots	Crossing number	Number of prime knots
0	1	9	49
1	0	10	165
2	0	11	552
3	1	12	2,176
4	1	13	9,988
5	3	14	46,972
6	3	15	253,293
7	7	16	1,388,705
8	21	—	—

Prime knots are traditionally labeled as N_k where N is the crossing number and k represents the number of the prime knot among those with the same crossing number. For example, the trefoil knot is 3_1, the figure-eight knot 4_1, and the three knots shown in Figure 12.19 are labeled 5_1, 5_2, and 5_3, respectively. The sub-order of knots with the same crossing number is not based on any particular scheme and has essentially been adopted from Tait's tables. Table 12.1 lists the number of prime knots for crossing numbers up to 16.

12.9 Knot Diagrams and Reidemeister Moves

It is very convenient to represent knots by schematic diagrams as in Figure 12.22. The knot is projected onto a plane and at each crossing, the under-strand is broken to distinguish it from the over-strand. A strand is the piece that goes from one undercrossing to the next. The number of strands is the same as the number of crossings (see Figure 12.22). However, there are of course many different ways of drawing a diagram for the same knot. Looking at two diagrams for the same knot, it may not at all be obvious that these are equivalent. How can we determine if two knot diagrams are equivalent, meaning that they are depictions of the same knot? In 1927, the German mathematician Kurt Reidemeister (1893–1971) and others independently demonstrated that if two knot diagrams represent the same knot, then one can be transformed into the other by a sequence of only three types of moves on the diagram. These manipulations are now called Reidemeister moves and they can be applied in both directions. Figure 12.23 shows the three Reidemeister moves.

The twist move (R1) twists or untwists a strand, the poke (R2) moves a strand completely over another and the slide (R3) moves a strand completely above or below a crossing. Figure 12.24 shows how a rather complicated looking knot can actually be revealed as the unknot by a sequence of Reidemeister moves.

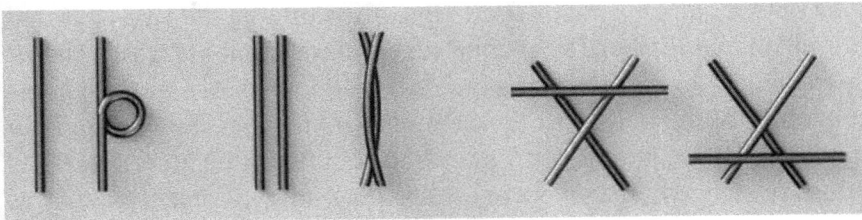

Fig. 12.23.   Reidemeister moves, from left to right: R1 (twist), R2 (poke), R3 (slide).

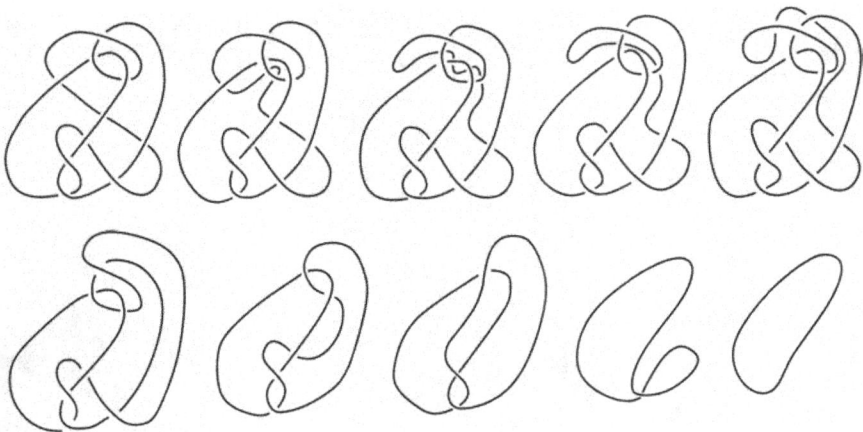

Fig. 12.24.   The unknot in disguise.

12.10 The Recognition Problem and Knot Invariants

In knot theory, determining if two knot diagrams represent the same knot is known as the recognition problem. An important special case of the recognition problem is the unknotting problem. This is the problem of algorithmically recognizing the unknot from a given knot diagram. The first algorithm to solve the unknotting problem was developed by the German-American mathematician Wolfgang Haken (1928–), who also devised some particularly complicated examples of how the unknot can be masked, also known as "Haken's Gordian unknots" (see Figure 12.25).

A practical way to generate a rather complex knot is letting a cat play with a ball of yarn for some time, then finding the two ends and joining them together. Untangling such a messy knot as far as possible corresponds to applying a sequence of Reidemeister moves in the diagram of the knot. With the help of Reidemeister moves, the knot diagram can be simplified until the knot can finally be identified. Now suppose that we are given two knot diagrams, perhaps looking as complicated as the one shown in Figure 12.25, and we want to find out whether or not these represent the same knot. If we can find a

Fig. 12.25.   Wolfgang Haken's "Gordian unknot".

sequence of Reidemeister moves transforming one diagram into the other, then two knots are the same. If the knots turn out to be composite knots, we may have to decompose them into prime knots. However, simplifying a knot diagram can be a very tedious procedure and we may not be able to identify the knots, considering that the number of prime knots with crossing number 16 already exceeds one million. Yet, if the knots are not equivalent, there is a chance to recognize this much quicker, without having to reduce their diagrams until we can identify them. This can be achieved by looking at knot invariants, an important tool for distinguishing knots. A knot invariant is a quantity or property which is defined for each knot and which is the same for equivalent knots. The crossing number of a knot, which we defined as the minimum number of crossings in any diagram of the knot, is an example for a knot invariant, while the actual number of crossings in a particular diagram of a knot is, of course, not a knot invariant (as we can always add more crossings with the twist or poke move).

12.11 Tricolorability of a Knot

Another knot invariant is the so-called tricolorability of a knot. A knot is called tricolorable, if each strand of the knot diagram can be colored with one of three colors, such that the following conditions are satisfied:

- **C1:** At least two colors must be used, that is, not all strands can have the same color.
- **C2:** At each crossing in the diagram, the three incident strands must either have all the same color or all different colors.

Clearly, the unknot is not tricolorable. The trefoil knot 3_1 is tricolorable (see Figure 12.26), but we could draw other knot diagrams representing the trefoil knot, with additional crossings, for example, the one shown in Figure 12.27 with five crossings. How can we show that the property of tricolorability is independent of the representation of the knot, and therefore, a property of the knot itself, rather than of it its diagram? Since two knot diagrams represent the same

Fig. 12.26. Tricolorability of the trefoil knot. Image by Jim.belk — Own work, Public Domain.

Fig. 12.27. Trefoil knot.

knot if they are related by a sequence of Reidemeister moves, we just have to show that tricolorability is preserved under each of the three Reidemeister moves.

Clearly, the first Reidemeister move does not change tricolorability of a knot diagram: If we add a twist to a strand, which is part of a larger diagram, we just keep the same color for both strands (see Figure 12.28). Conversely, if we start with a twisted strand, then both strands must already have the same color because of condition C2 (since two of the strands at the crossing are actually the same strand and must, therefore, have the same color).

Let us now look at Reidemeister move R2, the "poke". If we introduce two new crossings onto two separate strands that have different colors assigned, then we can use the third color for the short strand between the two crossings, so that at each crossing the three incident strands have three different colors, as required by condition C2 (see Figure 12.29). The other direction is clear, since in order to

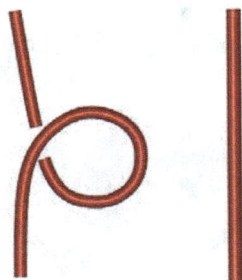

Fig. 12.28. Tricolorability is preserved by Reidemeister move R1.

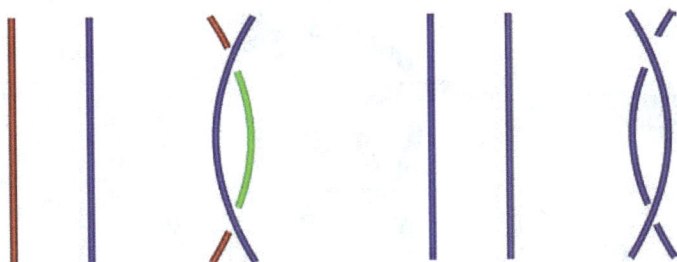

Fig. 12.29. Tricolorability is preserved by Reidemeister move R2.

satisfy C2, the two crossings at the overlap must already have been colored with three different colors or with one single color (see Figure 12.29). In the latter case, removing the two crossings with R2 and leaving the color as is, will preserve tricolorability of the whole diagram. Finally, if we start with two strands of the same color, we also use this color for all three strands after the poke.

To show that move R3, the "slide", does not affect tricolorability, we have to consider four possible coloring situations, one of which is shown in Figure 12.30. In each case, sliding the top strand from one side of the crossing to the other can be done without destroying tricolorability of the knot diagram.

Since tricolorability of a knot diagram is preserved under Reidemeister moves, it is indeed a knot invariant. If one knot diagram for a particular knot is tricolorable, then so is every other possible knot diagram for the same knot. On the other hand, if a knot diagram is not tricolorable, then this is true for any equivalent knot diagram as

Fig. 12.30. Tricolorability is preserved by Reidemeister move R3.

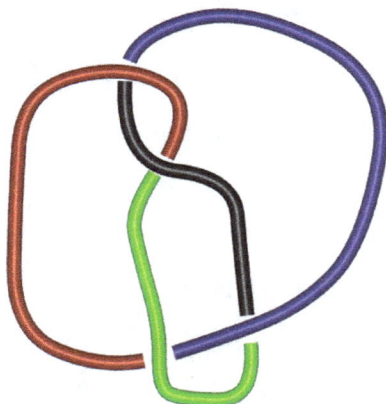

Fig. 12.31. The figure-eight knot is not tricolorable.

well. For example, the Gordian knot shown in Figure 12.25, is not tricolorable, since it is equivalent to the unknot, which is not tricolorable. Another example for a non-tricolorable knot is the figure-eight knot 4_1. In the diagram shown in Figure 12.31, it has four strands. Any two of the four strands meet at some crossing. To satisfy rule C2, all three strands meeting at a crossing must have the same color or all different colors. If for any crossing in the diagram of the figure-eight knot, the three strands had the same color, then all strands would be forced to have the same color, violating condition C1. Similarly, if each of the strands meeting at one crossing had a different color, then all four strands would be forced to have distinct colors.

To conclude, if we have two knot diagrams, one of which is tricolorable and the other is not, then we know that these diagrams represent different knots. In particular, if a knot diagram is tricolorable, then it must be a non-trivial knot (since the unknot is not tricolorable). Tricolorability is one of the simplest knot invariants, distinguishing only between two categories of knots: those which are tricolorable and those which are not. However, there are more refined knot invariants, for example the Jones polynomial, which is a certain function that can be assigned to each knot. All prime knots with up to nine crossings have distinct Jones polynomials, but there exist distinct knots with higher crossing numbers sharing the same Jones polynomial.

12.12 Applications of Knot Theory

Knot theory is not only an active field of research in pure mathematics, it also has applications in theoretical physics and even in molecular biology. It has been discovered that in the process of DNA replication inside our cells, DNA strands tend to form closed loops, some of which are knotted. Enzymes interacting with these knotted DNA strands can break them up and change their knottiness. In physics, knot theory plays a role in quantum field theory, the modern theory describing sub-atomic particles and their interactions. Although in quantum field theory, there is not a direct correspondence between different prime knots and elementary particles, as William Thomson had once imagined in his beautiful hypothesis, this still somehow closes the loop of knot theory, perhaps turning itself into a knot.

Chapter 13

Kinematics — The Geometry of Motion

We are so accustomed to living in a three-dimensional world that we do not generally give any thought to the more abstract aspects of moving around in our surroundings. We simply have the feeling of being able to enjoy unencumbered movement in every imaginable direction (ignoring the effects of walls and gravity for the moment), and this level of awareness is quite sufficient as far as it concerns our own mobility.

Things start to require a higher degree of sophistication once we start to think about constructing machines. The controlled movement of mechanical objects requires some concrete planning, and that, in turn, compels us to develop a more precise method of dealing with the options of movement in three dimensions and the limitations placed upon them by specific mechanical processes.

In this chapter, we aim to take a closer look at some of the more interesting geometric aspects of the movement of machines in three-dimensional space. After some introductory comments on the basic concepts, we will consider some specific examples of the geometric details involved in producing machines that can move freely in the pre-determined ways required by their intended applications. For example, we would expect a combustion engine, meant to power an automobile, to have an intrinsic geometry quite different from that of a robot arm built to manipulate objects outside of a space station. The examples we have chosen here are meant to convey some of the breadth of the geometric content required to master the construction of such objects.

13.1 Degrees of Freedom

An unrestrained rigid body has six *degrees of freedom* (*DoF*) for movement in space. As illustrated in Figure 13.1, it can be rotated around each of the three axes x, y, and z, and it can also be translated (which means to move straight in a single direction) in the directions of each of these three axes. Any movement in space can be considered to be composed of these six parts.

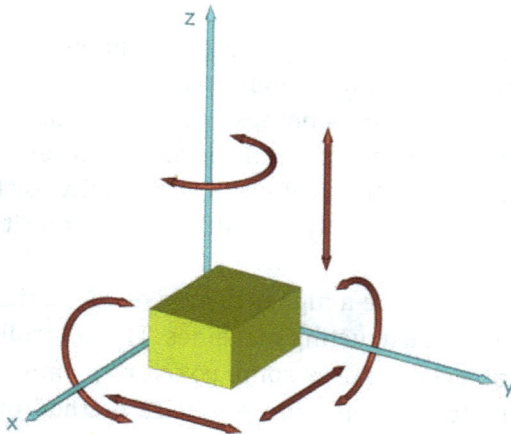

Fig. 13.1.   Degrees of freedom.

If we connect two or more rigid bodies, their DoF can be decreased by their interaction. For example, a swinging door is fixed to the door-frame with some revolute joints. Thus, it has one degree of freedom: DoF = 1, because we can only rotate it around one axis. If we remove the door from the frame, it has six degrees of freedom, DoF = 6, because we can rotate and translate it in each direction. If we connect it to the frame again, it loses five DoF. In that case, we say that we are introducing *kinematic constraints*.

As a first example of this, two rigid bodies can be joined to another as a *spherical pair*, as shown in Figure 13.2. The centers of the two spheres are kept identical at all times in this coupling. As we can see, each of the bodies can rotate freely relative to the other with respect to the three axes x, y, and z, but is not able to translate along them, as they are hindered by the other body. Thus, three degrees of freedom are lost, and the body has a degree of freedom: DoF = 3.

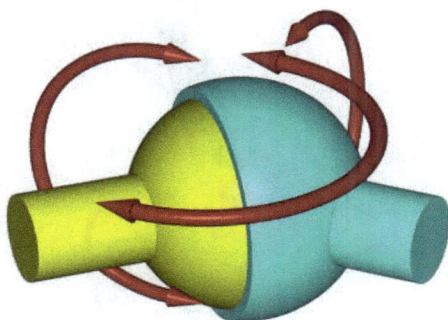

Fig. 13.2.   Spherical pair.

As another example, we observe a co-axial pair of cylinders, shown in Figure 13.3, in which we have the option of rotating each cylinder relative to the other with respect to their common axis and translating with respect to the same axis. This leads to a degree of freedom of DoF = 2.

There exist many more such constellations, but we do not have to develop a separate rule for each of them. Once we understand the principle behind them, we can calculate the DoF for all mechanisms.

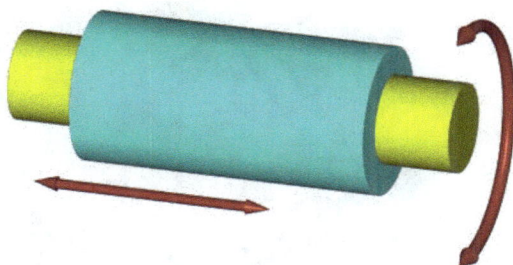

Fig. 13.3.   Cylindrical pair.

If we take a look at the mechanism shown in Figure 13.4, we immediately recognize that it has DoF = 1, because we can only rotate the blue bar relative to one axial direction. We call such a constellation a *revolute pair*.

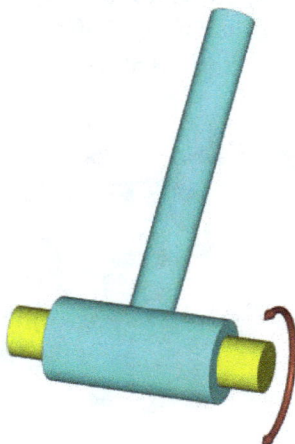

Fig. 13.4.   Revolute pair.

Let us now take a look at a mechanism consisting of more than two rigid bodies, as shown in Figure 13.5, and try to calculate the DoF for one of its parts.

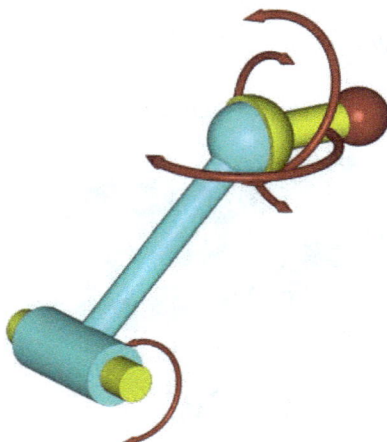

Fig. 13.5.   Three rigid bodies.

Considering the DoF of the red endpoint, we obtain DoF = 4. We arrive at this number in the following way. First of all, we can count the rotation of the revolute pair as one DoF. To this, we add the three possible rotations of the spherical pair, yielding four DoF in total.

13.2 Gruebler's Equation, Robot-Arms and the Stewart-Gough Platform

Now that we have made these preliminary observations. We are ready to consider a somewhat more concrete example. Let us say we want to construct a robot arm that can move a full cup to every position under the constraints given by the range of the mechanism, without tipping it over. In order to do this, we will have to include enough moving parts for the DoF of the mechanism to be DoF = 6. An example of such a robot arm is shown in Figures 13.6 and 13.7.

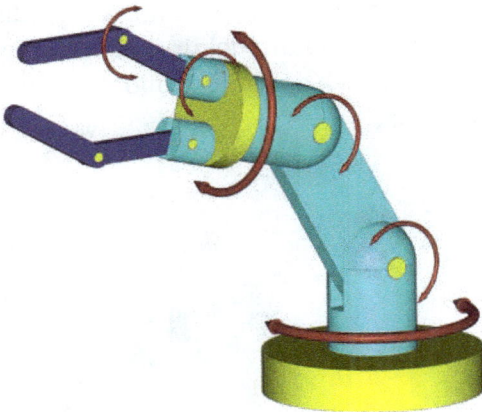

Fig. 13.6.   Robot-arm with DoF = 6.

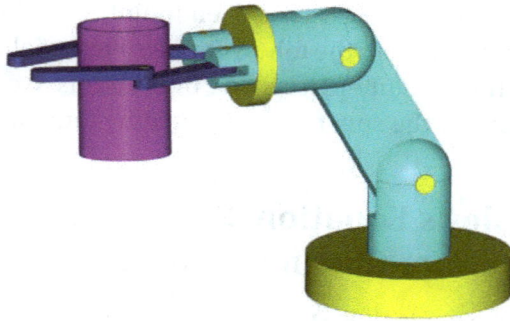

Fig. 13.7.   Robot-arm holding a cup.

For some robot arms, a DoF of less than 6 may be enough. The orientation of the tip of a welding robot doesn't matter, for instance, and for practical purposes it is sufficient for the tip to be able to reach any point in a certain range without the direction to which it points being of any real practical importance. Usually, robot arms are designed with three main twisting axes between the actuators (the moving bars) to control the position of the effector (the tip we wish to move to a specific position), and depending on the use of the robot, the effector may have an additional 1, 2 or 3 joints to gather the needed DoF. A model of such a welding robot is shown in Figure 13.8.

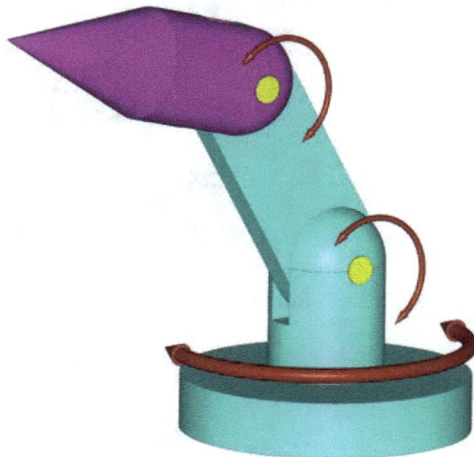

Fig. 13.8.   Welding robot.

As a side comment, we should be aware that we also have to consider its workspace, that is to say the region of all the positions to which we are potentially able to move the effector, if we actually want to build a practical robot. Depending on the robot's combination of joints and the dimensions of its actuators and its effector, this workspace can be box-shaped or kidney-shaped, for instance. This is an important consideration for practical applications, but goes beyond what we are considering here.

If we use spherical joints in designing a robot with DoF = 3, we need fewer systems (i.e., robot arms) than we do if we use cylindrical joints (with DoF = 2) or a revolute pair (with DoF = 1). If we take a closer look at our own bodies with this in mind, we can calculate the relative DoF of our fingers, arms, toes and so on. If the DoF exceed the number 6, we have more possibilities to reach each position in space (within the range of the system, of course). It is easy to see a practical application of this: if we concentrate, we can, for example, hold the tip of one finger in the same position while we move the rest of our arm around.

The sub-topic of *inverse kinematics* deals with the calculations required to determine how to move the end of a robot arm into the position required by some practical application. Unfortunately (and perhaps not too surprisingly), the calculations that need to be made to solve problems of this type are not at all easy! Fortunately, we do not always have to do all of these calculations when we design the machines we need. Geometric modeling with Computer Aided Design (CAD) is a great tool to help us design and avoid mistakes in the planning, programming and construction of a robot. If we first construct a virtual three-dimensional model in a computer program (using the principles described in Chapter 14), we can do all the necessary testing on the screen before we actually build a physical model, and all of this can be done without actually calculating the exact movements in advance.

Along with *serial kinematics*, in which actuators are connected by joints in series, as seen in the robot arms we have been looking at until now, we also have *parallel kinematics* at our disposal. An example for this is a hexapod robot, as shown in Figure 13.9.

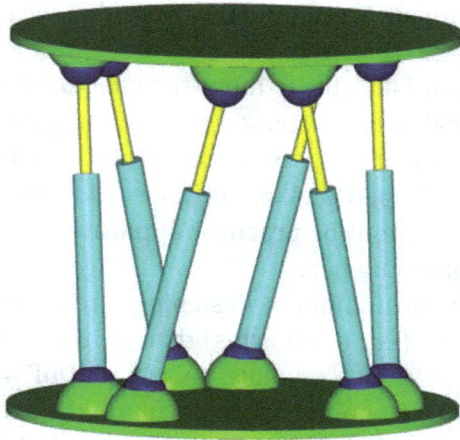

Fig. 13.9.   Illustration of a hexapod robot.

This type of mechanism is also either Known as a *Stewart platform* or a *Gough platform*, named after the two British engineers D. Stewart and V.E. Gough, who invented it in the mid-20th century. Six linear axes move a platform, as shown schematically in Figure 13.9. The platform has DoF = 6 (as will be explained below) and can, therefore, be moved into every position within the physical range of the system. This type of mechanism is commonly used in driving simulators and also in some amusement park rides. They are especially also used for flight-simulators, like the AXIS ATR 72-500 FFS, shown in Figure 13.10. The company AXIS is located near the city of Graz (Austria) and was founded in 2004 by engineers and pilots. Their spectrum of products ranges from training devices to full flight simulators.

Now that we have some understanding of the principle of calculating the DoF of a *kinematic chain* (i.e., rigid solids in systems connected by joints), we are going to turn our attention to an equation that allows us to calculate the DoF of each such a kinematic chain.

What we are referring to here is *Gruebler's equation*, an expression for a kinematic chain consisting of n systems and m joints with the DoF $f_i(i = 1$ to $m)$. This useful result was developed by the German

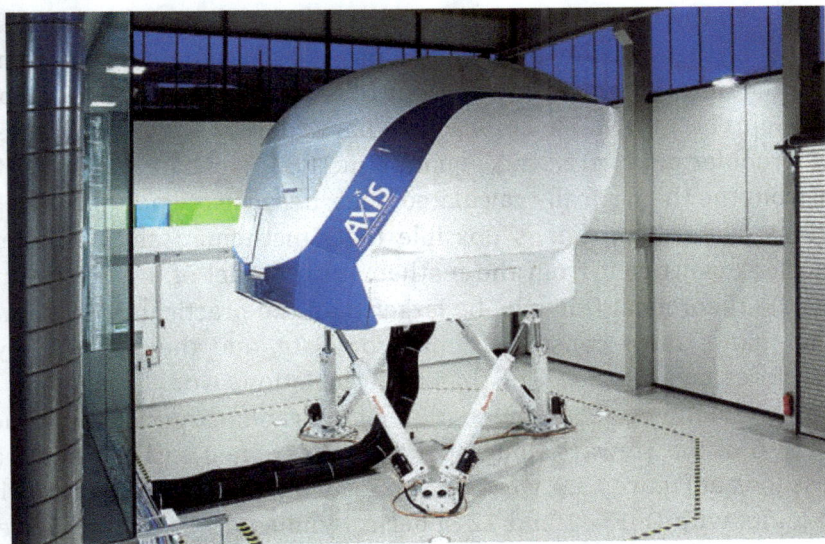

Fig. 13.10.   AXIS ATR 72-500 FFS © Croce&WIR.

engineer Martin F. Gruebler (1851–1935), and this is what it looks like

$$f = 6(n-1) - \sum_{i=1}^{m}(6 - f_i).$$

We shouldn't let this equation frighten us. The capital sigma (Σ) is simply a symbol for the sum; it is a shortcut for adding the terms $6 - f_i$ (six minus the DoF) for every single joint. (It is just a shorter way to write $(6 - f_1) + (6 - f_2) + \cdots + (6 - f_m)$.)

How can we establish this equation? If we study a mechanism consisting of n systems (we are referring to the parts of the mechanism that are not joints as *systems*), we choose one of the systems as fixed. (Perhaps we could choose the rack to which the other systems are connected as our reference.) This means that we have $n - 1$ movable systems left with respect to this fixed one. If we imagine all the joints to be removed, each of these systems has a DoF = 6 with respect to the fixed one. For the description of all possible constellations of the system we now have $6(n - 1)$ parameters. The joint with the index i reduces the number of free parameters by $6 - f_i$, for each joint $i = 1, \ldots, n$. This is exactly what Gruebler's equation states.

Gruebler discovered this equation in 1883. The true DoF can differ from the theoretically calculated ones, because of special constellations. For instance, it is possible that (mathematically) complex solutions can result from the mathematical model of a mechanism, and the theoretical DoF may be larger than the practical DoF. On the other hand, some constraints may correlate, and the system may, therefore, have a higher DoF than the calculations would suggest.

We will now give a rough outline of the explanation for the value DoF = 6 of the Stewart–Gough platform mentioned earlier. Taking a close look at Figure 13.9, we count a total of $n = 14$ systems, 6 cylindrical joints with DoF = 1 and 12 spherical joints with DoF = 3. Hence, using Gruebler's formula, we obtain DoF = 12

$$f = 6(14-1) - \sum_{i=1}^{m}(6-f_i) = (6 \cdot 13) - 6(6-1) - 12(6-3)$$
$$= 78 - 30 - 36 = 12.$$

It turns out that we have a redundancy of 6 DoF. If we change the six spherical joints (with DoF = 3) either on the bottom or on the top of the mechanism to cardan joints (which are hinges with DoF = 2), we get a value of DoF = 6 for our platform, as shown in Figures 13.11 and 13.12.

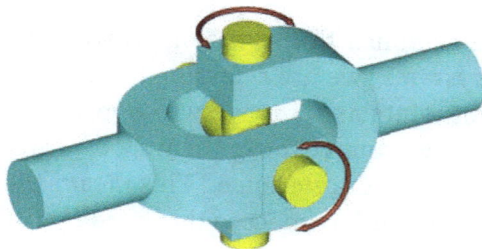

Fig. 13.11.   Cardan joint.

Fig. 13.12.   Stewart–Gough platform.

Let us now investigate another special mechanism, a link chain consisting of six systems connected by six revolute joints. Figure 13.13 shows a virtual model of this so-called 6R link chain (6 revolute joints). If the constellation of the joints is chosen in a very special way, it has a theoretical DoF = 0, but it moves nevertheless, with DoF = 1. Such mechanisms, which, in theory, shouldn't be able to move at all, actually can, and are called *over-constrained*.

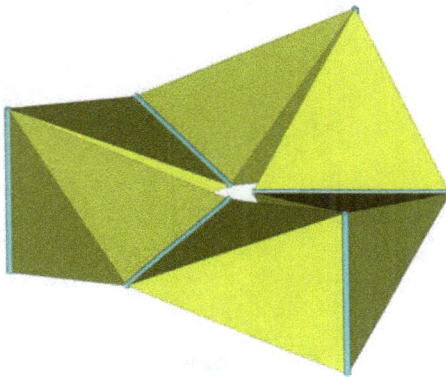

Fig. 13.13.   Link chain (6R).

Let's calculate its theoretical DoF, using Gruebler's formula. We have six systems and all the joints are revolute joints with DoF $= 1$. We, therefore, have the following:

$$f = 6(n-1) - \sum_{i=1}^{m}(6-f_i) = 6(6-1) - 6(6-1) = 0.$$

A complete proof would go beyond the scope of this book, but even without one, it is interesting to consider the conditions that need to be fulfilled for our special mechanism to be over-constrained. Raoul Bricard (1870–1943), a French mathematician found the following conditions for a 6R link chain to have DoF $= 1$:

- Any two adjacent axes $a_{i,i+1}$ and $a_{i+1,i+2}$ of the revolute joints cross at a right angle.
- Perpendicular minimal transversals l_i form a closed hexagon.
- The following condition holds for the side lengths l_i of the hexagon:

$$l_1^2 - l_2^2 + l_3^2 - l_4^2 + l_5^2 - l_6^2 = 0.$$

Figure 13.14 shows, for example, the axes a_{34} and a_{45}, the transversal l_4, which is the perpendicular minimal transversal for a_{34}, and a_{45}. The first condition says that if we translate a_{34} along l_4, until it crosses a_{45}, then the two axes cross in a right angle.

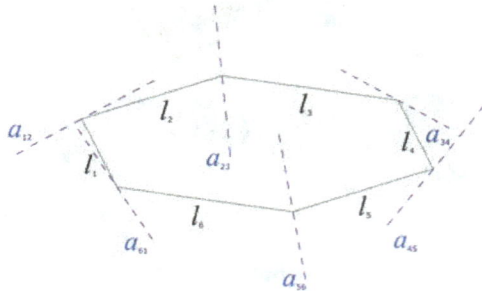

Fig. 13.14.   Overconstrained link chain (6R).

Fig. 13.15.   Paper model of a link chain (6R).

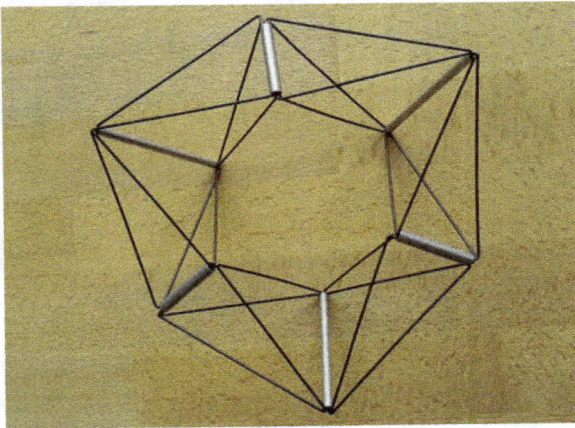

Fig. 13.16.   Wire model of a link chain (6R).

We can choose a very special case of a link chain that fulfills all three of Bricard's conditions, which is shown in Figures 13.13 and 13.14. This special constellation is achieved when the transversals form a regular hexagon and six congruent tetrahedra have connecting adjacent axes. We can easily build models of such link chains (Figures 13.15 and 13.16). For example, for the construction of the paper model shown in Figure 13.15, we have to construct six nets of congruent tetrahedra, assemble them and then join them pairwise with duct tape. The easiest way to get tetrahedra that fit the constellation shown in Figure 13.14 is to cut them out of congruent cubes and scale them vertically by the same factor (Figure 13.17). We can either use CAD-software to construct the nets of those tetrahedra (see Chapter 14) or calculate the lengths and draw the nets by hand — by taking the same considerations into account.

The lengths of the edges of the tetrahedra may also vary (with respect to the conditions of Bricard, described above), as shown in Figure 13.16. We can have fun playing with the resulting mechanisms, turning them around and around.

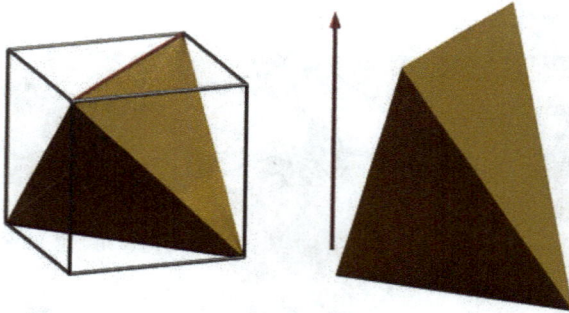

Fig. 13.17.   Construction of fitting tetrahedra.

13.3 A Fascinating Application: The Turbula® Mixer

As a next step, we observe that our very special link chain has the extraordinary property, which is that opposite axes always stay in a common plane when we rotate the chain. This means that we can cut the mechanism in half and investigate its kinematic properties. Figure 13.18 shows half of the wire model shown in Figure 13.16, while Figure 13.19 shows half the virtual model shown in Figure 13.13.

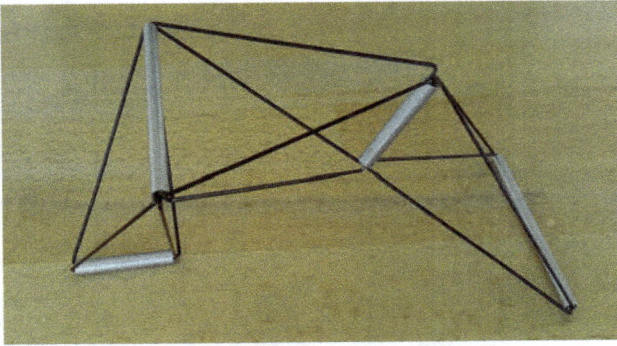

Fig. 13.18.   One half of the link chain.

Fig. 13.19.   The cut chain.

Taking a closer look at this model gives us an idea for a new mechanism with an exciting property. What if we fix the two joints at the ends of the chain to a base by taking two vertical revolute joints through the mid-points of the axes, as schematically shown in Figure 13.20 (the blue arrows are pointing towards those two joints)? We can imagine something such as using pins to attach them to the base, as we can see in Figure 13.21.

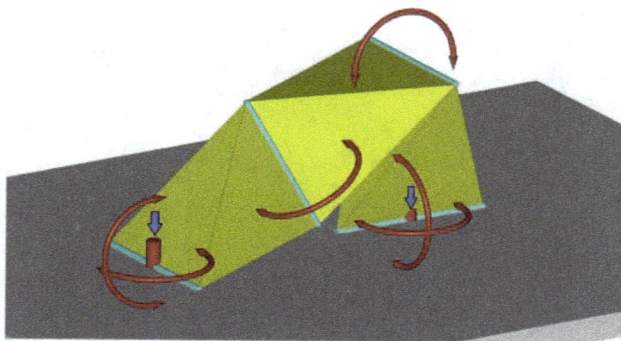

Fig. 13.20.   Idea of a new mechanism.

Fig. 13.21.   Handcrafted model.

If we calculate the DoF of this new mechanism, we get a theoretical value of DoF = 0 (see below). However, because of the special constellation of our initial 6R link chain (fulfilling the Bricard conditions), which we have cut in half to get our new mechanism, it still moves with DoF = 1.

$$f = 6(n-1) - \sum_{i=1}^{m}(6 - f_i) = 6(4-1) - 2(6-1) - 2(6-2) = 0.$$

If we play around a little bit with a model of this sort of mechanism (as the one shown in Figure 13.18) and focus on the system in

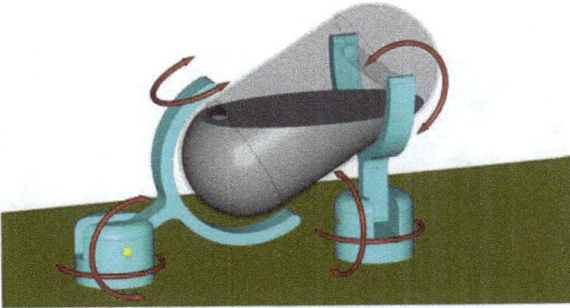

Fig. 13.22.   Mixing.

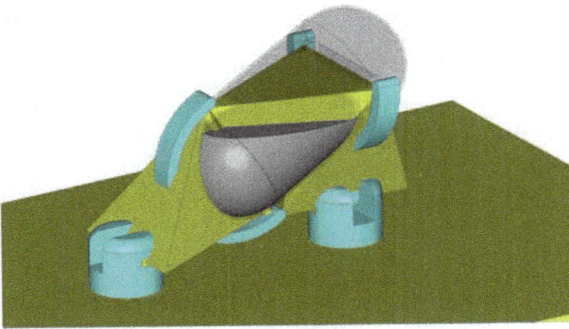

Fig. 13.23.   Link chain (6R) and Turbula® mixer.

the middle, we notice that it moves around in a strong manner. We say that this section is *well shaken*. If that middle section were to be filled with different substances, we can easily imagine that they would be mixed together very well. This observation was made by the German mathematician, engineer, philosopher and artist Paul Schatz (1898–1979), who invented the *Turbula*® mixer in 1960, and is used for the homogeneous *mixing* of powdery substances with different specific weights and particle sizes. This is a nice application of our theoretical approach to over-constrained mechanisms in kinematics!

In Figures 13.22 and 13.23, we can see how our link chain is embedded in the mechanism.

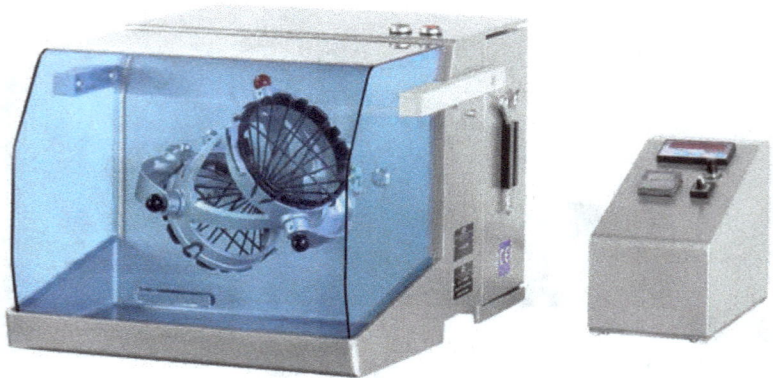

Fig. 13.24.   Turbula® mixer T2F.

Figure 13.24 shows a Turbula® mixer by Glenn Mills Inc.[1] This company manufactures Turbula® mixers in three different sizes, and each mixer can be constructed with a special coating for the production of foodstuffs or pharmaceuticals.[2-5]

13.4 Involute Gears

Having considered some interesting types of couplings that we can use to create machinery, we are ready to focus on how we can actually get such machines to move. Any kind of mechanical machinery relies on some form of transmission of power, and by far the most common tool to do this is a gear. The simplest types of gears can be flat (that is, cylindrical) and simply transfer one rotation to another, as is the case in the internal mechanism of a mechanical watch, as illustrated in Figure 13.25.

[1] 220 Delawanna Avenue, Clifton, New Jersey 07014, USA.

[2] Gfrerrer, A. (2008). *Kinematik und Robotik*. Graz: Institute of Geometry, TU Graz.

[3] Hesse, S. and Malisa, V. (2010). *Robotik — Montage — Handhabung*. München: Carl Hanser Verlag.

[4] Bottema, O. and Roth, B. (1990). *Theoretical Kinematics*. New York: Dover Publications.

[5] Zhang, Y., Finger, S. and Behrens, S. *Introduction to Mechanisms* [online] https://www.cs.cmu.edu/~rapidproto/mechanisms/chpt4.html [accessed on 01.05.2020].

Fig. 13.25.   Gears of the kind we would expect to find inside a mechanical watch.

In principle, one wheel could just roll while pressing on another, with the rotation of the first transmitted to the second by virtue of the friction between the two. If the wheels are made of some appropriate substance, like rubber, for instance, this will certainly work, but not with any high degree of precision. With a system of this type, there is inevitably some slippage, as well as a gradual degrading of the surfaces that are in constant contact. In order to avoid this, gripping teeth can be added to the rotating wheels. An elementary illustration of this idea is shown in Figure 13.26.

Simple spur-gear systems like this can certainly be used to avoid slippage, but they are not very sophisticated. Their main disadvantages are immediately obvious. The stress of each tooth is always on its outside corner, causing constant wear on that one spot. Furthermore, such systems will have uneven power transference, slippage of the corners on the flat sides of the opposing teeth and small gaps between the intervals in which successive teeth mesh with their counterparts on the opposite sides. Surely, we can do better.

What properties would we ideally want a pair of gears to have? One possibility is that we would like the transfer of power to be as

Fig. 13.26.   Very simple gears with teeth.

constant as possible, with as little loss of power to friction as we can obtain. In other words, it would be good if we could find a way in which to form the teeth of our gears to roll on each other without rubbing at all, and to simultaneously keep pressure from one gear to its partner oriented in an unchanging direction. This may seem like a big challenge, but as it turns out, it is not difficult to design teeth with these properties, and gears with such teeth are, in fact, the most commonly used in practical mechanical applications.

The curves we will require for the edges of the teeth in this context are *circle involutes*. We can think of such curves as being created in the manner illustrated in Figure 13.27.

Imagine the string wrapped around a can, as we show in Figure 13.27. As the string unwinds in a clockwise direction, with the marker kept taut at all times, the marker describes the curve shown

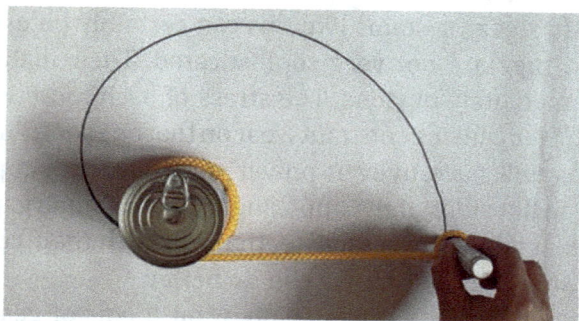

Fig. 13.27.   Drawing a circle involute by unraveling a coiled string.

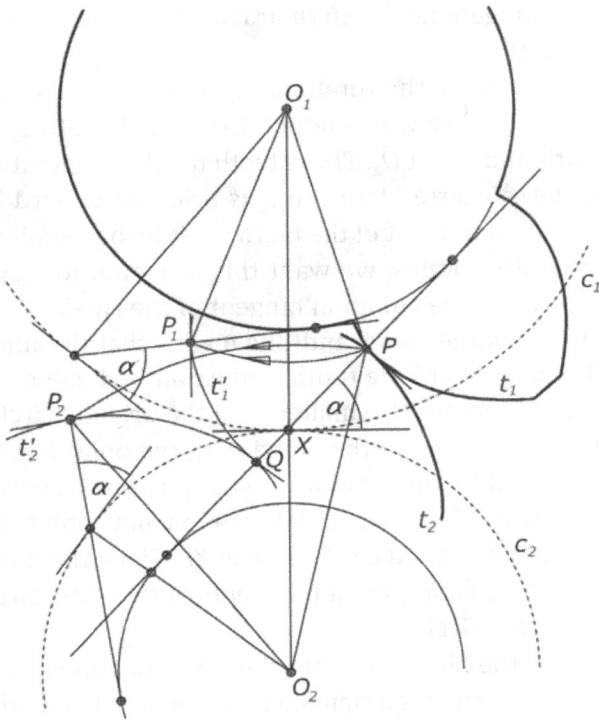

Fig. 13.28.   Involutes with the properties we require from gears' teeth.

in Figure 13.27. As we can see, the position of the marker is at all times at a point on a tangent to the circular can, whose length is equal to the length of the part of the string we have unravelled up to that point. Among other things, this means that the tangent of the curve we have drawn in the point where the marker is placed at this moment is perpendicular to the taut string, that is, to the tangent of the circle.

In Figure 13.28, we can see how the fulfillment of the conditions we require of our gears' teeth leads us to the curves shown.

Furthermore, as we show in Figure 13.28, we start with a system in which a circular disc c_1, with center-point O_1, rotates in a clockwise direction. The rotation of this disc is transferred to a counterclockwise rotation of a second disc c_2, with mid-point O_2. To start with, these discs are tangent at X with their common tangent line at X perpendicular to O_1O_2.

In order to improve the rotational transfer, we introduce a tooth t_1, connected to the disc with midpoint O_1, and a tooth t_2, connected to the disc with mid-point O_2. These teeth touch at a point P in such a way that the force exerted by t_1 on t_2 is oriented toward X. In other words, the common tangent of the teeth must be perpendicular to the line XP. As the disc rotates, we want this direction to stay constant. The angle α between the common tangent of the two initial discs and the line XP remains the same, and this means that the lines perpendicular to the tangents of the tooth t_1 must all be tangents of a common circle with midpoint O_1, smaller than the original circle c_1.

As the tooth t_1 rotates to the position t_1', the point P rotates to P_1. At the same time, the tooth t_2 rotates to the position t_2', with the point P rotating here to P_2. The teeth now touch in a new point Q, and XQ is a tangent of the same small circle as was XP. Since the lines perpendicular to the tangents of t_1 are all tangents of the same circle, t_1 is an involute of this very circle.

If we look at the situation from the point of view of the bottom circle c_2, we see that the arguments we have just made hold in exactly the same way. This means that t_2 is also an involute of a circle, namely the circle with midpoint O_2 and tangent $XP = XQ$.

While all of this may seem a bit confusing, taking a good look at Figure 13.29 should help clear things up a bit.

In Figure 13.29, we see the same construction as in Figure 13.28 in each of the two parts. In Figure 13.29(a), the gear (with one solitary tooth) is superimposed, giving some idea of the result of the construction. The arrows show the directions in which the two gears rotate. In Figure 13.29(b), we see the same configuration after the part of the rotation used in the construction of Figure 13.28. It is well worth studying these pictures before reading on. This is really a case in which a picture is worth a thousand words.

Taking a closer look at the way these involutes are connected to the rolling action of the two circles with which we started out, we see that the teeth mesh perfectly. There is no energy lost to friction (at least in principle), as the teeth just roll off each other. Also, if we space the teeth in an appropriate way, as illustrated in Figure 13.30,

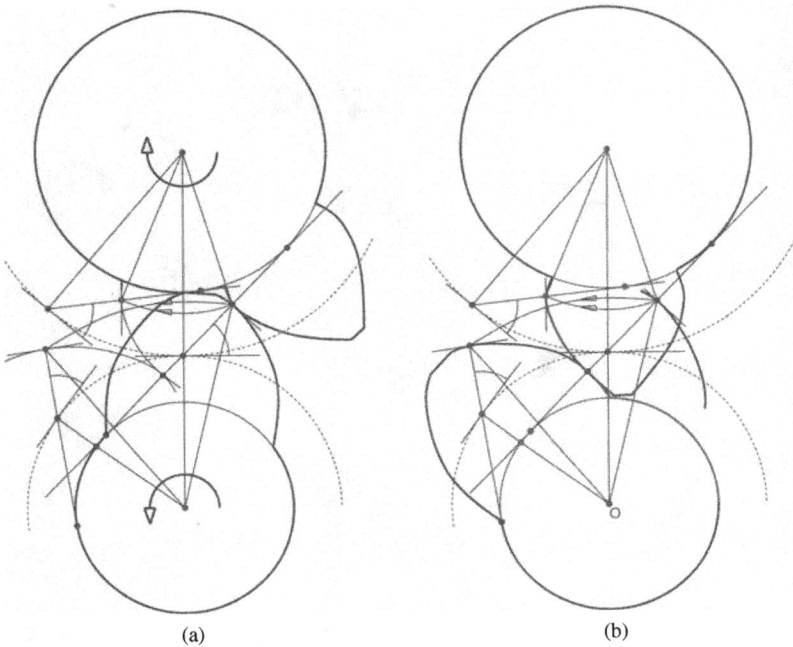

(a)   (b)

Fig. 13.29.   One tooth on each of the two gears from Figure 13.28.

we can arrange it such that the transfer of the pushing action from one pair of teeth to the next pair is gradual, with two pairs of teeth briefly pushing simultaneously. For this reason, we can arrange these teeth so that the transfer of power is continuous, with no gaps.

Now that we know a little bit more about how gears can mesh to transfer rotation to a parallel axis, we can start to think about what we can do to transfer rotational movement to a rotation with an axis in some other direction. In order to do this, we can use a *bevel gear*.

As we can see in Figure 13.31, bevel gears are shaped like serrated sections of right circular cones. The vertices of the cones are cut off, but if we imagine how the cones would be extended, we see that the vertices of the two cones of the meshing bevel gears would coincide. In Figure 13.31, the axes of the cones are at right angles, and this is by far the most common angle used for bevel gears. However, it is possible for such gears to be made with any acute angles between the axes, and we can then transfer rotation to any direction with such gears.

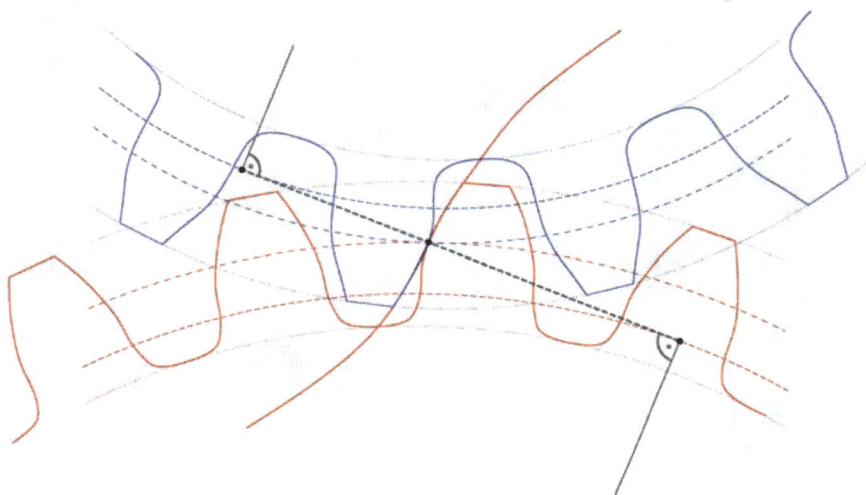

Fig. 13.30.   Involute gears with multiple teeth.

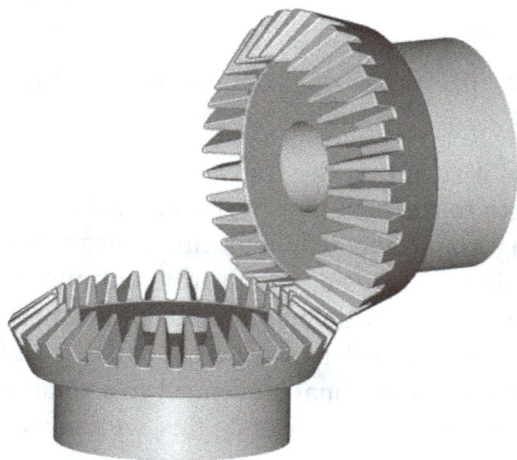

Fig. 13.31.   Simple bevel gear.

Ideally, we would like to make such bevel gears with the same properties as the simple flat gear we can create using circle involutes. This would mean using the spatial counterpart of the circle involute, which is the so-called *spherical involute*. (Think of rolling one sphere on another, the way we rolled one circle on another in two dimensions to create the circle involutes.) In practice, this is generally not the form that is used, because of practical manufacturing considerations. As it turns out, a close approximation (the so-called *octoid*, the precise definition of which is much too technical to be included here) has a number of practical advantages, both in manufacturing and with respect to some specific issues of bending stress and contact stress. The path of contact in such a bevel gear is not quite a straight line, as it would be for teeth modeled on a spherical involute, but slightly bent.

These are only a few of the more geometric aspects of creating machinery that can move precisely in three-dimensional space. As any mechanical engineer will be more than happy to tell you, there is a whole world of fascinating things to be discovered in the workings of any man-made creation that moves. If you have ever taken any mechanical gadget apart and inspected the geometric connections, you might be enlightened by what we demonstrated in this chapter, even though putting it back together could be quite challenging.

Chapter 14

The Art of Designing Three-Dimensional Models

One of the basic problems of computer graphics is the representation of three-dimensional shapes. These virtual shapes are then assembled into complex photo-realistic scenes that are usually referred to as Computer Generated Imagery (CGI). CGI has become an important industry through the interest of large film studios in creating artificial scenes, characters, buildings, and landscapes. Realistic spaceships, futuristic cities and scary monsters have become a familiar sight in the movies.

14.1 Introduction

The process of creating realistic-looking virtual objects involves a variety of highly-specialized techniques. Typically, three-dimensional objects are scanned or recorded photogrammetrically and processed further using methods of computer-based modeling. Computer animation and simulation of physical processes often gives the scene a realistic, or sometimes even supernatural, dynamic. Photo-realistic rendering creates a natural look with sophisticated methods. This rendering uses mathematical algorithms to simulate surface properties and lighting, possibly even tracing all light rays from the (virtual) camera back to the (virtual) light sources, taking into account all types of reflections, coloring, and refraction.

A data-structure in a computer that represents a real object is usually called a three-dimensional model, and the process of creating these models is called three-dimensional modeling. The large family of software programs that have been created especially for this purpose is called "CAD software", where the letters CAD stand for Computer-Aided-Design. Once a virtual object has been successfully created, modern CAD software allows us to display the object on screen and to manipulate it. One can turn it around and view it from all sides and one can zoom-in to inspect the details. Even though it is displayed by a computer, it is like holding the object in your hand and playing with it, and modern virtual-reality systems try to convey exactly this experience. In that respect, three-dimensional models are very much different from two-dimensional drawings, which typically do not allow to change the view dynamically.

Important applications of CAD can also be found beyond the entertainment industry. Medical imaging uses three-dimensional models of organs and three-dimensional representations of the interior of a body to reveal hidden structures, as a help for diagnosing diseases, and for planning surgery. Architects use three-dimensional models of buildings and interiors to convey design ideas. Moreover, three-dimensional models have become an important tool for engineering. Technical objects can be created virtually, tested and improved virtually. Nowadays, it is even possible to assign physical properties to virtual objects, such as to measure forces occurring in mechanical joints. So, one can already simulate, test, and improve an object before actually manufactures it. This procedure typically saves a lot of money in technological development.

In this chapter, we will focus on fundamental techniques of creating three-dimensional models using CAD software, and we will exemplarily use these techniques to create some nice virtual objects. Moreover, we will discuss different printing methods to get our models out into the real world.

14.2 Creating Virtual Three-Dimensional Models

For a computer, the data that describes a three-dimensional model is typically a collection of points and polygons (a so-called "polygonal mesh") that represent the surface of the object — very much like the vertices and faces of a polyhedron. The polygonal surfaces, which consist of planar pieces, are often improved by smoothing techniques (e.g., by using so-called "spline-curves") in order to better approximate smooth surfaces.

An early example is the Utah teapot, which is depicted in Figure 14.1. It was created in 1975 by the British computer scientist Martin Nevell who found this object among his wife's tea set. He entered by hand the points defining the surface. The shape has become an iconic shape in computer graphics, and it has been used ever since for demonstrating new methods of creating surface textures, lighting effects, and rendering techniques, etc.

While this example shows that the points that define an object or surface can be entered manually into the computer, this is not always a practical method, even if a (very high quality) three-dimensional scanner helps to automate the process. Here, we will discuss a method called "constructive solid geometry", that is supported by modern CAD software. This is a very useful method of simple three-dimensional modeling that is based on easily comprehensible principles.

Fig. 14.1.   Utah teapot.

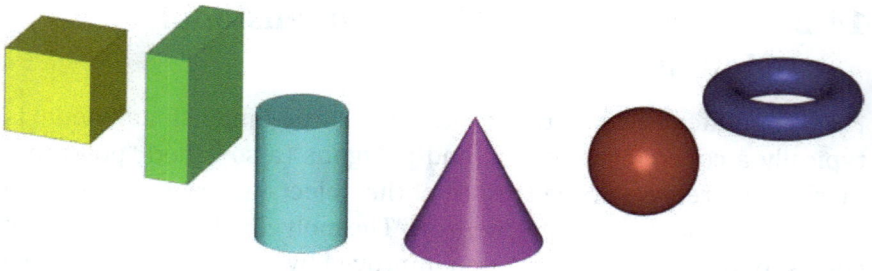

Fig. 14.2.   Basic forms for geometrical modeling.

The main idea is to combine simple objects into visually complex ones by applying a set of basic operations. The primitive shapes that form the basis of constructive solid geometry are predefined within any CAD software program. A collection of these basic shapes is shown in Figure 14.2, where, from left to right, these solids are the cube, the cuboid, the cylinder, the cone, the sphere, and the torus. Depending on the CAD software, there might be additional pre-defined shapes, but these basic solids are always part of the toolbox.

Each CAD program now offers some basic operations for manipulating these shapes. The details of their implementation may depend on the software, but the fundamental principles are always the same and they will be described later in this chapter. We will see, how these principles can be used in order to create visually-complex forms out of the basic solids.

The operations for manipulating three-dimensional models with CAD software can be basically divided into four groups:

(1) Congruence transformations changing the location and orientation of a model.

(2) Scaling transformations changing the proportions of a model.

(3) The so-called Boolean operations, combining several objects in a variety of ways.

(4) The operations loft, sweep, revolve and the creation of freeform-curves and freeform-surfaces.

14.3 Congruence Transformations

Let us first discuss the congruence transformations. In three dimensions, a congruence transformation has the same meaning as in two-dimensional geometry. Any translation, rotation, reflection or any composition of these operations, is called a congruence transformation. It is an operation that changes the location and the orientation of a body but does not change its shape. Figure 14.3 shows the application of a congruence transformation to a cube. The *translation* shifts a yellow cube on top of another cube without changing its orientation in space. In the second image, the yellow cube gets *rotated* about one of its vertices. Finally, the operation of *reflection* in a plane creates a mirror image behind the plane, as we see in the last picture of Figure 14.3.

To sum up, we can notice that we can use congruence transformations for placing and/or duplicating objects. Furthermore, we can combine transformations. If we, for example, uniformly rotate an object around an axis and at the same time uniformly translate it in the direction of the axis, we get a screw displacement.

We can easily construct more complex forms, using our basic forms by placing and transforming them cleverly. For example, we can create a rhombic dodecahedron — one of the Catalan solids — by placing six appropriate pyramids on a cube as shown in Figure 14.4(a).

If we choose half the edge-length of the cube as the heights of the pyramids, pairs of two neighboring faces of the pyramids form a rhombus. All in all, this yields to 12 rhombi — thus, we get a rhombic

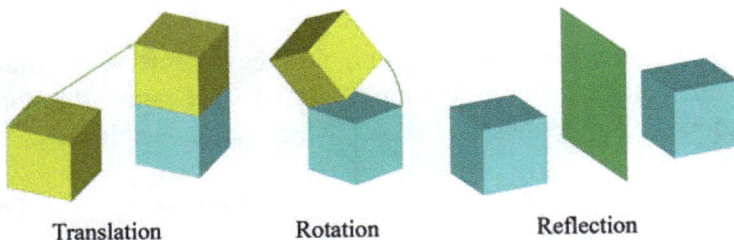

Translation   Rotation   Reflection

Fig. 14.3.   Congruence transformations.

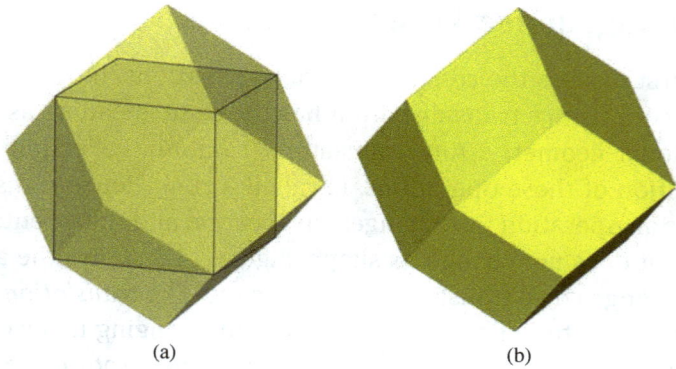

(a)   (b)

Fig. 14.4.   Creating a rhombic dodecahedron.

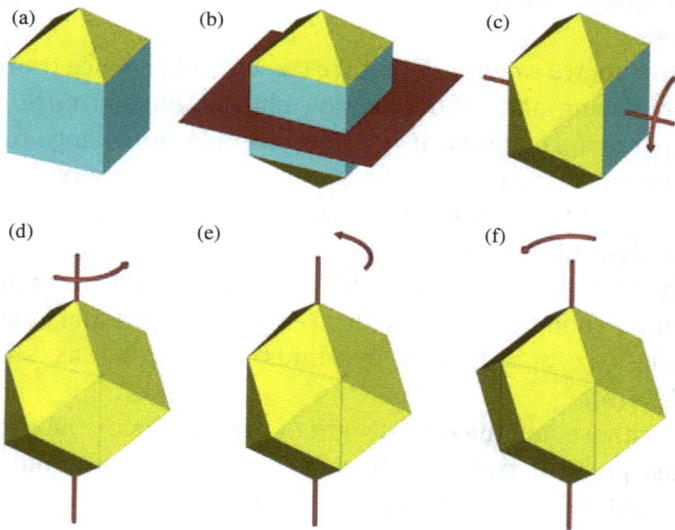

Fig. 14.5.   Application of congruence transformations.

dodecahedron. If we want to combine the cube and the pyramids to really one solid, as shown in Figure 14.4(b), we can use the Boolean operation *union* for the seven solids, which will be explained further below.

We will now clarify the procedure as to how we can construct a rhombic dodecahedron using congruence transformations, as shown in Figure 14.5. We can take a cube, and place a pyramid with half the height of the cube on one of its faces. Then we can use congruence transformations to generate five more pyramids with the same shape, standing on the other faces of the cube. There are many possibilities, as to how we can carry out this construction. We will describe one possible strategy for modeling the object. After placing the first pyramid on the top-face of the cube (see Figure 14.5(a)), we can reflect and copy it to get the opposite pyramid with the top of it facing downwards (Figure 14.5(b)). If we rotate (and copy) our starting pyramid by 90° around an appropriate axis, as shown in Figure 14.5(c), we get the pyramid standing on the front-face of the cube. We can now rotate and copy the obtained pyramid three times around the shown vertical axes to get the three missing pyramids (look at Figures 14.5(d)–14.5(f)).

14.4 Application: Modeling a "Shark Wheel"

Let's now immerse ourselves more deeply into the world of three-dimensional modeling by analyzing how to create a very special wheel virtually. One might even say that in 2015 the wheel has been reinvented by the American mathematician David Patrick, who accidentally discovered the *shark wheel*, which is shown in Figure 14.6.

Skateboarders found out that this new wheel displays many performance advantages compared to the traditional round wheel (a cylinder) — it enables one to have low rolling resistance, but also to have enough traction on soft terrains. We can make a virtual model of this new sort of wheel using the tools above.

After geometrically analyzing the wheels shown in Figure 14.6, we can find three red components of the same shape for each wheel. If we take our time and carry-on analyzing, we find out that each of the three red components consists of six quarter-tori, or colloquially: quarter-donuts. Now we know which basic forms we can use for modeling our wheel. The next question is how we can put the parts together correctly.

Let's first think about how to create one of the three red chain-links. We can build up a model by taking a quarter-torus and copying it by using the proper congruence transformations. Let's follow in

Fig. 14.6.   Skate board with shark wheels.

sequence the pictures shown in Figure 14.7. As a first step we create a quarter-torus, using the toolbox including the basic forms (picture 1). Then we reflect and copy it (picture 2) and after that we rotate the duplicated torus by an angle of 90° in the displayed direction (picture 3). Now, we are able to iterate the last two steps with the next torus. If we continue to reflect, copy and rotate, we get one chain link (picture 4). If we put a cube around the object, as shown in picture 5, we are able to intensify the impression of its geometrical structure. We add the space diagonal of the cube, which fits through the hole of our chain link (picture 6) and then we turn our object, until we look at it in direction of this diagonal (picture 7). We now can see, why this form could be used as a wheel — because it is *round*. We can approximately fit a cylinder, with the space diagonal as its axis, into the chain (picture 8), rotate the figure, translate and copy the chain link twice. What we get at the end is our shark wheel (picture 9).

If we want to make a model of a shark wheel, we just need two wooden rings, a saw and some glue. We cut the two rings in half and then again in half. Thus, we get eight quarter rings. We take six of them and glue them together as shown in Figure 14.8. If we give it a spin, we can enjoy watching this wheel roll!

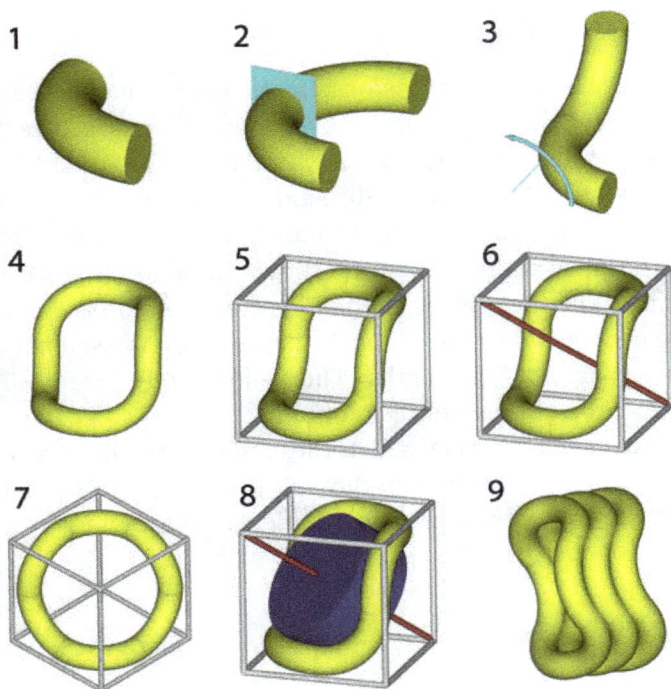

Fig. 14.7.   How to create a shark wheel.

Fig. 14.8.   Building a real model.

14.5 Scaling Transformations

Let's now discuss scaling transformations, which change the proportions of a model. Scaling an object by a specific factor is a very useful tool to develop new shapes from basic forms. In Figure 14.9, we can see how a cube is scaled in the direction of the red arrows. Here the scaling factor is 0.5, and a scaling factor of less than one means compression, because all lengths in the shown direction (Figure 14.9(a)) are multiplied by the scaling factor. This compression is symmetrical about the center of the cube. Thus the cube is compressed in the given direction by halving all lengths. The yellow body (Figure 14.9(b)) shows the shape of the resulting cuboid.

While scaling a figure by a factor less than 1 results in a compression of the figure in the chosen direction, scaling by a factor greater than 1 results in a stretching. Of course, scaling by factor 1 has no effect at all.

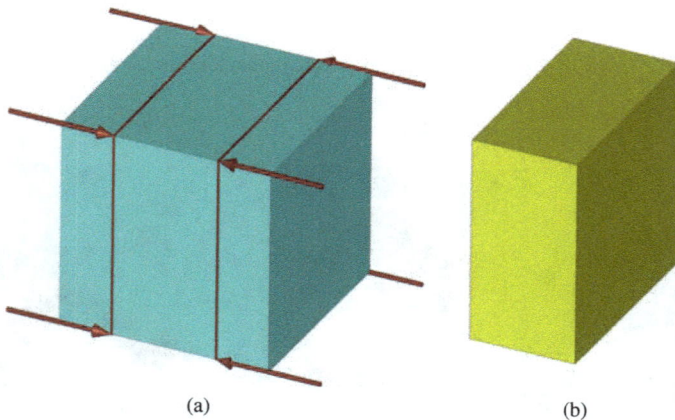

(a)   (b)

Fig. 14.9. Scaling a cube in one direction.

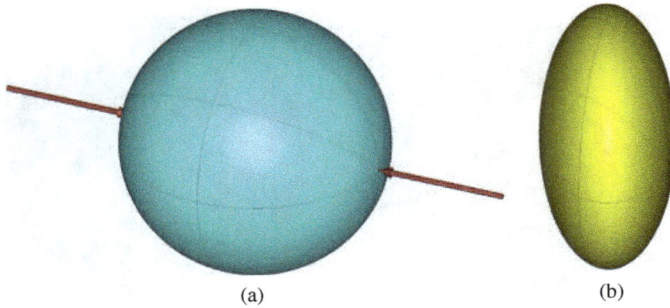

(a)   (b)

Fig. 14.10.   Scaling by a factor 0.5 turns the sphere into an ellipsoid.

In Figure 14.10, we show the effect of a scaling transformation on a sphere. This gives us a new kind of object, which is called an *ellipsoid*. Figure 14.10 shows the result of scaling a sphere by the factor 0.5 (which means a compression) in the direction of the red arrows. Again, all distances in the direction of the red arrows are getting compressed symmetrically to the center of the sphere.

Conversely, we can take the ellipsoid from Figure 14.10(b) and scale it by factor 2 in the same direction as before. This results in a stretching of the object, because the scaling factor is greater than 1. If the scaling factor is 2, this means that all distances in the chosen direction are getting doubled. In Figure 14.11, this procedure is applied to the ellipsoid obtained in Figure 14.10. All distances in the body that were previously halved are now doubled again. Hence, if we apply this transformation to the ellipsoid, it will bring back the sphere we originally started with.

If we scale an object with the same factor in all of the three axial directions *x*, *y*, and *z* of a given coordinate system, it will get smaller (if the factor is less than 1) or bigger (if the factor exceeds the number 1), but it will keep its shape. Thus, the tool *scaling* is very useful, if an object is too small or too large compared to another — we can blow it up or shrink it. We can also make objects narrower, if we scale them in direction of the two horizontal axes, which we will demonstrate later in this chapter by using the example of the Klein bottle (Figure 14.46).

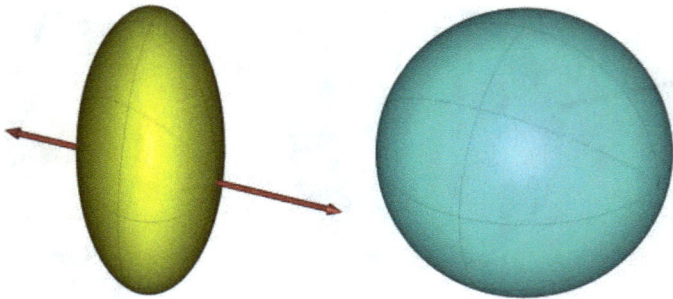

Fig. 14.11. Scaling the ellipsoid by a factor 2 in the direction of the arrows gives back the sphere.

14.6 Boolean Operations

Let's take a look at our third group of operations for the manipulation of three-dimensional models. The English mathematician, George Boole (1815–1864), laid the foundation for what we refer to as *Boolean Operations*. Boole's research in logic and in algebra led to results with important impacts on mathematics, physics and computer science. He invented his own algebra by formalizing logic into a symbolic language. In 1854, Boole published a monograph, *An Investigation of the Laws of Thought on Which are Founded the Mathematical Theories of Logic and Probabilities*, where he showed how the rules of logic can be represented by algebraic operations with symbols. For this reason, Boole invented an algebra where the symbols were limited to the values 0 and 1, so everything fit together. The usefulness of Boole's algebra in computer science is due to the use of the binary system in computer programming. Computers basically work on the two states of electric tension: *on* (1) or *off* (0). In geometry we apply Boole's operations to solids instead of logical statements: The use of the Boolean operations *union, subtract* and *intersect* allows us to create further forms.

Let us first explain how the Boolean operations work on simple, plane figures. We choose a triangle A and an overlapping circle B and interpret each of them as the set of points lying within their bounds, and we choose three sample points P, Q, R within those figures. (see Figure 14.12).

We can now find three different regions in this figure. One region is the left part of A, where the point P is located; one region, in the middle of the figure, where the point Q is located, and the region at the right side of B, where the point R is situated. Point P only belongs to the set A, point Q belongs to A and B and point R just belongs to B, as shown in Figure 14.12.

Now, let's apply the Boolean operations to our figure. First, we perform the addition (in this context we speak of *union*) of A and B. As we would naturally expect, this means to take all points of A and all points of B together. The result of this operation is shown in Figure 14.13(a). It is the set of all points that belong either to A, or to B or to both. Therefore, the points P, Q and R all belong to the union $A \cup B$.

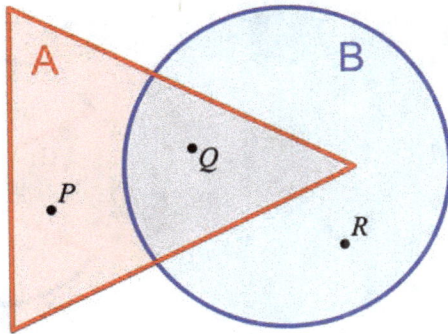

Fig. 14.12.   Two overlapping sets A and B.

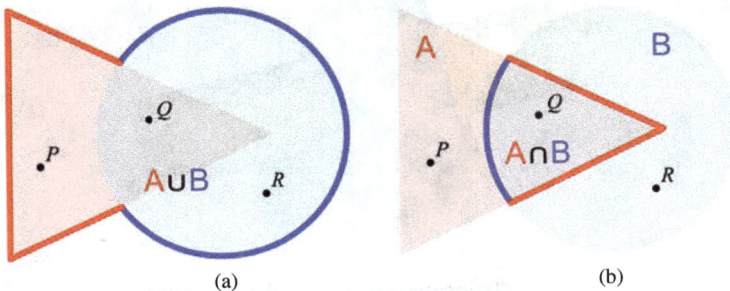

Fig. 14.13.   Union $A \cup B$ (a) and intersection $A \cap B$ (b) of two sets A and B.

The Boolean operation intersection leads to the result shown in Figure 14.13(b). The result of this operation, written in symbols A ∩ B, is the set of points, which belong to A and also belong to B. Q is the only one of the three points P, Q, and R, which belongs to this intersection.

The subtraction is a non-symmetrical operation. We know that from calculating with numbers (the calculation 3–2 has a different result than 2–3). Figure 14.14 shows the results for A\B (A minus B) in (a) and B\A (B minus A) in (b), respectively.

After our investigations of applying the Boolean operations on two-dimensional figures, we now want to study them as a very important tool for three-dimensional modeling. Analogous to the two-dimensional case, we interpret three-dimensional solids as sets of points in space. Let's start with a truncated cone (set A) and a cylinder (set B) as shown in Figure 14.15.

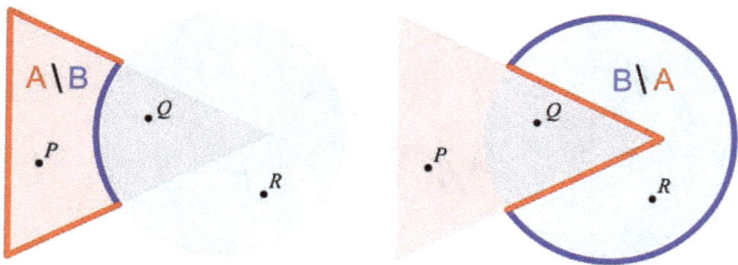

Fig. 14.14.   Subtraction of regions: A\B (a) and B\A (b).

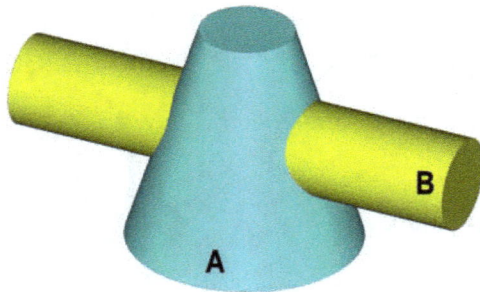

Fig. 14.15.   Truncated cone (a) and a cylinder (b).

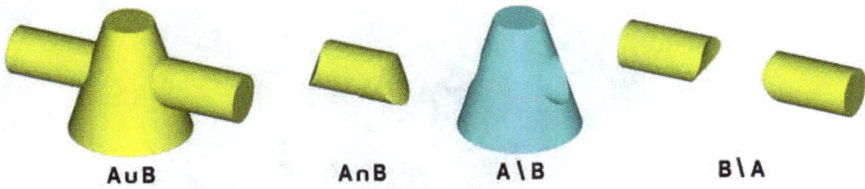

Fig. 14.16. Boolean operations: Union A ∪ B, intersection A ∩ B, subtraction A\B, and subtraction B\A.

Figure 14.16 shows the results of the Boolean operations applied on the two solids.

In the first picture in Figure 14.16 we see, what happens, when the two given solids get *united, forming the region* A ∪ B. All parts of both figures are stuck together. The second picture shows the result A ∩ B after *intersecting* the cylinder and the truncated cone. Just those parts of the figures remain, that belong to both of them. The result of the *subtraction* A\B in the third picture shows that the parts of the truncated cone, which also belonged to the cylinder, have been eliminated. In the last picture we see the result of the subtraction B\A, when we take those parts away from the cylinder, which also belong to the truncated cone. We can also notice that the subtraction is the only non-symmetrical Boolean operation. It can, for example, be used to make a hole in an object by subtracting a cylinder.

We have already seen an example for the use of the Boolean operation "union" above, when constructing a rhombic dodecahedron (Figure 14.4). To give an example for the power of the operation *intersect,* we use it for modeling the shape of the die with rounded vertices, which is shown in Figure 14.17.

Many (but not all) dice look like the one shown in Figure 14.17. If we want to create a virtual model, we have to analyze its form. It's a hybrid of a sphere and a cube. The sphere has its center in the center of the cube and touches all the edges of the cube. Let's now use the power of the operation *intersect.* The pictures in Figure 14.18 show, how this works. In Figure 14.18 (a), we can see the cube and the sphere stacked together. The operation intersect eliminates all parts of the cube, which do not also belong to the sphere (red parts in

Fig. 14.17.   Real die.

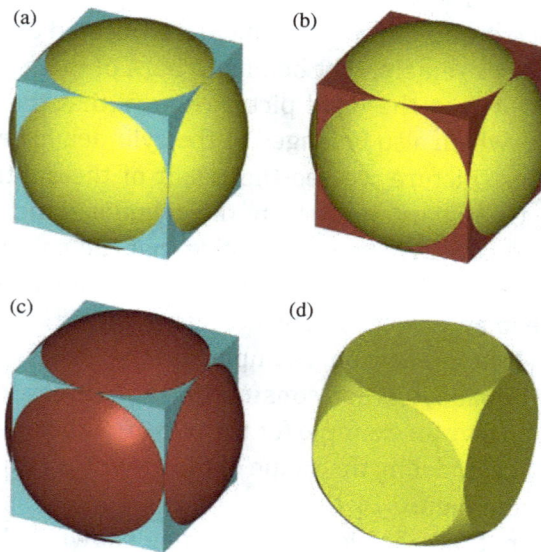

Fig. 14.18.   Modeling the form of a die.

Figure 14.18(b)) and furthermore it eliminates all parts of the sphere, which do not belong to the cube (red parts in Figure 14.18(c)). In Figure 14.18(d), we see, what is left after the elimination of all the red parts.

Fig. 14.19.   Adding the numbers.

Fig. 14.20.   Virtual die.

Now we have found the base-form of our die. We can add the numbers by cutting them out with spheres or cylinders, using the Boolean operation *subtract,* as shown in Figure 14.19.

In the same way we can generate the rest of the numbers to develop a proper die (Figure 14.20).

14.7 Application: Creating an Ambiguous Cylinder

Let's now use the Boolean operations for modeling something even more fascinating — let's make a so-called ambiguous cylinder, an object for which Kokichi Sugihara, a Japanese mathematician, won the

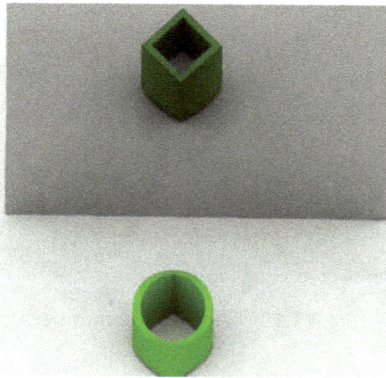

Fig. 14.21.   Ambiguous cylinder illusion: A "cylinder" and its image in a mirror.

second price of the 12th Best Illusion of the Year Contest 2016. Kokichi Sugihara's research areas involved the mathematics in visual perception and also included Voronoi diagrams (see Chapter 8). In Figure 14.21, we can see a virtual created picture of this illusion. The object in front of a mirror seems to be a (circular) cylinder, but its reflection in the mirror behind the object shows a square object, namely a prism. Because of its illusory appearance as two different objects the ambiguous cylinder got his name.

If you put an ambiguous cylinder in front of a mirror and you take the right perspective, you will see a cylinder, but its reflection appears to be a prism. How can we construct such a model? The first consideration is that we want to see a cylinder from our viewpoint and prism as its reflection. Figure 14.22 demonstrates the situation: the red lines in direction of the red arrows represent our looking direction. If we are looking at the object and at the reflection of it from a larger distance, we can assume the lines of sight as parallel. Considering the law of reflection (angle of reflection = angle of incidence), we have to start with an object like that shown in Figure 14.23.

If we are looking at the cylinder, we should also see a prism, if we are looking into the mirror (Figure 14.24(a)). Our object should be both cylinder *and* prism. How such an object may appear, is shown in Figure 14.24(b), which shows the same object as in Figure 14.24(a).

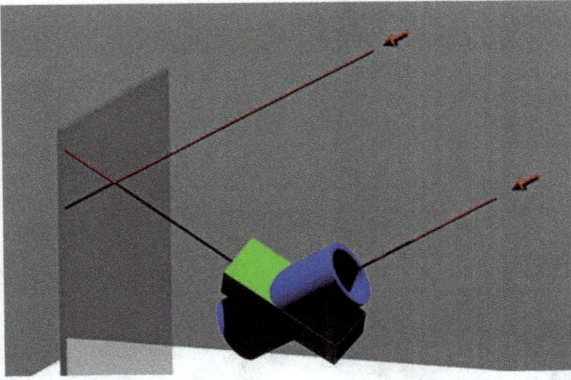

Fig. 14.22.   Looking at the object and into the mirror.

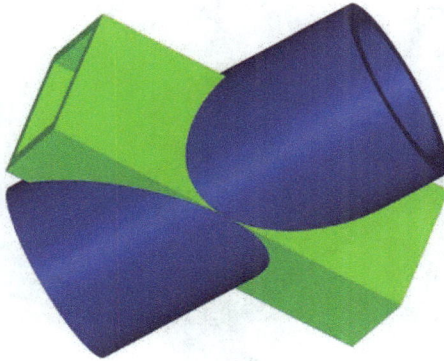

Fig. 14.23.   Intersection.

Figure 14.22 shows that the constellation of the cylinder and the prism in Figure 14.23 allows us to see the top of a cylinder (a circle) and furthermore the top of a prism (a square) in its reflection. The geometric Boolean operation for the logical *and* is the *intersection* (see Figure 14.25). We want to see the circle *and* the square from our viewpoint, thus we can intersect the two objects from Figure 14.23, because we do not need the rest of the two objects.

If we look at it from above, we can see some kinds of waterdrops and the symbol of infinity (Figure 14.26).

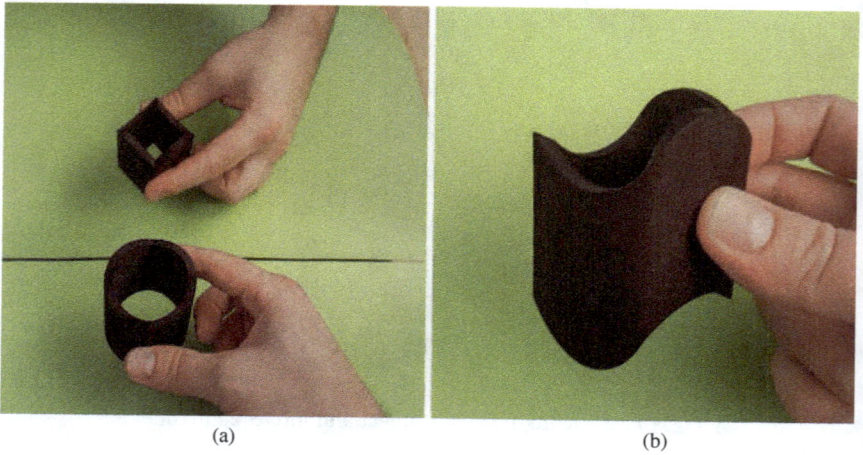

(a)   (b)

Fig. 14.24.   An ambiguous cylinder and its reflection, and the same object, shown from another perspective.

Fig. 14.25.   Object after intersection.

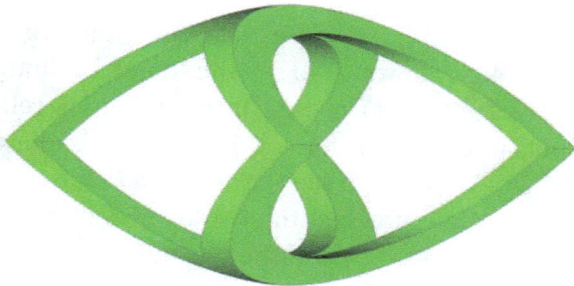

Fig. 14.26.   Interesting view.

What a nice result! But weren't we searching for an ambiguous cylinder?

If we put the object, shown in Figures 14.25 and 14.26, in front of a mirror (Figure 14.27), we can see that we have made our first step to success — now we have to take care of the rest of our object.

We can build a cylinder around our figure by *sweeping* (this tool will be explained a bit later) a line along the shown red curves (which are elliptic arcs), as we see in Figure 14.28.

After that we thicken the cylinder (Figure 14.29). *Thickening* means, in very simple terms, to make a solid wall out of a surface. CAD software programs offer a tool where we can choose the direction (in/out) and the thickness of the wall.

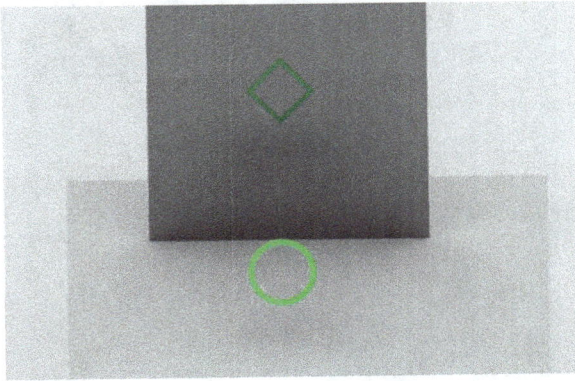

Fig. 14.27.   Circle and square.

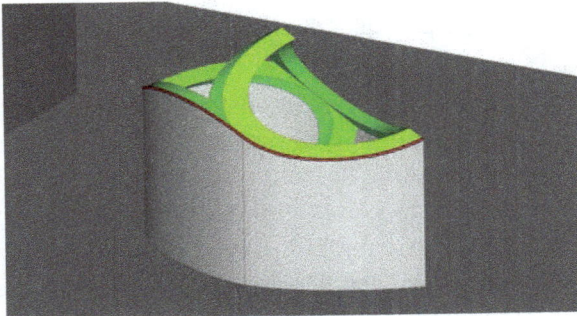

Fig. 14.28.   Building the cylinder.

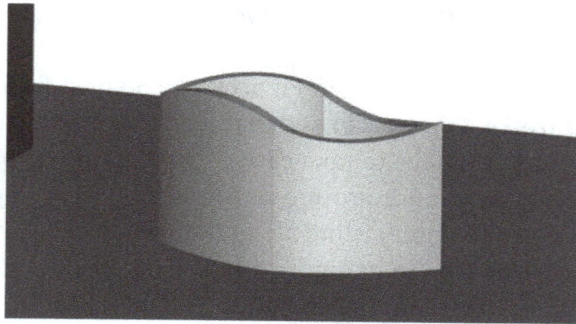

Fig. 14.29.   Thicken the wall.

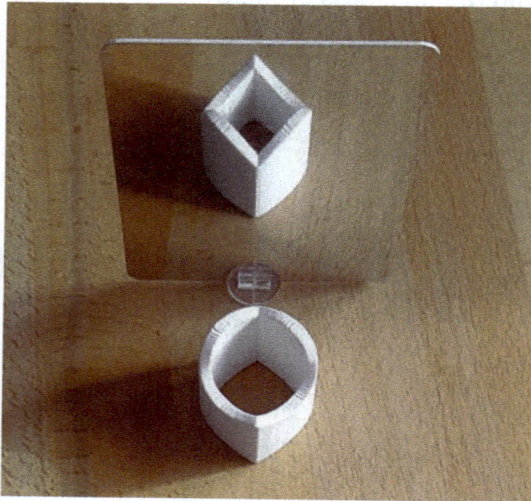

Fig. 14.30.   Illusion in real-life.

If we have access to a three-dimensional printer, we can test our object in real-life. Figure 14.30 shows a three-dimensional printed ambiguous cylinder standing in front of a mirror.

The illusion of the ambiguous cylinder works better, if the bottom of the object is not level and flat but curved in exactly the same way as on the top — to get this object, we can cut it parallel to the boundary on the top. Unfortunately, then the cylinder drops in real life and so we have to hold it (Figure 14.24). Figure 14.31 shows, what the object

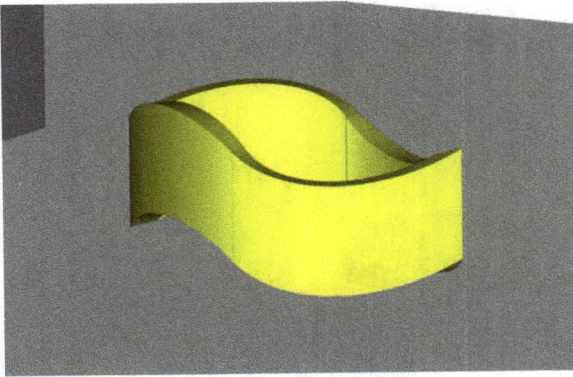

Fig. 14.31.   Ambiguous cylinder with cut bottom.

Fig. 14.32.   Virtual illusion.

looks like and then looking at Figure 14.32, we can observe, how the illusion works (virtually).

14.8 Loft, Sweep, and Revolve

Let's briefly discuss three further important types of modeling-strategies offered by CAD software: *loft*, *sweep* and *revolve*. *Loft* creates a connecting surface between two given profiles. In Figure 14.33(a), we see two surfaces (a circular disk and a hexagon),

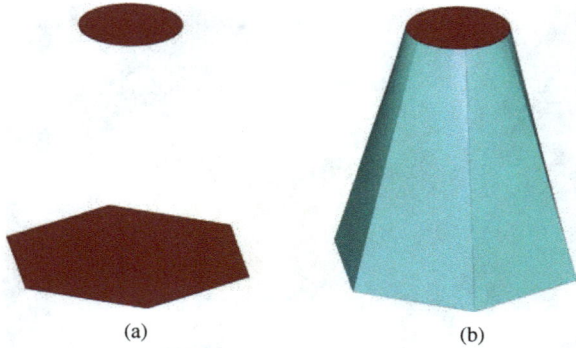

(a)   (b)

Fig. 14.33.   Loft.

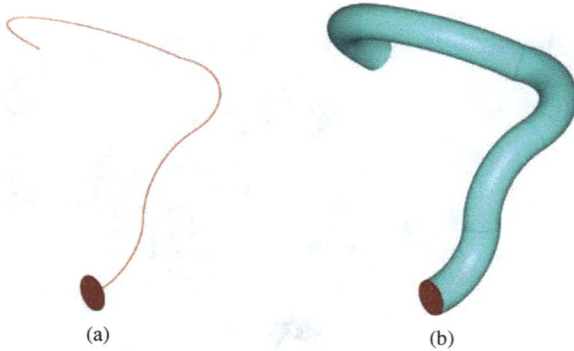

(a)   (b)

Fig. 14.34.   Sweep.

and in Figure 14.33(b), we see the computer-generated surface that provides a smooth transition between the two shapes.

The operation *sweep* enables us to create a new surface by moving a certain profile along a given path. Figure 14.34(a) shows a curve in space and a circular disk. We can imagine that the space curve is realized by a piece of bent wire. Let us further imagine that the disc has a hole in the middle and that it is threaded onto this wire. When we move this disc along the wire, it creates a figure in space, a "hose" that exactly follows the path of the bent wire. This is exactly what the sweep operation does. It creates the blue "hose" shown in Figure 14.34(b).

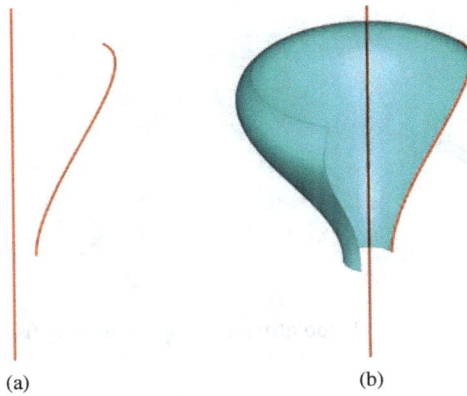

(a) (b)

Fig. 14.35.   Revolve.

Using the operation *revolve* we can easily create a surface of revolution. Figure 14.35(a) shows a vertical axis of rotation and a curve that serves as a cross-sectional profile. Figure 14.35(b) shows the surface obtained by rotating the profile through an angle of 180° around the given axis.

Without getting into too much detail, we also have to mention the important modeling-tools for creating *freeform curves* and *freeform surfaces*. They can be used to generate more abstract forms. Roughly speaking, there exist two types of freeform curves, called *interpolating* and *approximating* freeform curves, respectively. These two types are shown as the red lines in Figure 14.36.

Freeform curves are determined by so-called *control polygons*, shown as blue polygonal lines in Figure 14.36. The shape of the freeform curves can be manipulated by varying the vertices (*control points*) of the control polygon. The two types of freeform curves are distinguished by the way they are determined by the control points. Interpolating freeform curves pass through the given set of control points, as shown in Figure 14.36(a). Approximating curves do not pass through the control points, (except through the start- and endpoint) but approximate the control polygon, as shown in Figure 14.36(b).

Freeform curves have their origin in the field of computer-aided design. A designer is well supported by the great variety of appealing

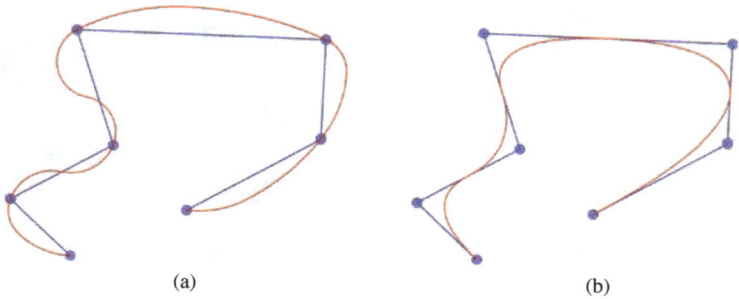

(a)   (b)

Fig. 14.36.   Interpolating and approximating freeform curves.

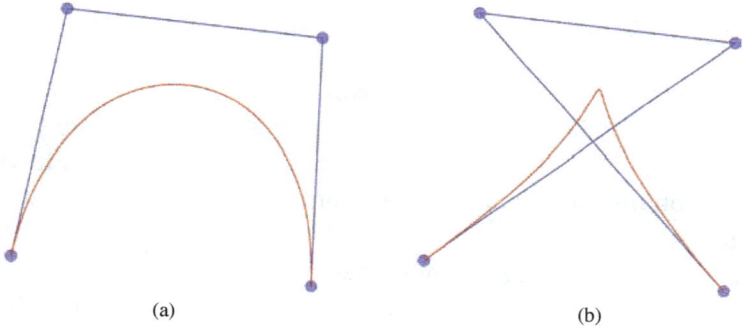

(a)   (b)

Fig. 14.37.   Two Bézier-curves.

shapes that can be easily created just by manipulating a small set of control points. The French engineers Pierre Bézier (1910–1999), who worked for Renault, and Paul de Faget de Casteljau (1930), who worked for Citroën, discovered approximating freeform curves and surfaces in the 1960s. Mathematically, freeform curves were easy to describe and they were first applied to create and describe curved shapes for the automobile industry. Today, these curves are called *Bézier curves,* and the algorithm to calculate them is called *the algorithm of De Casteljau.*

In Figure 14.37, we see two Bézier curves (red) fitting the given control polygons. Each of them is defined by four control points (blue). Bézier curves can be easily described mathematically with polynomials. Without going into further depth, we can note that the degree of these polynomials is always one number lower than the number of control points. Thus, we can assume the two Bézier curves, shown in Figure 14.37, to be of degree 3.

A disadvantage of Bézier curves is, that if we vary one point, the whole curve changes its shape. Therefore *B-spline* curves got invented. They consist of Bézier curves (of lower degree), which fit together with smooth transitions. Figure 14.38 shows a B-Spline curve of degree 2, consisting of Bézier curves of degree 2 (Figure 14.38(a)) and the Bézier curve of degree 3 (Figure 14.38(b)), taken from Figure 14.37.

We can manipulate the shape of a freeform curve just by moving the control points. Figure 14.39 shows, how the curves are changed (from the red curve to the black curve) if just one of the control points

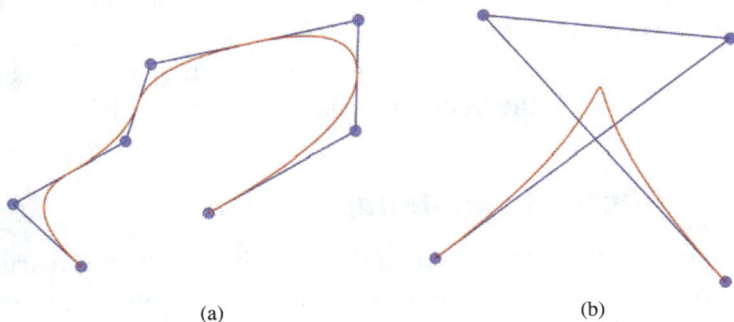

(a)   (b)

Fig. 14.38.   Freeform curves.

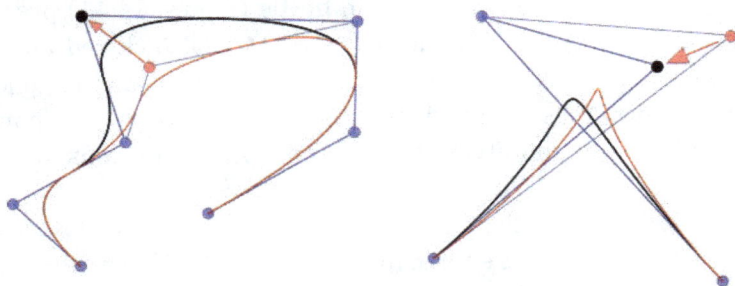

Fig. 14.39.   Variation of freeform curves.

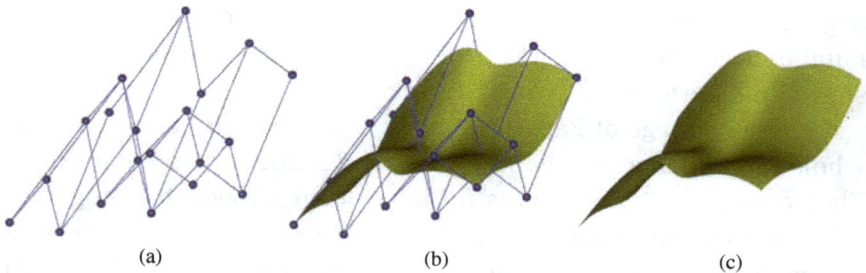

(a)   (b)   (c)

Fig. 14.40. Creating a freeform surface.

is shifted from the red position to the black position, as indicated by the arrow.

The idea of generating freeform surfaces is the same as for freeform curves, but in the former case we start with a three-dimensional net of polygons. This net of polygons is used to create and manipulate the freeform surface in three dimensions. In Figure 14.40, we can see a net of polygons on the left side (a), the net and the fitting freeform surface in the middle (b) and the freeform surface itself on the right side (c).

14.9 Application: Modeling a Klein Bottle

We will now demonstrate some of the modeling strategies discussed in the previous sections to build something very fascinating: a virtual model of the Klein bottle, the one-sided surface invented by the German mathematician Felix Klein (1849–1925). We have already discussed the Klein bottle in detail in Section 10.5.

If we analyze the form of a Klein bottle (Figure 14.41) precisely, we find the two curves shown in Figure 14.42 (thick red and blue curves), which determine the object. We can, for example, generate two B-Spline curves of third degree approximating the shape of the Klein bottle by modifying the related control polygons, until the curves fit.

Now we can rotate the red curve around the black axis (we use the modeling-tool *revolve*, as shown in Figure 14.35) and use the function *sweep* by moving a circle along the blue curve (compare this to Figure 14.34). When we embark on this path, we need to assure that

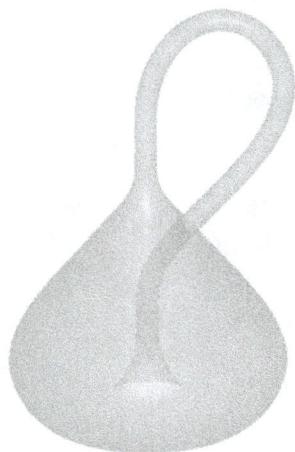

Fig. 14.41.   Virtual model of the Klein bottle.

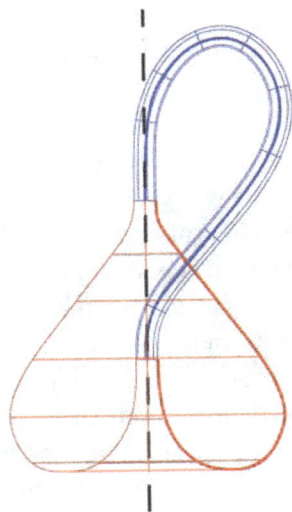

Fig. 14.42.   Two determining freeform curves.

the transition between the two surfaces is tangential, and also that the distance between the endpoints of the curves is equal to the radius of the circle used for sweeping. In Figure 14.43, we see the two sections circled, where we have to connect the two surfaces.

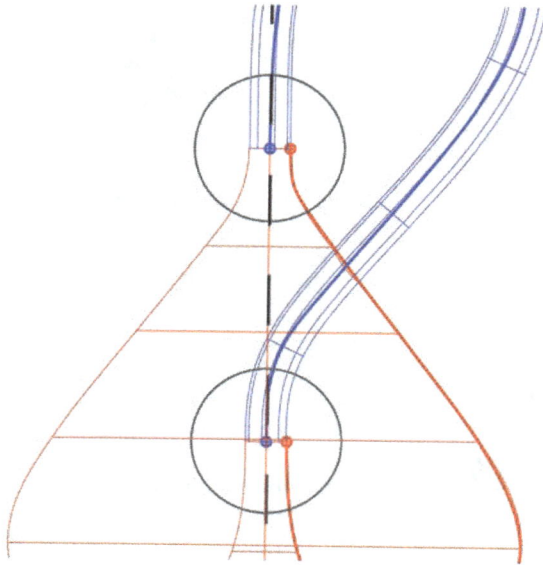

Fig. 14.43.   Transition between the two surfaces.

We make a hole into the red surface by *trimming* (a tool used to manipulate surfaces, which works similarly to the manipulations of solids by using the Boolean operations) it with the blue one. Essentially, we cut a hole from the red surface using the blue one as cutting tool. Figure 14.44 shows the surfaces after trimming. Finally, we join the two surfaces and get the Klein bottle shown in Figure 14.41.

If we would like our bottle to be narrower, we can scale it in the directions of the two horizontal axes by a factor less than 1 (Figure 14.45). We could also scale it just in the direction of the vertical axis — by a factor greater than 1 (this would lead to a larger model).

When we finally created our virtual three-dimensional model, we can choose different formats for exporting our object. We can export two-dimensional files or three-dimensional files. Before deciding on which type of file we need, we first must think about how — and with which kind of machine — we want to produce our model.

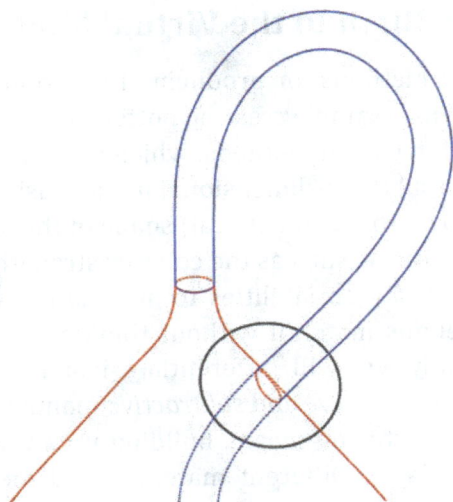

Fig. 14.44.   Rotate and sweep.

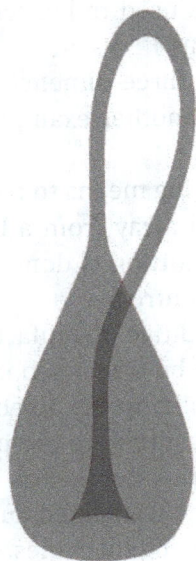

Fig. 14.45.   Narrow Klein bottle.

14.10 Giving Birth to the Virtual Model

There exist many methods for producing three-dimensional models in our real world, for example creating pottery, folding paper or bending wood. We will focus on methods, which allow us to produce real models out of virtual three-dimensional models using machines. Our focus will depend on the (geometrical) shape of the object and not on the physical appearance, such as the color or strength of it. For example, we will search for possibilities to automatically produce parts made of homogenous material without thinking about how to join them. Consequently, we will differentiate just between two basic types of production: *additive* and *subtractive* manufacturing.

Additive manufacturing means building up an object by adding material layer by layer. Different materials and methods are used depending on the costs, physical requirements of the object or the required necessary precision. Examples of current additive manufacturing methods are three-dimensional printing (a three-dimensional printer works basically like an automatic hot glue gun) and VAT Polymerization (UV light is used to harden the object layer by layer out of a photopolymer resin).

If we make a simple three-dimensional sketch with a three-dimensional pen, we have another example for additive manufacturing, see Figure 14.46.

Subtractive manufacturing means to construct three-dimensional models by cutting material away from a block (such as of wood or aluminum). Normally this cutting is done by CNC- machines (CNC = Computerized Numerical Control).

If we want to use additive manufacturing, we need a three-dimensional file, which can be translated into a map of the paths the machine has to follow to build up the object. If we want our object to be cut out, we need a file containing the paths for the milling head. Depending on the software, there exist different types of three-dimensional files that can be translated automatically into the paths for both types of manufacturing machines.

A kind of mixture of additive and subtractive manufacturing would be to slice the object (software can do this with the

Fig. 14.46.   Simple three-dimensional sketch.

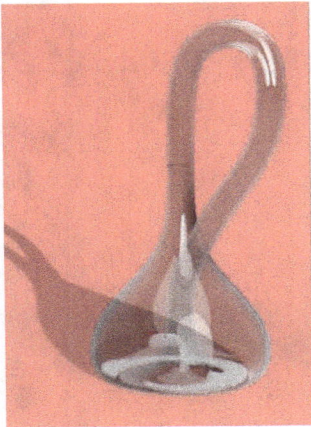

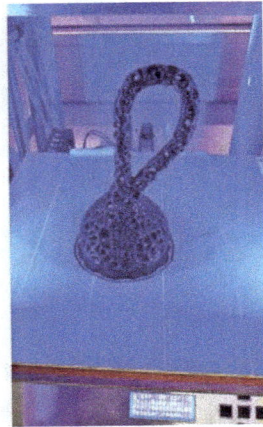

(a)   (b)   (c)

Fig. 14.47.   Virtual Klein Bottle, Voronoi-tessellation, and the real model in a three-dimensional printer.

virtual model), to cut the slices with a laser-cutter, and to subsequently assemble the slices. In this case, we just need a two-dimensional file of the parts of our model to instruct the laser-cutter.

We can see an example of a virtually generated model of a Klein-bottle shown in Figure 14.47(a) and a picture of it after a

Voronoi-tessellation of its surface in Figure 14.47(b). For the realization of a Klein bottle by means of three-dimensional printing (Figure 14.47(c)) we benefit from the Voronoi-tessellation, because we can look inside the model, even if the material is not transparent.

In this chapter, we have obtained an overview about fundamental principles of designing geometric three-dimensional models. We have seen, how we can use those principles for virtually generating some fascinating objects, such as Shark wheels, an ambiguous cylinder and a Klein bottle. Additionally, we have discussed different ways of giving birth to virtual models.

Chapter 15

A View of a World with More Than Three Dimensions

Four-dimensional "hyperspace" is a well-known motif in science-fiction literature. The general idea is that one could take a shortcut through hyperspace to quickly cover large distances in normal space, and thus, be able to move faster than light. As exotic and esoteric as the idea of hyperspace may seem, higher-dimensional spaces are nothing special in mathematics, where they even represent a very common topic. Multi-dimensional or even infinite-dimensional spaces are taught at an early stage in advanced mathematics courses at universities. The reason is that even in the description of very simple, everyday systems, more than three dimensions are needed to capture the variety of possible behaviors.

15.1 Higher-Dimensional Spaces in Mathematics

What does the dimension of a physical system actually mean? It is the number of its degrees of freedom, and is best described by how many independent quantities are required to uniquely determine its status. Let us consider, as an example, a very simple "system", such as a single point object on a plane, and we only want to describe its location. We typically do this with respect to a coordinate system, that is, we introduce an x-axis and a y-axis, orthogonal (perpendicular) to each other, and describe the location of the point by two coordinates, x and y (see Figure 1.6). The two coordinates can be varied independently, and this is expressed by the statement, that two-dimensions are required for the location of a point in the plane. In a similar way, we would need three independent coordinates to specify the location of a point in space (see Figure 1.8). A point in space, therefore, has three degrees of freedom, and the set of all its possible locations (x, y, z) is three-dimensional.

Now, as a next, somewhat more complex, system we can describe the configuration of two points in a plane. We would need two numbers (x_1, y_1) to describe the location of the first point and two other numbers, say (x_2, y_2), to describe the location of the second point. Hence, any configuration of the two-point system in the plane would require the specification of four coordinates, (x_1, y_1, x_2, y_2), and each of these four coordinates can be varied independently (see Figure 15.1). Therefore, we say that the set of all possible configurations of two points in a plane forms a four-dimensional space. The location of a point in this four-dimensional space is given by the coordinates (x_1, y_1, x_2, y_2), which describe a unique configuration of *two* points in the two-dimensional plane. In this mathematical sense, the word "space" simply means the set of all possible lists of four of coordinates, that is,

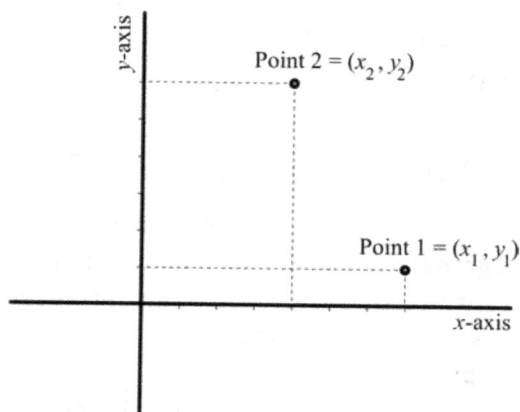

Fig. 15.1.   Four coordinates (x_1, y_1, x_2, y_2) describe a configuration of two points.

a set of "points" for which four coordinates are necessary to describe them.

Now, the path to further generalizations is obvious. For example, the combined positions of two points in *three*-dimensional space is given by six coordinates, since each of the two points requires three coordinates to be described. Hence, the set of all the configurations of a two-point system forms a six-dimensional space.

15.2 The Physics of Space–Time

Very often it is claimed that time represents the fourth dimension. Indeed, if one wants to describe not only the location of an event (e.g., taking a photo), but also the time this event took place, one would need a total of four numbers. Three numbers describe the spatial coordinates (x, y, z) and another number, t, is needed to describe the time of the event. Events are, thus, characterized by four independent coordinates which are considered as the coordinates of a point (x, y, z, t) in a four-dimensional space. This space is commonly called the "space–time continuum", or Minkowski space, named after the German mathematician Hermann Minkowski (1864–1909).

One might think that combining a point in space with a point in time into a four-dimensional entity is just a mathematical trick. In physics, however, Minkowski space is actually the very basis of the theory of relativity. It is a fundamental discovery of the German-born physicist Albert Einstein (1879–1955) that any change in the velocity of an observer has to be described by a kind of rotation in space–time, where space and time coordinates can merge into each other. Space- and time-coordinates are, therefore, closely linked and must always be considered together. In 1908, Minkowski stated that "Henceforth, space by itself, and time by itself, are doomed to fade away into mere shadows, and only a kind of union of the two will preserve an independent reality."[1] In physics, the temporal dimension is, thus, treated in a similar way as the spatial dimensions. Einstein's

---

[1] Minkowski, H. (1908). *Raum und Zeit*. Lecture given at the 80th Naturforschertagung in Köln.

general theory of relativity even identifies gravity as a phenomenon related to the geometry of the space–time continuum. By Einstein's theory, however, the notion of distance between points (the metric) as well as the notion of orthogonality of two directions in space–time is quite different from what we would expect by analogy from our familiar three-dimensional space. And that's not all. Modern physical theories such as superstring theory require even more dimensions — these are hypothetical extra dimensions being rolled up into tiny closed loops that are not observable under normal circumstances. Therefore, whatever the nature of hyperspace might be according to physics, it is of limited use for what we are going to demonstrate in this chapter. We will, however, have an opportunity to discuss Einstein's cosmological model of three-dimensional space towards the end of the chapter.

15.3 Geometry in Four Dimensions

Let us take a glimpse of a geometry in four dimensions, where the fourth dimension can be considered as an independent *spatial* direction, perpendicular to the other three directions. Unfortunately, we cannot really imagine a four-dimensional space in this sense, because our imagination is firmly rooted in our three-dimensional world. However, we can recognize many aspects of four-dimensional geometry by thinking in terms of analogies to three-dimensional geometry.

In three-dimensions, we can move left–right, back–forth, and up–down. For the fourth dimension, we need a new direction, which is perpendicular to all the three other directions, in the same sense, as up–down is perpendicular to both the directions left–right and back–forth. We cannot point in that new direction, because we are confined in a three-dimensional world. Just to have a name for it, the British mathematician and eccentric Charles H. Hinton (1853–1907) suggested to use the ancient Greek words "ana" (meaning "up to") and "kata" (meaning "down from") as an additional pair of words to describe the motion along the fourth axis.

The British author Edwin A. Abbott (1838–1926) (whose middle initial strangely also stands for "Abbott") wrote the book entitled *Flatland*, where he tried to demonstrate higher-dimensional thinking in analogies and wrapped it all up in an interesting story. From the perspective of a hypothetical being confined to a two-dimensional world, our third dimension must be in the same sense incomprehensible as for us is the fourth dimension. A flatlander can move left and right or back and forth, but up and down would be unknown and unthinkable.

15.4 Thinking in Analogies

A lot can be learned from Abbott's idea of thinking in analogies. For example, a one-dimensional straight line can separate Flatland into two halves, which we denote by "Leftland" and "Rightland" in Figure 15.2(a). Thinking of this line as a barrier, there is no possibility for a flatlander to go from one side to the other without breaking through the line. In the same way, a two-dimensional plane would divide our 3-space into two separate parts, one part "above" the plane, and one part "below". We denote these two regions of space by "Upland" and "Downland" in Figure 15.2(b). By analogy, we conclude that our three-dimensional space divides the four-dimensional hyperspace into two separate parts, one part is "ana" and the other is "kata" from our world.

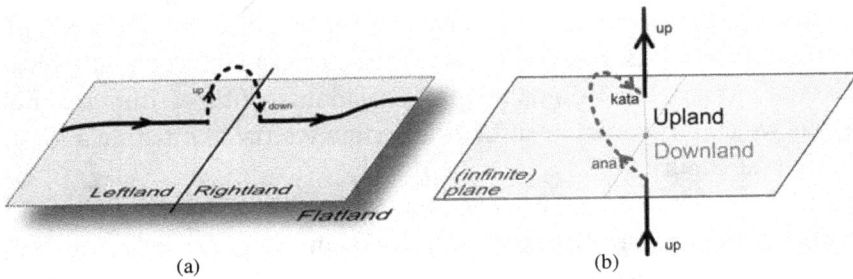

Fig. 15.2.   Dividing Flatland by a line and three-dimensional space by a plane.

Figure 15.2(a) shows a method of surmounting the barrier separating Leftland from Rightland. One can get to the other side of that line simply by using the third dimension, that is, by going "up" where there is no obstacle, and then returning "down" to Flatland on the other side of the line. Clearly, this would be impossible and would seem like a miracle for any being confined to Flatland. By analogy, a four-dimensional being could always go on the other side of a barrier in 3-space, just by going around the obstacle in the fourth dimension, that is, by going temporarily a little bit "ana" from our space, as indicated in Figure 15.2(b). A two-dimensional plane cannot separate hyperspace, no more than a line can separate our 3-space into two parts.

By analogous reasoning, any closed line (for example, a circle or a square) in Flatland has an inside and an outside. From within Flatland, it is impossible to get to the inside without breaking through the boundary line, but you always could "dive in" from the third dimension, that is, from above or below. Likewise, a closed room in our world could be entered without difficulty from the fourth dimension, that is, from ana or kata. The inside of a closed box might be inaccessible to us, but it would be open and visible for a four-dimensional being in very much the same way, as the interior of a square is open and visible for us.

15.5 Möbius strip in Flatland

Recall the Möbius strip, a one-sided surface that is obtained by gluing a rectangular strip of paper into a ring with a half-turn (see Figure 10.2). A Möbius strip needs three-dimensional space for its realization and it is, therefore, very difficult for a Flatland resident to correctly imagine a Möbius strip. As residents of three-dimensional space, we run into similar difficulties when we try to imagine a Klein bottle (see Figure 10.23).

We could try to explain a Möbius strip to a resident of Flatland in the following way. Consider a rectangular strip as in Figure 15.3, diagram 1. You may consider the rectangle to consist of an elastic material that can be bent and stretched and squeezed at will. Now, a Möbius strip is created by gluing the two ends of the strip together in such a

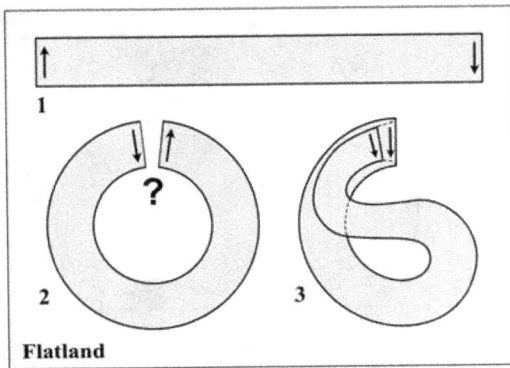

Fig. 15.3.   Flatland-analog of a Klein bottle — the Möbius strip in two dimensions.

way that the two arrows can perfectly overlap each other — in the same direction. At a moment of reflection, the resident of Flatland will realize that it is obviously impossible. If we bend the strip to form a ring, the two arrows will always be in opposite directions (see Figure 15.3, diagram 2). But if we allow the strip to penetrate itself, we can pass one end of the strip through the side wall and, thus, join it from "behind" with the other end, as shown in Figure 15.3, diagram 3.

Actually, there is no way to represent a Möbius strip in two-dimensions without self-intersection; so, Figure 15.3, diagram 3 is perhaps the best we can do to explain a Möbius strip to a resident of Flatland. In three dimensions, of course, the side of the strip is not a barrier and can be avoided by lifting the strip a little out of Flatland before joining the ends, as shown in Figure 15.4. With the help of a third dimension, a Möbius strip can be created as a smooth closed strip without any self-intersection.

In the same way, we can understand that a Klein bottle can be created without self-intersection by making use of an additional fourth dimension. In Chapter 10, we saw how we could join the two ends of a tube with its ends in opposite orientation. We found that this was impossible without letting the surface penetrate itself, as we can see in Figure 15.5. But if we embed the Klein bottle in four-dimensional space, we just lift the tube, before it penetrates its own side wall, a little bit in the fourth dimension. That is, in Figure 15.5, we bend the

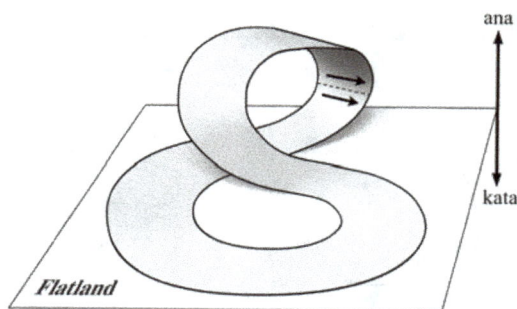

Fig. 15.4. A Möbius strip can be created without self-intersection by making use of an additional dimension.

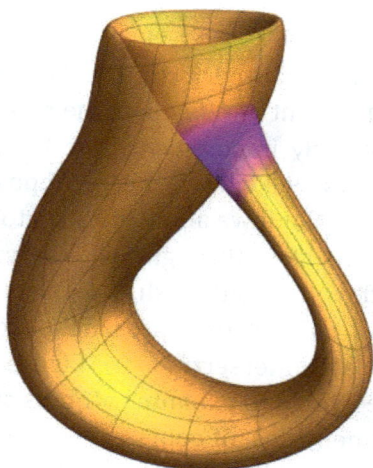

Fig. 15.5. Klein bottle — a one-sided surface with self-intersection.

violet part of the Klein-bottle in the ana- or kata-direction, just a little to avoid the self-intersection. This gives a smooth surface without any self-penetration in four dimensions. Only its three-dimensional projected image would appear to intersect itself.

15.6 A Four-Dimensional Shape: The Hypercube

Simple geometric figures can also be defined in four-dimensional space. But how can we try to imagine a four-dimensional solid?

We could never take it in hand to feel its shape and understand it as we would with a shape in three dimensions. Again, we have to make use of the analogy to the lower-dimensional world of our everyday experience. In order to learn how to represent a four-dimensional object in our three-dimensional world, we can use what we experienced in Chapter 3 about the representation of three-dimensional objects on a two-dimensional drawing plane.

Consider, for example, the image of a cube on a two-dimensional page of this book, as in Figure 3.2. The third dimension is typically represented by drawing lines at an angle to the existing lines of a two-dimensional object. Here, we start with a square, the analog of a cube in two dimensions. At each vertex of the square two mutually-perpendicular edges meet. Such a vertex becomes the corner of a cube through a third edge that is perpendicular to the other two edges. In a two-dimensional representation, this third edge will be an oblique line. And this is, in fact, the easiest way to draw a three-dimensional cube: Just draw two squares slightly offset but parallel to each other and connect their vertices through oblique lines, as shown in Figure 15.6. The oblique lines are all parallel and have the same length. The resulting image is the parallel projection of a three-dimensional cube onto a two-dimensional plane, as described in Chapter 3.

The three-dimensional cube has eight vertices, twice as many as the two-dimensional square. The number of its edges is 12. (Two times four for the two squares and the four oblique line segments connecting the squares.) The faces of the cube are two-dimensional squares, and there are six of them.

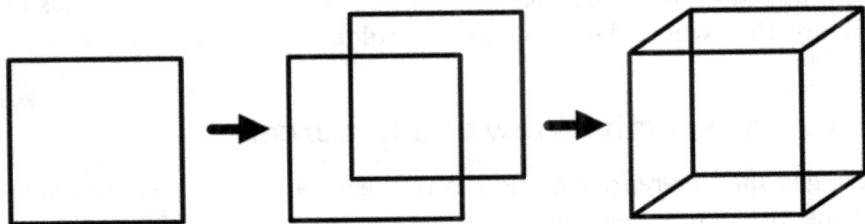

Fig. 15.6.   Two-dimensional representation of a three-dimensional cube.

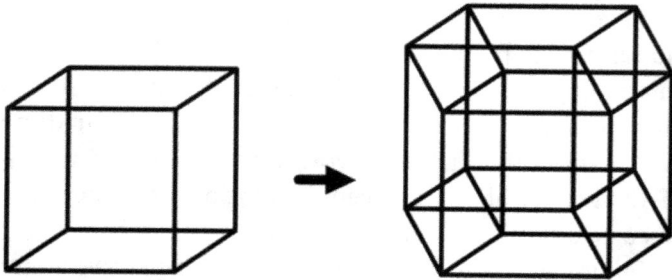

Fig. 15.7.   Representation of a four-dimensional cube.

Now let us proceed by analogy and create an image of a four-dimensional "hypercube", also called a *tesseract*. We start with a cube in three dimensions. Let's draw parallel lines in an oblique direction connecting vertices of our cube with the vertices of another slightly-offset cube, as in Figure 15.7. We, thus, obtain the three-dimensional image of a four-dimensional cube (and, of course, we had to display this three-dimensional object two-dimensionally on our book page).

The four-dimensional hypercube has twice as many vertices as the three-dimensional cube that we started with. Thus, it has 16 vertices. The number of edges is easy to determine. It is the number of edges of the two three-dimensional cubes (12 + 12 = 24) plus the number of oblique lines (8) that join the adjacent corners of the two cubes. Thus, we obtain 32 edges.

A square in two dimensions is composed of four one-dimensional sides. A cube in three dimensions is composed of six two-dimensional squares, which form its surface. A hypercube in four dimensions consists of eight three-dimensional cubes. We can see parallel projections of these cubes in Figure 15.7. It might be easier to see these cubes in Figure 15.8, where they are drawn in color.

15.7 Perspective View of a Hypercube

We get another representation of the tesseract, if we form the analogue of a two-dimensional perspective view of a three-dimensional cube. (We described this in Chapter 3, for example, Figure 3.13.)

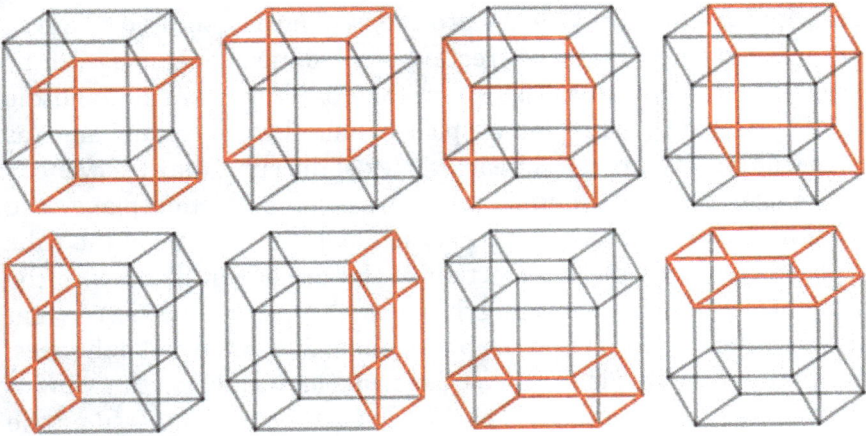

Fig. 15.8. "Surface-cubes" of a hypercube.

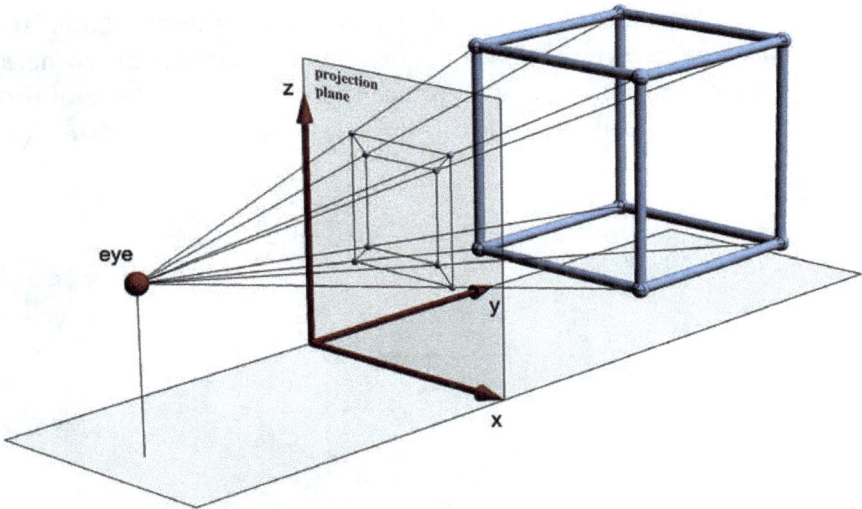

Fig. 15.9. Creating a two-dimensional perspective view of a three-dimensional cube.

We consider first the situation in three dimensions, as shown in Figure 15.9.

The observer looks with one eye from a fixed position through a "window" (called the *projection plane*) at the object behind the window. In Figure 15.9, the projection plane is indicated by a gray square. It is here, where we create the two-dimensional image of the three-dimensional cube. The eye of the observer is in front and the cube is located in the region of space behind the projection plane (as seen from the observer). For each vertex of the cube, the observer can now mark the corresponding position on the projection plane that, from his point of view, coincides exactly with the vertex of the cube (that is, he marks the intersection of the projection plane with the straight line of sight connecting the eye with the cube's vertex). In that way, the observer draws on the projection plane exactly what he sees. For the situation in Figure 15.9, this results in the perspective view of the cube, as shown in Figure 15.10. This is the image created on the projection plane.

Since the projection plane is parallel to the front side of the cube, the front, as well as the back, appears as a square in the projection. The back of the cube appears smaller than the front because it is farther away from the viewer. The oblique lines that connect the corners of the inner with the outer square are the perspective views of the cube edges that extend into the third dimension, that is, in a direction

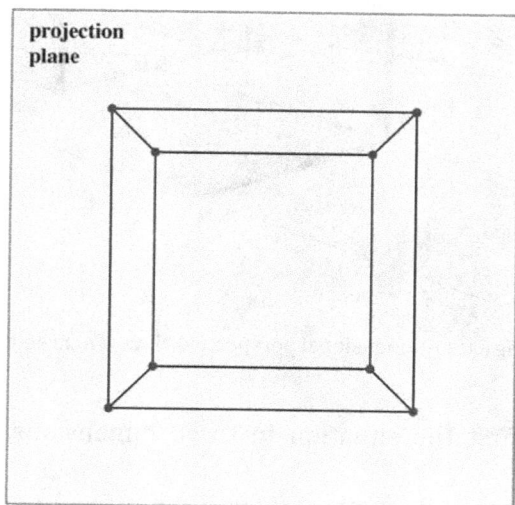

Fig. 15.10.   Perspective image of a three-dimensional cube seen from the front.

which is perpendicular to the projection plane. Note, that these oblique lines are all parallel to each other in three-dimensional space.

We can perform quite a similar procedure in order to obtain a three-dimensional perspective view of a four-dimensional cube. We think of the three-dimensional projection "plane" (mathematically, it is called a *hyperplane*), where the image is created, as "our" three-dimensional world. The ana-kata direction of hyperspace is perpendicular to our world, which we take as a projection plane for a four-dimensional observer. The three-dimensional world separates hyperspace into two regions. The eye of a four-dimensional observer is placed on one side of our world, while the four-dimensional hypercube is "behind" our world, as seen from the observer. The line of sight from the eye of the observer to a vertex of the hypercube is again a (one-dimensional) straight line, which intersects our world in precisely one point. This point is the three-dimensional perspective image of the hypercube's vertex. The observer can do this for all the 16 vertices of the hypercube and then connect the adjacent edges with straight lines that represent the edges of the hypercube. In that way, the observer would draw an image of the hypercube into our three-dimensional world for us to behold, in the same way that the perspective image of the ordinary cube was drawn onto the projection plane in Figure 15.10.

The resulting three-dimensional image of a hypercube is shown in Figure 15.11. This is a perspective projection of the hypercube into 3-space. In hyperspace, the hypercube is positioned in such a way, that two of its "surface cubes" are parallel to our world and hence appear undistorted in the projection which is visible to us. Therefore, we interpret this image as a small cube inside a large cube. The small cube is the "backside" of the hypercube, it is further away from the observer in the fourth Direction, and thus, it appears smaller than the front side. The other surface cubes are distorted by the perspective and appear as sort of truncated pyramids. But in four dimensions, all these "cubes" have the same size and shape.

The oblique edges that connect the outer to the inner cube are the perspective views of the edges that run in the direction of the fourth dimension, orthogonal to all directions of our world. They are, in fact, parallel lines in hyperspace — in the same sense as the oblique lines in Figure 15.10 are parallel to each other in three-dimensional space.

Whenever the hypercube is not parallel to the coordinate axes, its appearance in the perspective view can be very different from Figure 15.11. A four-dimensional rotation of the hypercube is shown in Figure 15.12.

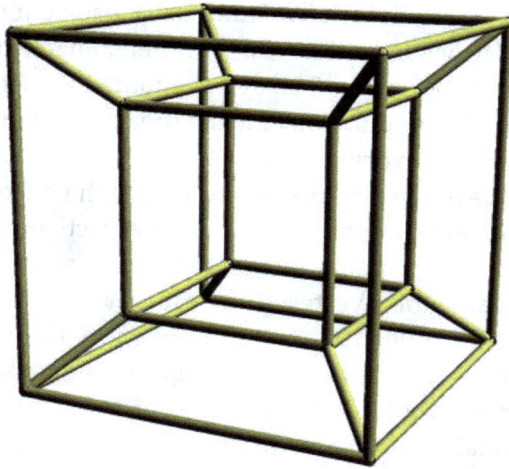

Fig. 15.11. Perspective projection of the four-dimensional hypercube into our three-dimensional space.

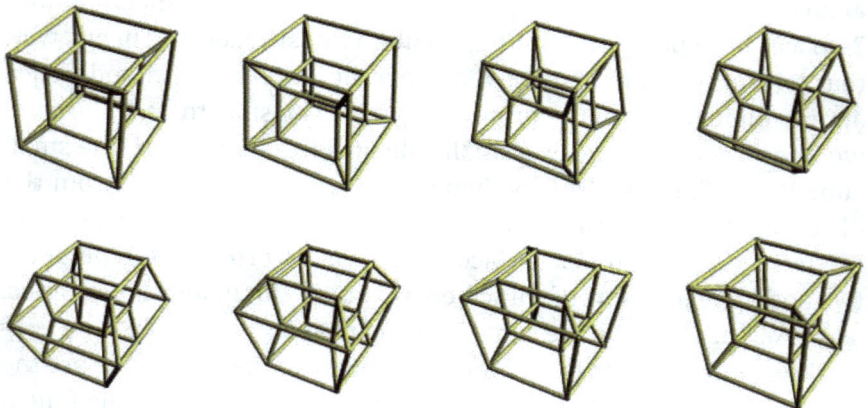

Fig. 15.12. Rotation of a hypercube in four-dimensional space.

15.8 The Hypersphere

The hypersphere in four dimensions is, in a sense, both simpler and yet more complicated than the cube. In three dimensions, the sphere is the set of all points that have the same distance from its center. Here we use the word "sphere" or, more precisely, "2-sphere" to refer to the two-dimensional surface. We distinguish this from the ball, which is the sphere together with its interior in three dimensions. This definition remains the same in four dimensions. But here the set of points that all have the same distance from a center point forms a shape with three-spatial dimensions. We call it 3-sphere or hypersphere and it is the "surface" of a four-dimensional ball.

Perhaps the easiest way to imagine a 3-sphere is again by analogy with the three-dimensional situation. In three dimensions, we imagine that a sphere is created in the following way. We use two circular discs of exactly the same size, which are made of a rubber-like elastic material. We place the two disks exactly one above the other and we glue their borders together, as shown in Figure 15.13. Then we inflate the structure by blowing air between the two circular disks. The upper disk is gradually transformed into the "Northern Hemisphere", the lower disk becomes the "Southern Hemisphere" and the circle along which the two disks are glued together becomes the equator of the sphere. The sphere is, therefore, obtained by continuously deforming two copies of circular disks. Thus, we can say that it is topologically equivalent to two circular disks whose edges (circles) have been linked with one another. We remind the reader that this idea of topological equivalence has already been discussed in Chapter 10.

Let us reconsider the procedure of inflating a sphere from the perspective of a two-dimensional resident of "Flatland" as we see in Figure 15.13. Imagine, that one disk is brought up to Flatland from below, the other from above, until they both touch Flatland in the same region and meet in such a way that the circles that form their boundaries are perfectly aligned. Flatland residents would see two circular disks miraculously occupying the same region. It would be impossible to create this situation within Flatland, unless one allows

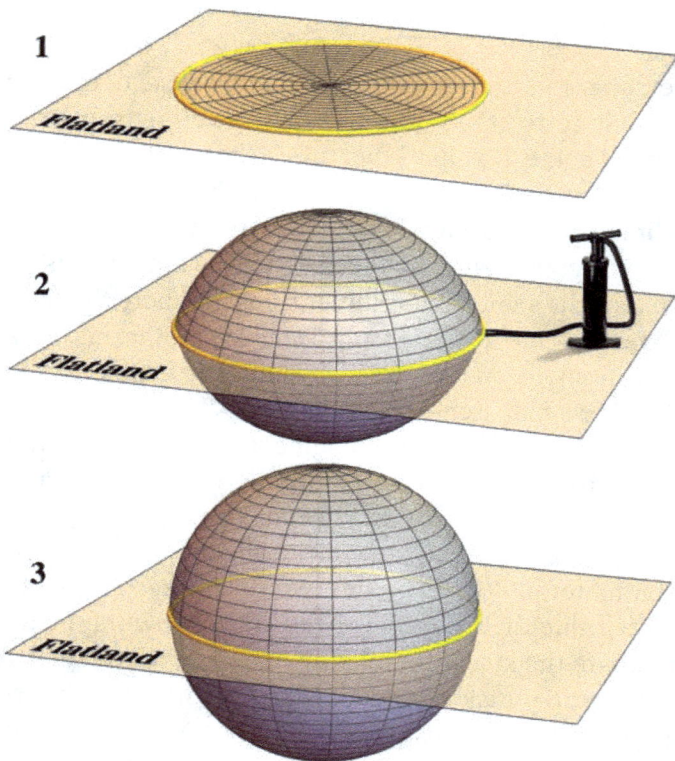

Fig. 15.13.   Inflating a sphere. Only the equator remains in Flatland.

the two disks to penetrate each other without being damaged. Next, the two disks are glued together along their circumference. When inflation starts, the upper disk is lifted up into the third dimension while the lower disk is pressed down. For the inhabitants of Flatland, this means that the interior of the disks suddenly vanishes and all that remains is the circle along which the two discs are glued together, that is, the equator of the three-dimensional sphere.

By analogy, we could imagine a procedure for creating a 3-sphere in four-dimensions, by starting with two three-dimensional rubber balls. Let us assume that it is possible to place these two balls in the same region of three-dimensional space, so that their spherical surfaces are perfectly aligned. Then we glue the two balls together along their entire surface. Inflating the hypersphere would now mean to separate the interior of the two balls by lifting one of the balls up (or "ana") from our space into the fourth dimension, while pushing the other ball down (or "kata"). Only the surface (an ordinary 2-sphere) would remain in our three-dimensional world. This sphere is the intersection of the hypersphere with our three-dimensional space and may be called the equatorial sphere. It is a two-dimensional surface that separates the three-dimensional hypersphere into two hemi-hyperspheres. The equatorial sphere is a cross-section through the hypersphere that cuts through it exactly in the middle.

15.9 Cross-Sections of a Hypersphere

By moving the sphere through Flatland, as shown in Figure 15.14, we can create a sequence of cross-sections, with the help of which we can further explain the shape of the sphere to a Flatland resident.

From Flatland's perspective, a point would appear as soon as the sphere touches Flatland from below (Figure 15.14, diagram 1). This point quickly widens into a small circle that grows larger the further the ball is pushed through Flatland. The circle has the largest circumference as soon as the plane cuts the sphere exactly in the middle (along the equator). After that, everything happens in reverse order

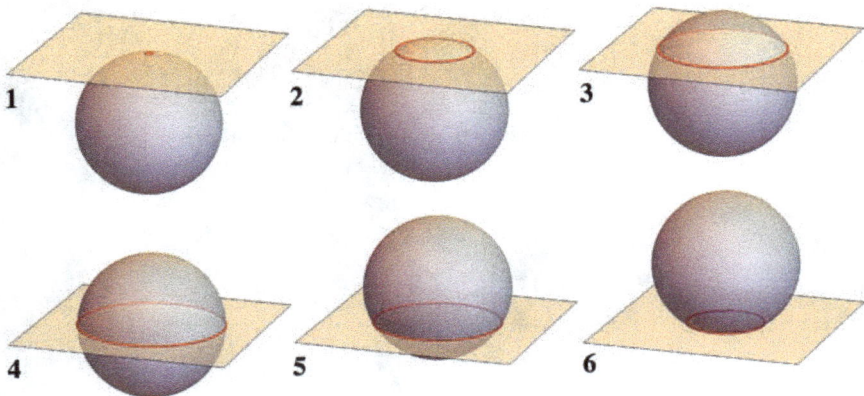

Fig. 15.14.   A sphere moving transversally through Flatland.

and the circle visible in Flatland shrinks again to a point that finally disappears when the sphere leaves Flatland in the upward direction.

The trail that the sphere leaves in Flatland is a continuous series of concentric circles, cross-sections of the sphere, as shown in Figure 15.15, diagram 1. These circles sweep the entire circular disk twice, once as they grow and once as they shrink. This shows us again that the sphere is topologically equivalent to two circular disks that are connected along their circumference.

If the ball also moves horizontally when diving through Flatland, one can spread the circles side by side as shown in Figure 15.15, diagram 2. This gives a better impression of the timing of this process.

By analogy, we can guess what we would see, if a resident of the four-dimensional space pushed a hypersphere through our space, say, from ana to kata. As soon as the hypersphere touches our three-dimensional space with one of its poles, we see a point appear that quickly widens into a sphere as the hypersphere moves through our space. The visible cross-section reaches maximum diameter at the equatorial sphere of the hypersphere, and then begins to shrink back to a point before it completely disappears. When the hypersphere is pushed through our space in an oblique direction in hyperspace, one would see a growing and shrinking sphere that moves horizontally, as shown by the sequence in Figure 15.16.

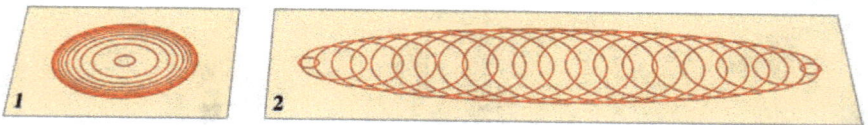

Fig. 15.15.   Cross-sections of a sphere in Flatland.

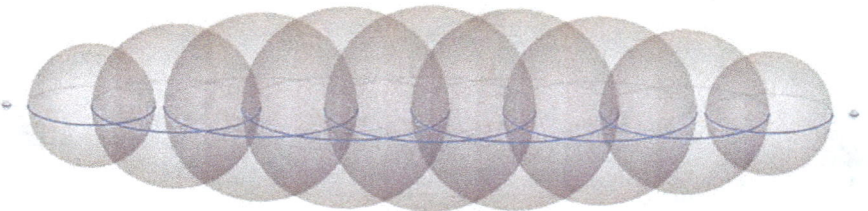

Fig. 15.16.   Cross-sections of a hypersphere in 3-space.

15.10 Stereographic Projection of a Hypersphere

It is possible to visualize the hypersphere with the help of a stereographic projection. We remind the reader that a stereographic projection was used in Chapter 4 to create a map of a hemisphere by using, as an example, the "North Pole" as a center for a perspective projection mapping the Southern Hemisphere onto a plane tangent to the sphere at the South Pole. This procedure is shown in Figure 15.17, where we indicate the projection rays for six points on the equator (see Section 4.17 for more details).

Assuming that the sphere has radius 1, the North Pole has the z-coordinate 1 and the South Pole the z-coordinate -1. Hence the point with coordinates $(0, 0, 1)$ is the North Pole and the South Pole has coordinates $(0, 0, -1)$. The projection plane, that is, the plane tangent to the sphere's South Pole, is the two-dimensional set of all points with coordinates $(x, y, -1)$.

Figure 15.18 shows the resulting image in the projection plane, this is the stereographic map of the lower or "Southern" Hemisphere. It is just a circular disk with the coordinate lines of the sphere.

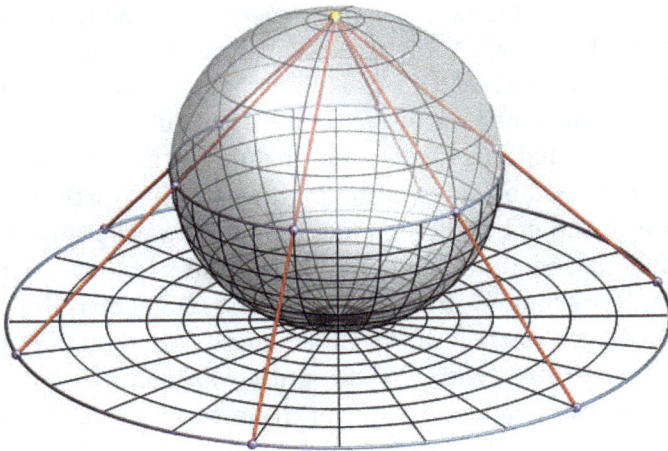

Fig. 15.17.   Principle of stereographic projection mapping a hemisphere onto the plane.

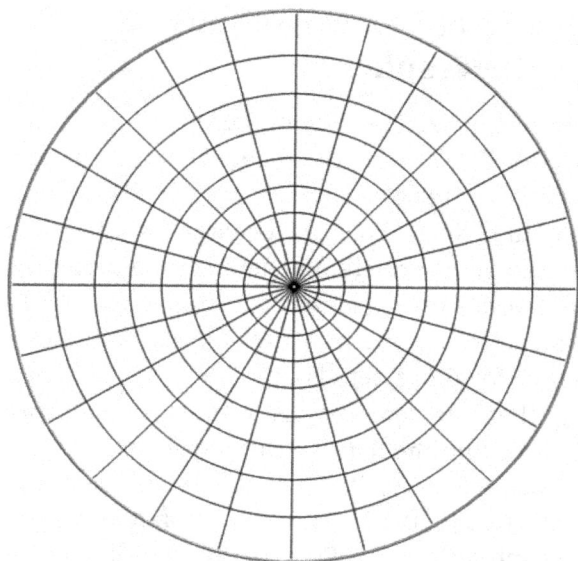

Fig. 15.18.   Image of the coordinate lines of a hemisphere under a stereographic projection.

The straight radial lines are the meridians and the concentric circles represent the parallels of latitude.

If we want to explain the 2-sphere to an inhabitant of Flatland, we could show this image — a two-dimensional map of the Southern Hemisphere. We could also try to explain that the radial lines are actually curved (but bent in a direction perpendicular to Flatland) and that they are called meridians, and, furthermore, that they connect the South Pole at the center of the image with the North Pole on the other side of the sphere. We could even enlarge the image area to include a larger part of the sphere, which also includes a part of the Northern Hemisphere. In the image, the Northern Hemisphere would surround the Southern Hemisphere. Unfortunately, a map of the entire sphere (with the exception of the North Pole) would require the entire Euclidean plane for its presentation, as discussed in Section 4.17. It is more convenient to show a separate map of the other (northern) hemisphere, obtained by projecting stereographically from the South Pole onto a plane tangent to the North Pole.

We could then point out, that the two images show two disks that have to be glued together along the equator in order to obtain the whole sphere — something that would be extremely difficult to imagine for our Flatlander.

The method of stereographic projection can be generalized to higher dimensions. We can, therefore, take a completely analogous view of the hypersphere. The "North Pole" (or "kata pole") of the hypersphere would be the center of projection, and we think of our three-dimensional space as being tangent to the South Pole ("ana pole"). We can perform a stereographic projection onto that tangent space in complete analogy to the situation in three dimensions. What would we see as the stereographic image of the hemi-hypersphere in 3-space? Of course, we would see a ball, and just as the center point in Figure 15.18 corresponds to the South Pole, the center of that ball would be the point where the hypersphere touches our three-dimensional space, that is, the ana pole.

In order to see some structure, we need to describe the 3-sphere in terms of coordinates analogous to latitude and longitude. We can, since the hypersphere is three-dimensional, define three angles, let's say a longitude, a latitude-1 and a latitude-2. The lines where the longitude is constant (the meridians) are radial rays originating from the center of the ball, like the radial rays in Figure 15.18. If we draw just these lines and perform a stereographic projection the image would be as that shown in Figure 15.19.

The stereographic projection of the hypersphere is a three-dimensional image of a hemi-hypersphere. It shows the pole in the center, with radial green meridians emerging from it. They are actually curved lines, but their curvature cannot be seen, as these lines are bent in the fourth direction, which is orthogonal to all directions of space.

15.11 The Hypersphere in Physics

Thus far, we have always considered the hypersphere as a geometric object embedded in a four-dimensional space, just as we perceive the 2-sphere as part of our three-dimensional space. The hypersphere is a three-dimensional space without a border.

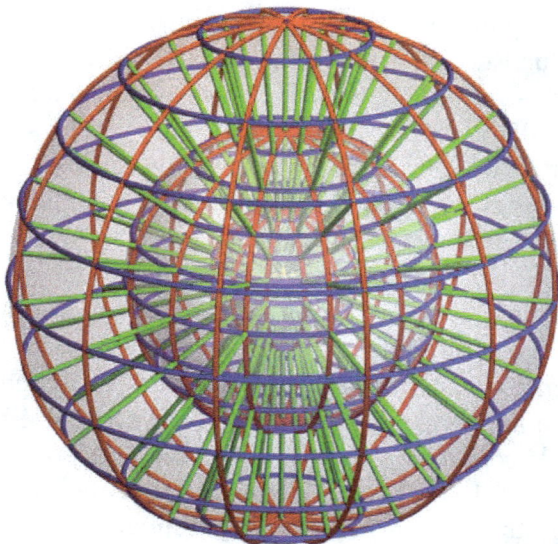

Fig. 15.19.   Stereographic projection of a hemi-hypersphere.

We could actually imagine living within a sufficiently large hypersphere. We could move in all three spatial directions and would not notice anything unusual unless we covered very long distances. This is certainly difficult to imagine, but the analogy to Flatland will help us again. Would a resident of Flatland even notice if he or she didn't live in a flat Euclidean plane, but actually lived on the slightly curved surface of a very large sphere? Probably not, because the difference to a really flat Flatland is not noticeable at short distances. Only if our Flatlander set off and kept going straight on and on, would he eventually return to its starting point after a long time and after circumnavigating its entire two-dimensional world. And it doesn't matter in which direction he started. For the Flatlander, his world has no limit and no borders, but it is, nevertheless, finite. Actually, this is exactly the situation in which we find ourselves on the surface of the Earth. At first sight, the surface of the Earth appears as a flat plane, however, only if we move far enough in one direction, will we eventually encircle the whole Earth and return to the starting point from the opposite direction.

Let us consider if our situation in the universe could be completely analogous to the situation of a flatlander on a spherical surface. Imagine that we actually live in a huge hypersphere. At short distances, we would have no indication that we would not be living in a normal, three-dimensional space. The curvature of our world would only become noticeable at astronomically great distances. But, if we were to fly straight ahead in a spaceship in any direction, we would, much to our surprise, come back to our starting point from the other side. The universe would have no limit and no border, but it would only be finitely large. Its volume would be finite.

The reader might be tempted to dismiss that as pure speculation, but it has a very real background. In 1917, in the course of the development of the general theory of relativity, Albert Einstein designed a cosmological model in which the three-dimensional space is a closed 3-sphere. Einstein's famous saying "Two things are infinite: the universe and human stupidity; and I'm not yet completely sure about the universe[2]" can be taken literally in this context.

According to the general Theory of Relativity, the presence of matter bends the space around it. Einstein, thus, explains gravitational force as a geometric deformation of space–time, which also causes a curvature of the three-dimensional space. This suggests that, seen as a whole, the universe could be a "curved space", like the three-dimensional surface of a hypersphere, of finite volume, yet, without three-dimensional borders. Einstein's cosmological theory involved the introduction of a so-called "cosmological constant", which he later dismissed again (he is said to have even called it the greatest blunder of his life). Nevertheless, the introduction of dark energy in order to explain observations of an expanding universe had effectively the same consequences as Einstein's cosmological constant. Hence, Einstein's theory of a closed universe has become highly popular again and cosmological models giving space the global structure of a hypersphere are still being discussed.

---

[2] Quoted from a personal conversation by Frederick S. Perls in his autobiographical book *In and Out the Garbage Pail*, Gestalt Journal Press, 1969.

15.12 The Hypersphere in Art

It is interesting that a cosmological world view that describes a closed spherical universe may have been conceived already in the Middle Ages. It has been repeatedly claimed that the Italian poet, Dante Alighieri (1265–1321), in the final portion of his *Divine Comedy*, had created a cosmological model that can be interpreted as a geometric description of a 3-sphere. Probably the first to point this out as early as 1925 was the Swiss mathematician Andreas Speiser (1885–1870).

After the *Inferno* and the *Purgatory*, the final part of the *Divine Comedy*, the *Paradiso*, sees Dante with his companion Beatrice — a symbol of divine love and grace — on the way through Heaven. He ascends through the nine heavenly spheres, passing the spheres of Moon, Sun, planets and fixed stars until he reaches the Primum Mobile, the ninth of the heavenly spheres, where he gets a glimpse of the Universe as a whole. Looking back, he sees the spheres of heaven surrounding Earth. Turning around, he sees with new and enhanced abilities of perception, the Empyrean, which is the paradise and spiritual heaven of the angels, arranged in nine concentric spheres around an unbearably bright center which symbolizes the essence of God. The structure of the Empyrean's angelic spheres surrounding God precisely mirrors the structure of the nine heavenly spheres surrounding Earth. The Empyrean surrounds the Primum Mobile and at the same time appears as an interior part of it.

In an article of 1975, the American physicist Mark A. Peterson explains, how Dante's description reminds him of a mathematician's description of a hypersphere. Two spherical semi-universes glued together along an equatorial sphere (the Primum Mobile) in order to form a mind-boggling new whole — the 3-sphere. We may admire each semi-universe, for example, by a stereographic projection as in Figure 15.19, where it becomes a three-dimensional ball with the hypersphere's pole at its center and is surrounded by the other semi-hypersphere. But the whole hypersphere is only obtained by gluing the two balls together along their whole surface, which gives a structure, which — due to its four-dimensional nature — is beyond our imagination. We cannot be sure that Dante really tried to describe the geometry of a hypersphere, but he certainly tried to conceive a consistent image of a heavenly cosmos that is beyond the understanding of ordinary humans.

Even the French artist Gustave Doré (1832–1883), who created perhaps the most impressive series of illustrations for Dante's work, did not try to represent Dante's sky as a whole in a single image. We show one of his fascinating visualizations in Figure 15.20. In this image, Doré concentrated on the representation of the Empyrean,

Fig. 15.20. Illustration to the Divine Comedy, Paradiso Canto 28 by Gustave Doré (1832–1883): Rosa Celeste: Dante and Beatrice gaze upon the highest Heaven, the Empyrean (1892) [Public Domain].

the mirror of the material sky, which represents half of a hypersphere, with the pole as a place of dazzling brightness, where God himself resides.

Peterson concludes his article with the words: "There are people who will never believe that physics is beautiful, and who perhaps still resent the way it demolished the medieval world view. They might be surprised to learn that modern physics can also illuminate the richness of the medieval imagination." The same could be said about the mathematics underlying and supporting physical theories. And the idea that Doré's picture could show part of a larger, four-dimensional structure might also add another dimension to the viewer's reception of this piece of art.

Index